Ethics of Chemistry

From Poison Gas to Climate Engineering

Other Related Titles from World Scientific

The Public Image of Chemistry
edited by Joachim Schummer, Bernadette Bensaude-Vincent and Brigitte Van Tiggelen
ISBN: 978-981-277-584-9

Nanotechnology Challenges: Implications for Philosophy, Ethics and Society
edited by Joachim Schummer and Davis Baird
ISBN: 978-981-256-729-1

Fundamentals of Patenting and Licensing for Scientists and Engineers Second Edition
by Matthew Y Ma
ISBN: 978-981-4452-53-3

The Ethics and Biosecurity Toolkit for Scientists
by Judi Sture
ISBN: 978-1-78634-091-7
ISBN: 978-1-78634-092-4 (pbk)

Design and Applications of Single-Site Heterogeneous Catalysts Contributions to Green Chemistry, Clean Technology and Sustainability
by Sir John Meurig Thomas
ISBN: 978-1-84816-909-8
ISBN: 978-1-84816-910-4 (pbk)

Ethics of Chemistry

From Poison Gas to Climate Engineering

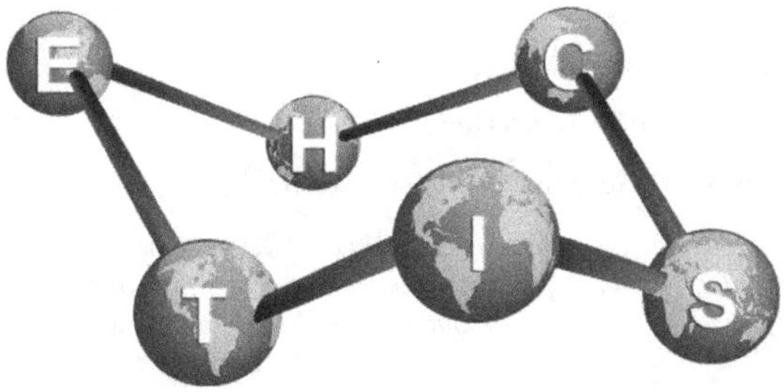

Editors

Joachim Schummer

Tom Børsen

Aalborg University, Denmark

World Scientific

NEW JERSEY • LONDON • SINGAPORE • BEIJING • SHANGHAI • HONG KONG • TAIPEI • CHENNAI • TOKYO

Published by

World Scientific Publishing Co. Pte. Ltd.
5 Toh Tuck Link, Singapore 596224
USA office: 27 Warren Street, Suite 401-402, Hackensack, NJ 07601
UK office: 57 Shelton Street, Covent Garden, London WC2H 9HE

Library of Congress Control Number: 2021930374

British Library Cataloguing-in-Publication Data
A catalogue record for this book is available from the British Library.

ETHICS OF CHEMISTRY
From Poison Gas to Climate Engineering

ISBN 978-981-123-353-1 (hardcover)
ISBN 978-981-123-354-8 (ebook for institutions)
ISBN 978-981-123-355-5 (ebook for individuals)

For any available supplementary material, please visit
https://www.worldscientific.com/worldscibooks/10.1142/12189#t=suppl

CONTENTS

Introduction

Ethics of Chemistry: Meeting a Teaching Need

Joachim Schummer and Tom Børsen

1. The Purpose of This Volume

Perhaps more than ever, chemical know-how is in demand for developing solutions to many global issues, including the protection of the natural environment, healthcare, nutrition for a growing world population, water treatment, energy production, waste treatment, recycling, and clean-up of environmental damages by former generations. Rather than just doing their isolated lab work, chemists are expected to engage with other disciplines and with society at large to work together on managing these issues. Of course, real-life solutions are never as simple as those for crossword puzzles. They always involve various advantages and disadvantages, improvements and drawbacks, and opportunities and risks to be discussed and balanced against each other. Thereby, chemists are inevitably involved in disputes about values. They would badly fail if they were not prepared to reflect on the values, develop and analyze moral and political arguments, build moral judgments, and perform responsible actions, all of which belong to the domain of ethics.

For many years, national and international organizations, such as UNESCO's World Commission on the Ethics of Scientific Knowledge and Technology (COMEST), have therefore recommended mandatory courses of ethics for all university students of science and engineering. Moreover, funding agencies and corporate associations increasingly require or assume that scientists are familiar with ethical issues, such as the European Responsible Research and Innovation program and the Corporate Social Responsibility agenda. While the necessary courses have been established in numerous countries and for various student groups, chemistry is still lagging behind, despite the particular importance to the discipline. One reason, or excuse, is the lack of appropriate course materials, because ethics of chemistry does hardly exist in academia, unlike, for instance, ethics of medicine, biology, computer science, engineering, and so on.

The main purpose of this volume is to fill that gap both by providing the necessary resources for teaching and by establishing ethics of chemistry as an academic field of future research. We do so by presenting a well-selected set of ethical cases that we have developed and co-edited over four years in a series of special issues of *Hyle: International Journal of Philosophy of Chemistry* (hyle.org). Our chapters are authored by the best available experts worldwide, many of which have a background both in chemistry and the humanities or social sciences. The cases are carefully selected so as to cover the most important ethical issues that have raised moral concerns in the past and have strongly shaped the public image of the discipline. They both define the scope of ethics of chemistry and narrate a cultural history that every chemistry student should be familiar with based on academic resources.

The objectives of using the book in university chemistry education is not only to transfer knowledge about ethical issues and events in chemistry, but also to facilitate the development of analytical skills and action-oriented competences so that ethical dilemmas can be identified and addressed by the next generations of chemists.

In this chapter, we introduce ethics of chemistry and guide the educational use of this volume, depending on purposes and resources. We first provide a brief outline of philosophical ethics by distinguishing it from best practice methodology or behavior, sometimes misleadingly called 'ethics', and then provide a brief introduction to ethical theories, followed by a discussion of their application to fields such as chemistry. Next we introduce all chapters by our typology of ethical issues according to the order of this volume. This is our recommended order of reading and teaching ethics classes to chemistry students. However, cases can also be considered and taught according to historical order, which provides a selective cultural history of chemistry; or by subject fields, which allows for integrating selected ethical issues into particular chemistry classes, for both of which we provide recommendations below.

Unlike the common theory-plus-experiment-based teaching, which depicts chemistry in radical societal isolation and improves technical skills, ethics courses enable students to act responsibly in projects that matters in society (Schummer 2018). They reestablish the long-neglected link to the humanities, and raise awareness of the simple fact, that chemical research is of the greatest societal importance, for better or for worse.

2. What Is Ethics?

2.1 Ethics of chemistry versus 'ethics in chemistry'

If a scientist fabricates data, wrongly assumes authorship of something that some colleagues have achieved before, or if senior scientists put their name first (or last, depending on the customs of the discipline) on the publication of work mostly done by junior scientists, or if someone denigrates the public image of science, the scientific community is usually upset. There is no question that all that is bad behavior. However, does it really concern general ethics?

While it is, in any regard, morally wrong to betray somebody else, like your colleagues and the general public with fabricated results, it is not so clear if the other issues scientists are usually concerned with are of ethical relevance. After all, lobbying for the image of one's profession is not an ethical issue but in the self-interest of the discipline, usually at the expense of other disciplines, nor is the question of who deserves most credit for a scientific achievement. Indeed most other societal fields do not care about intellectual priority claims of, say, business ideas, as long as there is no legal protection. And if they do so, like in art, copying the style of others is considered poor art rather than moral failure.

Many issues that scientists sometimes call 'ethical issues' are actually violations of methodological or epistemic norms. There are a wealth of methodological rules in science proposing, for instance, how to produce and interpret data correctly, how to publish and review scientific results, and so on. These so-called 'best practice' rules ensure that results are reproducible by anyone else, such that the scientific community can build further research on them. Any violation could harm scientific progress when others credulously rely on these results. Of course, they are upset then because they have wasted time and efforts, and because the scientific quest for knowledge was misguided. However, it depends on the specific circumstances whether such violation of methodological norms additionally violates ethical norms. While there is an overlap area between both topics to be discussed in Chapter 2, this volume is not about methodological issues but about ethical issues proper. What then is ethics of chemistry about?

There is a double meaning of 'ethics' as there is a double meaning of 'chemistry', like of any other science in the English language. In the vulgar meaning, 'ethics' just means the behavioral rules of a group, such as 'chemistry' means the regular behavior of molecules. In this sense, 'ethics in chemistry' describes the behavior of chemists and what they think one ought to do, which largely covers the aforementioned methodological norms. In the academic meaning, however, 'ethics' is that branch of philosophy that deals with moral values, ideally putting it on general principles that are valid for anyone at any time.

A chemical example may illustrate the difference: there is a 'chemistry' of two particular insects interacting chemically (subject matter level, the actual behavior of insects) and there is the academic field of pheromone chemistry that studies the chemical interactions between insects by various molecules (epistemic level, scientific knowledge about the insect behavior). Like academic chemistry studying substances and molecules, ethics studies moral values, norms, and judgments. Thus, ethics is not the behavior, but the study of the behavior. However, unlike science that aims at the perfect description, prediction, and explanation of its subject matter, ethical studies aim at normative ideas of what is morally right or wrong. These studies can be very general and abstract, similar to theoretical chemistry, or applied to particular fields of human activity, such as in ethics of chemistry.

What then are moral values, norms, and judgments? Moral values are general goals or goods that are morally valued and recommended to strive for or to support (as, for instance, justice, well-being, or human rights). If these values additionally meet ethical requirements detailed below, they are ethical values. Moral norms are both obligations to strive for moral values (*e.g.*, 'act so as to increase or reinforce justice, well-being, and human rights') and obligations to avoid counter-productive goals (*e.g.*, 'act so as to avoid injustice, harm, and human rights violations') also known as moral prohibitions. Moral judgments include the full range of normative statements, from blame and praise to moral recommendations pro and con, unlike scientific judgments that state whether statements are true or false. In ethics, moral judgments are usually about the actions of moral subjects, both individuals and corporations, distinguishing between doing morally right and doing morally wrong. They can come in advance as guidance (*e.g.*, 'it is morally right to do this or that') and in retrospect as verdict ('that was morally wrong'). However, moral judgments may also be about the general attitudes of moral subjects, which typically distinguish between virtues and vices and is then called 'virtue ethics' (*e.g.* 'this person is guided by justice'). The examples illustrate that values can to some extent be translated into general norms, judgments, and virtues and *vice versa.* If either of them forms the basis of an ethical system or theory, it is considered an ethical or normative principle.

Ethics of chemistry is thus the academic field that studies the moral values, norms, judgments, and virtues relevant to chemistry with the aim of providing moral guidance from a general ethics perspective. Because ethics takes, per definition (see below) an impartial stance, it may question, criticize, or confirm the moral attitudes of chemists – as individuals, corporations, or local or global scientific society – rather than taking them for granted. In ethics, like in any science, it does not matter what people believe at a certain time and place, but only what rational arguments can be provided to support or oppose a view.

2.2 A brief introduction to philosophical ethics

At the risk of over-simplification, a very brief introduction to philosophical ethics, or moral philosophy, may help setting the stage. Ethics, a major discipline of philosophy, has elaborated theories and normative principles to decide on ethical issues over many centuries. To begin with, it is useful to distinguish morality, which is what ethics is mainly about, from other normative realms (*i.e.*, realms of norms that say what one ought to do), in particular, customs and law, keeping in mind the distinction between the academic field and its respective objects of study. Ethics studies morality, while the academic field of law studies the laws, whereas social sciences, such as anthropology, describes the actual customs of particular societies.

Customs and law are both local regulations, enforced by various sanctions, from mild stigmatization to imprisonment, or even capital punishment in some countries. Morality differs from customs in that it refers to general principles that are independent of the particular group or society, whereas customs are based on the specific traditions of a culture. For instance, local customs greatly vary in their dress codes or rules of decency and politeness, hardly any of that can be justified across cultures. And some customs, like the widespread ancient religious practice of human sacrifice, are deeply questionable on ethical grounds. Law, which many mistakenly think is the most basic normative realm, requires ethical justification in parliamentary debates before its implementation, and can later be questioned and revised on ethical grounds at any time. Ideally, law harmonizes with morality at any time, but it may be based on misguided moral ideas as in many dictatorships. Moreover, law covers only those areas that the state can reasonably control and courts could possibly judge. For instance, criminal intent, although being pivotal to many legal judgments, frequently evades the judges' capacities of mind reading. Hence, morality, rather than being a residual category besides customs and law, is the most basic and most important one.

The very first principle of any ethical theory that aims at moral judgments, indeed its necessary condition or meta-principle, is impartiality, which can appear in different versions. It may just be the strict rule to disregard any partial interest and, instead, give any affected party an equal say. In the latter form, the principle is well established in democracy in the form of general elections, as well as in political philosophy (contract theories) and discourse ethics (which defines rules to solve conflicts in moral matters, such as that anyone has the right to propose rational arguments for one's own position). Or, it may demand considering the equal interest of anyone concerned, including oneself; or abstracting from one's own particular interests; or treating anyone by the same rules, regardless of heritage and social status (unlike the laws and customs in monarchic, feudal, and caste societies). In its oldest and most famous form, the so-called Golden Rule, it requires that one should treat others as one want to be treated oneself by all others, thereby dismissing the difference between me and them. The 18th-century-philosopher Immanuel Kant developed this principle into a meta-rule, that is, a litmus test to decide whether a rule is an ethical rule or not: only if I can reasonably want that my rule becomes a general rule for anyone, it is an ethical rule. (You might wish to betray others, but you certainly do not wish that 'anyone can betray others' becomes a general rule.)

The second principle of ethics says, do no harm to others and instead do good, the principle(s) of non-malevolence and benevolence. It can come as one combined principle, such that doing harm can be balanced by doing good and vice versa. Or it comes as two different principles, usually with priority for non-malevolence, as in law, because attempts at doing good frequently turns into unintended harm due to unforeseen adverse effects. Much of the classical philosophical ethics is about whether and how one can combine the first two/three principles into one. Many Kantian ethicists think that their own form of impartiality (s. a.) implies not-doing-harm, because nobody wishes to be the subject of being harmed. Utilitarian ethics, the other important tradition, unites impartiality (all people are equally considered) with the combined principle of benevolence and non-malevolence: act so as to maximize the happiness of all people, which is achieved by both doing good and avoiding harm.

Over the centuries, a variety of theories have been proposed in ethics, just as in science. They all have their pros and cons and mostly establish respectable positions for many areas, but they differ from each other in many regards of which we will introduce three important distinctions.

First, ethics may focus on different aspects of human action, both for providing moral guidance in advance and judgments in retrospect. One aspect, highlighted by consequentialist ethics, is the consequences of one's actions, such as when utilitarianism requires to maximize the happiness of all people. While scientists might take this to be the obvious approach, because it is based on familiar cause-effect relations of human action, they should consider this: if your action has unintended adverse consequences, your action is morally wrong, whatever your good intentions might have been. In contrast, deontological ethics (from Greek:

deon, duty, obligation) judges and recommends actions according to whether they are performed in accordance with moral duties, which cannot be defined in scientific terms of cause-effect relations. These duties certainly include avoiding harm and doing good. However, deontology is not an excuse for the naive good will, a major ethical problem, because it usually includes the duty to consider the best available knowledge for avoiding unintended harm.

Second, depending on their number of principles, ethical theories can be divided up in monist positions (a single principle) and pluralist positions (many principles). Utilitarian ethics is usually monist, which has the advantage of avoiding priority conflicts between different principles, whereas the Kantian meta-rule allows to derive numerous rules that might lead to conflicting recommendations and judgments. Moreover, most contemporary ethical theories include the principles of justice and human dignity that both cannot simply be derived from the principles of impartiality, benevolence, and non-malevolence. For instance, utilitarian ethics allows for the unequal distribution of happiness and harm among people, which might even include the torture and sacrifice of individuals, as long as the overall happiness rises. Once further principles are considered, they all need to be balanced against each other in every situation, which undermines the ideal of a simple algorithm for ethical judgments. That might explain why ethics is usually considered fuzzy. However, also science is fuzzy when we deal with real-world issues by models that necessarily include simplifications and approximations, rather than with mathematical simulations of idealized cases.

Third, while classical ethics was developed for local moral issues of the time, modern issues frequently affect a much broader range, both in terms of time, space, and species to be considered. In particular, technological changes, here and now, can have a global impact now and on future generations, and on nonhuman species. Thus, ethical theories differ in whether they consider only local people or also the global population, the current generation or also future generations, and only humans or also nonhuman biological species, as in environmental ethics. Many of the ethical issues discussed in this volume require considering a broader scope because chemistry has indeed an enormous impact across time, space, and biological species.

2.3 Applied Ethics: Top-down versus bottom-up

As mentioned previously, the general approach of philosophical ethics is not so different from that of theoretical science in that it tries to base its theories on a few principles. Ideally normative guidelines and judgments for a particular case are to be derived from the principles and the conditions of the case, just as scientists try to explain a particular phenomenon by applying basic theories or models to the specific context. There is a controversy about which and how many principles are needed in ethics, as there is an epistemological dispute in science if all natural phenomena can be explained by a single set of scientific principles, which theoretical physicists usually claim to be from physics in a reductionist manner.

Naturally, the epistemological distinction informs different educational approaches. According to the top-down or reductionist approach, education starts with first principles to be enriched only later by particular cases that are meant to illustrate the principles at work. In chemistry, you would start your first semester with quantum mechanics and do laboratory experiments only at the end of your study, in order to learn how to apply quantum mechanics to chemical phenomena. A corresponding ethics (of chemistry) education would begin with first principles and end up with particular cases (from chemistry) in so-called applied ethics. In contrast, the bottom-up approach, which is skeptical about a set of first principles fixed once and forever and about reductionism, at least in its current shape, starts with studying particular cases, in order to learn what matters and how to design useful distinctions, concepts, and models. Because that is how the experimental sciences, like chemistry, have successfully developed over centuries, we take this to be also the obvious approach in ethics of chemistry, both in research and education.

In philosophical ethics, skepticism of the top-down approach has a long tradition, reaching from Aristotle in the fourth century BC to the modern field of Applied Ethics (in capital letters!) that refuses to be just an application of first principles. That skepticism, as well as the bottom-up approach, is particularly strong in medical ethics. However, ethical principles can play an important role from the very beginning of exploring a new field such as ethics of chemistry. They guide our view in what matters from an ethical perspective, help introduce useful distinctions, and develop concepts and theories adjusted to chemical issues, without simply taking them as applied cases. They might even inspire us to modify or enrich them based on a deepened ethical insight of the cases. In a sense, they work as necessary instruments that we need at the start and that might be improved in the course of further research, just as chemistry has improved its models and instruments in the course of its history.

Even more so in ethics education, principle-based ethics helps sharpen important distinctions, crucial for any ethical reasoning, such as between descriptive and normative reasoning, moral arguments and rhetoric, good-for-us and morally good, and important-for-me and important in general. Discussions in Applied Ethics frequently become all too fast absorbed by personal interests, and ignoring those distinctions, and principle-based ethics is the most effective educational remedy against that.

Nonetheless, this volume favors a case-based or bottom-up approach to ethics of chemistry for various reasons. That is not only, as mentioned before, in accordance with the general epistemology of chemistry and the prevailing skepticism in Applied Ethics. Moreover, because we are at the very beginning to explore what really matters in chemistry from an ethics point of view, it is useful to start with the cases that have mostly stirred up public moral debates. There is the potential of learning something entirely new from the chemical cases because they have largely been ignored by philosophical ethicists. Furthermore, because there are a variety of different ethical theories, any top-down approach would prematurely and wrongly have to single out one theory for each and every case in chemistry. By contrast, this volume is authored by many different scholars, each one a specialist of the respective case, without following a single ethical theory.

From an educational point of view, the diversity of authors and viewpoints also provides students with a realistic view of the ethical issues and the different possible stances. What matters in ethics education is not the indoctrination of one particular view, as in dictatorships. Instead, ethics education aims at developing various moral capacities, including the ability to develop one's own moral position based on ethical arguments.

3. Cases of Chemistry According to Ethical Topics

There are various ways of ordering the cases and, correspondingly, of using the cases in class. Because our primary aim is to provide material for ethics teaching, this volume orders the cases by ethical topics. That has the additional advantage of structuring the field, that is, ethics of chemistry, right from the beginning, to provide guidance for further academic research or let others challenge and modify our still provisional ordering. Of course, there are entirely different options, such as ordering the cases by historical timeline or by chemical fields, which we will discuss in Sections 4 and 5. And there is, of course, the possibility of selecting individual cases according to specific interests, by choosing the most prominent cases or by using the table of contents along with the subject index at the end of this volume. However, for our purpose it is useful to divide the case in five groups according to the dimensions of the ethical issues.

3.1 Misuse and misconduct

The first group, *Misuse and Misconduct*, covers cases where the responsible actors clearly violate ethical norms, which they might do knowingly or unknowingly. Evil actors do this with intention, and in retrospect they are usually easy to identify and to blame, but these are the least interesting cases from an ethical point of view. Others do that apparently without intention but by sheer ignorance of ethics or of the dimension of their doings. Indeed, as we will see, in many cases of that group, actors misunderstand their ethical responsibilities and confuse it with other norms, such as patriotism or obedience to employers or governments. Others try to excuse themselves later by arguing they had not known the consequences of their actions, although ignorance or naivety is of course not an excuse per se. Instead, there is an ethical duty, of scientists as of anyone else, to inform oneself about the possible consequences of one's actions according to the best available knowledge. Naivety is no excuse but a major source of ethical misconduct and widespread harm. That is perhaps the most important lesson from this type of case.

Our volume starts with a fictitious case of scientific misconduct that Janet D. Stemwedel wrote for this occasion, drawing from various prominent historical examples, including the Sezen/Sames case from 2002 (Chapter 2). In her study, we meet a PhD student who cannot reproduce the experimental data published by a former member of the research group. As the case develops, the reader gradually realizes that data fabrication has been involved. Stemwedel analyzed her fictive case with the help of Muriel J. Bebeau's strategy for ethical decision-making. As the analysis progresses, various dilemmas faced by a number of involved people unfold, from the PhD student to the supervisor, to colleagues and university officials, illustrating the ethical complexities of typical real cases of scientific misconduct.

In the second case study, Joachim Schummer narrates the history of chemical weapons research, development, and deployment in WWI with focus on Fritz Haber in Germany (Chapter 3). The case shows how for the first time in history scientists engaged on a large scale in weapons research, established an academic-industrial-military-governmental complex, and created an unprecedented arms race that would become a model for the Cold War era. In his ethical analysis, he argues that chemical weapons research, which is widely conducted up to the present day, is morally wrong according to all major ethical theories, despite a battery of widespread but misleading excuses. The questions then arise as to why chemical societies do not condemn it in their codes of conduct, and why they instead still today consider the main weapons researchers of WWI role models for a younger generation.

The third case asks if the research, development, and production of chemical weapons can all be characterized per se as misuse of science (Chapter 4). Stephen M. Contakes and Taylor Jashinsky use the two examples of Louis Fieser, the inventor of napalm during WWII, and Dow Chemical, one of the producers of napalm during the Vietnam War. Based on Just War theories, their analysis is focused on the issue of whether the actors knew about and supported the actual military mass deployments of napalm against civilians, which by all standards count as severe war crimes. While they find Fieser, as WW II developed, increasingly guilty, they see Dow at least struggling with ethical issues and betrayed both by US politicians and the military through misleading information.

3.2 Unforeseen local consequences

The second group of cases, *Unforeseen Local Consequences*, is more intricate and distinguished from misuse particularly by the level of available knowledge at the time. While misuse disregards the best available knowledge of probable consequences, in this group the available knowledge is, at the time of making crucial decisions, indeed limited, such that the adverse consequences are unforeseen, but not totally unlikely. The ethical questions then arise if sufficient precautionary measures had been applied and how the actors respond to early alerts of potentially disastrous effects. Cases in this group typically include the three kinds of hazards of chemical industries derived from their production, their products, and their waste.

The first study deals with the worst chemical industrial disaster ever since, the leakage of methyl isocyanate in Bhopal, India in 1984 (Chapter 5). While the plant was meant to contribute to a 'green revolution' that should increase agricultural productivity through the use of pesticides, the disaster killed thousands of people and injured hundreds of thousands. Based on a detailed analysis of the historical events, Ingrid Eckerman and Tom Børsen scrutinize the responsibilities of the various actors, including the company Union Carbide Corporation and the governments of India and the local state of Madhya Pradesh. They argue that fundamental ethical values were violated and draw lessons on how future industrial catastrophes can be avoided.

The next two chapters deal with different aspects of unforeseen consequences of industrial chemicals. Klaus Ruthenberg investigates the early phases of the Thalidomide scandal in Germany that resulted in the births of thousands of malformed newborns and stillborn babies over several years (Chapter 6). In the late 1950s the company Grünenthal praised Thalidomide (brand name 'Contergan') as a 'wonder drug' that enables a good night's sleep without any side effects. However, their results were based on questionable scientific practice. Early warnings were dismissed for economic reasons. In his ethical analysis, the author discusses how the most important players in the case complied with four ethical principles: respect for autonomy, non-maleficence, beneficence, and the principle of justice.

The third case returns to chemical war efforts by investigating the notorious herbicide Agent Orange that the US military used as a strategic weapon on a grand scale during the Vietnam War to destroy possible hideaways of their enemy (Chapter 7). The industrial wartime manufacturing of the herbicide left considerable but unnoticed amounts of extremely toxic dioxins in the product, which killed or harmed a large part of the civil population of Vietnam, apart from the intended devastating environmental impact. Claus Jacob and Adam Walters use the case for discussing the responsibilities of various actors, from the chemical inventor and the industrial manufacturers to the actual user of the chemical, for the unforeseen harm.

In the fourth case of unforeseen local consequences, Ragnar Fjelland discusses the long-term effects of a former chemical waste disposal site in New York State, on which a suburban town called Love Canal was built (Chapter 8). He argues that citizen knowledge should be taken seriously in such cases. The official experts used scientific models and statistical methods with underpinning assumptions that did not capture the distribution of health problems among the citizens. Whereas the citizens themselves joined forces with other scientists to develop a model that incorporated their local knowledge. The author argues that scientists can fruitfully involve and collaborate with citizens, and should do so if they adopt the Precautionary Principle. They must not oversell their conclusions and should inform the public about uncertainties.

3.3 Global and long-term influences and challenges

Our third group of cases extends the theme of unforeseen adverse consequences toward global and long-term effects. Rather than with some local effects, we are now dealing with decisions that might affect the future of humanity or even the entire earthly environment for living beings. The fact that we find here many prominent cases illustrates the extraordinary power of chemistry on our global and future well-being or lack thereof. That role of chemistry has intuitively been acknowledged by a growing environmental movement since the early 1960s, but badly neglected by professional ethicists thus far, which has left chemistry students largely alone. They are still facing simplistic black/white schemes: either a condemnation of chemistry overall by a growing public opinion, or the promise of a rosary future by their potential employers.

Rather than just prohibiting chemical products right away because of possible adverse consequences, it is ethically advisable to balance benefits and harms, develop precautionary measures at an early state and invest in chemical research that develops less harmful substitutes, involves less harmful processes, redirects entire production chains, recognizes possible hazard as early as possible, and avoids environmental damage through recycling technologies.

Once considered the perfect remedy against both insect-borne diseases such as malaria and insect pests in agriculture, dichlorodiphenyltrichloroethane (DDT) proved to have numerous adverse effects on human and nonhuman health across the globe, most prominently pointed out already in 1962 by Rachel Carson's *Silent Spring*. In our first case of this group, Tom Børsen and Søren Nors Nielsen provide a general ethical framework to evaluate both the positive and negative aspects of industrial chemicals, which they apply to DDT in order to develop a balanced view (Chapter 9).

Next, Alistair Iles, Abigail Martin, and Christine Rosen examine the potential public health risks of Bisphenol-A (BPA), a precursor to and residual component of many plastics (Chapter 10). The compound may cause harm due to its endocrine disruptive effects, according to some experts. Also newly developed substitutes for BPA might have similar unknown adverse effects. The chapter critically discusses standard risk assessment procedures and industrial chemicals regulation in the United States. In their ethical analysis of how to deal with such uncertainties, the authors point out prototypical positions of key actor groups – managers/corporations, chemists/designers, and regulators/legislators – by distinguishing between three different regimes: deontology, consequentialism, and technocracy.

Another classic in its own right, polyvinyl chloride (PVC) is one of the oldest and still most widely produced plastics, despite various health and environmental issues. Alastair Iles and coworkers investigate its full production chain, from feedstock to consumer products such as toys. They illustrate how the historically entrenched production can be changed at various steps and by various actors to achieve a safer chemical world (Chapter 11). The case study of PVC also provides an introduction to Green Chemistry and shows how this approach can integrate ethical principles into the daily practice of applied chemistry.

Unlike compounds that can be produced in factories, elements have to be extracted from limited natural resources such as by mining, which frequently comes with huge local environmental damage. The present mining of rare earth elements, required for materials in numerous high-tech applications, unequally distributes benefits and harms in both space and time. It favors consumers in rich countries at the expense of the local population at mining sites, and it favors the present generation at the expense of future generations who might have no more access to these resources. Abigail Martin and Alastair Iles discuss the complex ethical issues that arise if dealing with global and intergenerational justice at the same time, which all chemists should be aware of if they consider applied research projects involving rare earth elements (Chapter 12). The issues invite chemists to research both substitutes for rare earth elements in high-tech materials and effective recycling processes that would reduce the demand for mining.

The two remaining chapters both discuss famous cases of chemical involvement in atmospheric and climate science. The first one, by Joachim Schummer, is on researching possible hazards before they actually occur, of which the most prominent case of chemical hazard foresight is Molina and Rowland's 1974 prediction of stratospheric ozone depletion by human-made chlorofluorocarbons (CFCs), which would have threatened most terrestrial life (Chapter 13). In this story, a postdoc project would trigger a revolution in global environmental politics, culminating in the Montreal Protocol of 1987 that banned CFCs worldwide. While Molino and Rowland are undoubtedly moral role models, the chapter discusses if and to what extent scientists have a special moral duty of hazard foresight based on their particular intellectual capacities.

As scientific evidence for global warming through the large-scale emission of carbon dioxide, methane, CFCs, and other substances had grown and international politics appeared unable to cope with that, atmospheric chemist Paul Crutzen in 2006 first broke a taboo by suggesting chemical methods for engineering the climate. Crutzen's suggestion prompted a vivid, but largely isolated, debate by ethicists that failed to address chemists. Dane Scott provides a balanced assessment of the various pros and cons of climate engineering with a focus on chemical methods of atmospheric carbon dioxide removal and answers the questions if today's chemists ought to engage in climate engineering research and which ethical conditions should be considered (Chapter 14).

3.4 Challenging human culture

Not all ethical issues of chemistry are a matter of life and death, or directly concern physical harm. Some arise from the way science questions, undermines, or ignores traditional ways of thinking and valuing, thereby challenging human culture. These issues, which cannot be studied by toxicological or other scientific methods, because they concern values such as justice, liberty, quality of life, or human dignity, are frequently overlooked by scientists, although they can raise very heated debates and potentially radical aversion of science. Others are so subtle and enduring that they might develop a long-lasting negative attitude toward scientists because of their potential ignorance of these issues. The chapters of this section address various cultural and societal aspects each typical of chemistry.

Based on a three-step model for making ethical judgments, Klavs Birkholm discusses the recently tremendous rise of psychotropic drugs as chemical tools for human enhancement (Chapter 15). While the promise of 'enhancement' sounds good at first glance, it always comes with downsides. First one should recognize the risks posed by a chemical drug to individuals, then assess the possibility for misuse, and finally imagine the cultural consequences of widespread drug use. The author identifies three potential cultural/societal consequences of psychotropic drugs: a medicalized culture that suppresses individuality, the fading quest for social recognition as an important driver of societal developments, and chemical enhancement becoming a societal norm that undervalues natural qualities.

In the second chapter, Joachim Schummer investigates societal and ethical issues of creating artificial life, from premodern times to current Synthetic Biology (Chapter 16). His case in focus is Craig Venter's 2010 announcement of having produced the first self-replicating cell, echoed by a worldwide media accusation of Venter 'playing God', which chemists in various forms have faced ever since. While the scrutiny of ethical arguments behind that accusation reveals meager results, the author argues that Venter himself provoked the public reaction in order to attract media attention, which raises the question of how to responsibly interact with the public on ethical issues.

A third issue that has caused heated debates in society is whether one can have intellectual property rights on human gene sequences (Chapter 17). While this suggests to some the disturbing notion that one can own parts of the human body, for a chemist, DNA is just a molecule. Saurabh Vishnubhakat takes the lawsuit of Myriad Genetics (2010-2013) to discuss the underlying scientific, legal, and ethical issues of patenting DNA. Rather than providing a simple answer, he argues for a procedural approach that mediates between specialist and generalist views and interests.

3.5 Codes and regulations

The final part moves to a more general level by discussing cases of how to rule the behavior of chemists. As was mentioned before, there are two other normative realms besides ethics, customs and law, which ideally harmonize with ethics. The first one is established in the form codes of conduct of chemical societies, and the second one as chemical regulations.

While the regulation of (the production, trade, and use of) chemicals is legally binding for anyone in a country, here exemplified by the European regulation (Registration, Evaluation, Authorization and Restriction of Chemicals [REACH]), the codes of conduct (exemplified by that of the American Chemical Society [ACS]) are not, not even for chemists. However, chemistry students should be familiar with both the codes and legal regulation of chemicals in their country, and they should be aware of the differences by which normative framework come into force. Legal frameworks are (in democratic society) decided by parliaments and thus express the current majority will of the population. Codes of conduct, at best, express the majority view of members of a local professional society. In contrast, ethical guidelines are supposed to be based on generally accepted ethical principles regardless of locality, nation, profession, and personal moral preference. Ethics thus is the very basis on which both local legislation and professional codes can possibly be judged, and thus justified or criticized. In other words, if you do not like some parts of your professional code or regulations, you need to refer to a normative framework of ethics to justify your criticism There is nothing else beyond.

In the first chapter of that part, Jeffrey Kovac reviews the codes of conduct by the ACS, arguably the oldest chemical society to establish such a code, from the 1965 'The Chemist's Creed' to the current version of 'The Chemical Professional's Code of Conduct' and various supplementary guidelines including on publication ethics (Chapter 18). By looking at the underlying ethical values, he compares the ACS codes with those by the British and German chemical societies and the more recent international codes, the 'Hague Ethical Guidelines' and the 'Global Chemists' Code of Ethics'. He argues that a future revision of the ACS code should put more emphasis on research integrity and on societal and environmental issues.

Incidentally, none of these codes has been set up or revised with advice from an ethicist or ethically trained chemist, to the best of our knowledge, nor is there much known about democratic procedures for approval by members. More than anything else, the way that chemical societies have produced their 'ethics' codes of conduct reflects the actual distance between the disciplines of chemistry and ethics.

The final chapter takes REACH, the European regulation of marketing chemicals, as a case to illustrate how ethical ideas have actually influenced legislation (Chapter 19). Jean-Pierre Llored demonstrates that, after curative and preventive environmental policies of the past, REACH has been informed and guided by the precautionary principle. Rather than providing a fixed set of interdictions, the regulation tries to face the uncertainty of the future with precautionary measures to be regularly revised.

4. Cases of Chemistry According to the Cultural History of Chemistry

The cases of this volume are selected so as to ideally form a canonical set that we think all chemistry students should be familiar with. Although future research will certainly explore further cases, the present set arguably comprises the most important historical events and processes that have shaped the public view of chemistry ever since the early 20th century. Whereas public media every now and then provides documentaries on these issues, usually on anniversaries, these cases are nowhere part of the standard curriculum of chemistry. As a rule, chemistry students are informed about the cultural history of their own discipline at best by public media, which illustrates the poor state of chemistry education.

Hence, this volume can also be read as a cultural history of the 20th and early 21st century. If the cases are historically lined up, it becomes more obvious why public aversion of chemistry arose during the 20th century. There were earlier strong resentments immediately after WWI, based on the large-scale deployment of chemical weapons (Chapter 3), which triggered literary fantasies of a chemical apocalypse (Schummer 2021). Chemistry's engagement in warfare continued to be important, from the atomic bomb and napalm in WWII, to Agent Orange and napalm in Vietnam War (Chapters 4 and 7), each contributing to public unease and mistrust. However, the most forceful impact began in the early 1960s when concerns about the adverse global impact of insecticides on the biosphere, such as DDT (Chapter 9), largely triggered the environmentalist movement. At the same time, the worst-ever pharmaceutical tragedy by the drug thalidomide, promoted worldwide as a remedy against 'morning sickness' of pregnant women, resulted in thousands of miscarriages and deformed babies (Chapter 6). These and other cases carried the general message that chemical products have an all-devastating potential of mass poisoning, from warfare and industrialized agriculture to pharmaceuticals.

Further cases made people aware that not only the chemical products but also their impurities (Chapter 7) as well as side-products and waste carry a deadly threat for the environment if not carefully processed. Sea water had been polluted with mercury for much of the 20th century, which the chemical industry had used as catalyst for the production of acetaldehyde in Japan (Minamata) and elsewhere (Chapter 11). Soil and groundwater became spoiled by uncontrolled dumping places of chemicals around the globe, some of which later turned into ground for greedy land development, such as in Love Canal, New York State (Chapter 8). Air pollution, a regular companion of industrialization, reached a new dimension when CFCs moved from the lower atmosphere in the local area to the global stratosphere, threatening all life on the earth by ozone depletion (Chapter 13). Moreover, apart from the unintended adverse effects of the products and waste, also the chemical production process can be extremely hazardous, as the worst-ever industrial disaster of chemistry in Bhopal, India, made obvious in 1984 (Chapter 5), as well as many previous and later cases. To complete the picture, because chemical production ultimately depends on natural resources, it is worthwhile to look also at how these resources are originally gained and at what costs, as the environmental damage of mining rare earth elements has increasingly illustrated (Chapter 12).

The full life cycle of industrial chemicals thus reveals many different hazards that have shaped the bad public image of chemistry over the 20th century. However, that image is one-sidedly focused on the chemical industry of the past and neglects the numerous chemists who have contributed to minimizing environmental harm. They did so by developing instruments and models to measure and predict environmental pollution, frequently being the first ones to raise public awareness of these issues (Chapter 13). They have worked hard to develop alternative products and processes that could replace the harmful ones of the past, guided by ideas of Green Chemistry (Chapters 11 and 10). And they are the only ones who research and develop methods to clean up the mess, from decontaminating dumping places to the large-scale binding of CO_2 in climate engineering (Chapter 14).

In recent years, chemistry increasingly became involved in cases that do not fit the simple scheme of doing/preventing/undoing harm through chemicals, or at least the public awareness has shifted to a broader scope of ethical issues and values. Global environmental issues, such as climate change and the exploitation of limited resources, have taught us that every harm and its remedy affect different populations and different generation unequally, which raises the issues of global and intergenerational justice for which there is no technical fix (Chapters 14 and 12). Recent attempts to create life in the laboratory are not seen as threats to one's health but as ways to undermine received cultural values (Chapter 16). Similarly, the public debate on whether a sequence of human DNA can be protected by patent law or not does not refer to physical harm but to a possible erosion of values (Chapter 17). Moreover, as chemistry has the potential of changing the human mind through drugs, both temporarily and enduring, it undermines the notion of what a normal human being is if enhancement drugs become regularly available (Chapter 15).

The various ethical issues and scandals of the past had an impact on the norms and regulations of chemistry. Both chemists and governments responded. Chemical societies tried to counteract the growingly bad public image of chemistry by issuing and revising codes of conduct, starting with the ACS (Chapter 18). Governments increasingly regulated the production, commerce, and use of chemicals by developing ever more restrictive and, in view of uncertainty, more flexibly adjustable regulations based on the precautionary principle, as did the European Union (Chapter 19). Readers of this volume are invited to check for themselves whether these normative schemes by chemical societies and governments have sufficiently coped with the growing challenges at a given time, or if and when additional guidance would have been required by ethics proper.

5. Cases of Chemistry According to Fields of Chemistry

The most prominent ethical cases, summarized above, cover almost all fields of chemistry, although there have been different emphases at different times.

Chemical warfare research was initially dominated by *physical and inorganic chemistry* during WWI, before organic and biochemist took the lead with the research of nerve agents, napalm, and other weapons (Chapters 3 and 4). There is a strong link between organic warfare chemistry and *agricultural chemistry*, because the same substance classes that were investigated as potential nerve agents appeared to be also promising candidates for insecticides for pest control, and vice versa. While that indeed enabled, along with fertilizers, an agricultural revolution by fighting famines through strongly increased productivity, it also produced devastating environmental disasters, from bio-contamination by widespread use of DDT for pest control to the terrible accident of the Bhopal plant that was producing another insecticide (Chapters 9 and 5). Last but not least, the herbicide that the US army employed as a strategic weapon in the Vietnam War, named Agent Orange, and that poisoned thousands of civilians because it contained toxic impurities, was a severe failure of the organic synthesis and product control (Chapter 7).

Pharmaceutical chemistry had been suspect to many people ever since, if only because of religious reasons. However, its potentially disastrous effects by lack of careful advance studies became evident by the thalidomide scandal in the early 1960s (Chapter 6). Fortunately, strict standards of clinical studies have been established since then before the official approval, which some politicians and their submissive administrators are trying to undermine only in the current Covid-19 pandemic. Other issues have risen since then because of the efficacy and power of chemical drugs in modifying the human mind, because that also modifies our understanding of human normality (Chapter 15).

Geochemistry, particularly *atmospheric chemistry*, along with *instrumental chemistry*, has been a major player in recent environmental matters. It even established itself, out of a combination of *physical chemistry* and meteorology, to understand the human impact on the atmosphere (Chapter 13). Climate change poses an entirely new challenge in not only contributing to understanding the global phenomena, but in posing the ethical question whether climate change should be managed by chemically manipulating the atmosphere (Chapter 14).

Moreover, the classical field of *inorganic chemistry*, which was originally about studying which elemental combination has what property, beyond the realm of organic chemistry, has become of pivotal ethical importance. If we take it, in its applied orientation, as the study of the earth economy of the chemical elements, it explores the possibilities and problems of any elemental substance flow, which includes the mining, recycling, and dumping of chemical elements. Once that dimension becomes obvious, every research decision on which elemental composition to explore further and for whatever research project must take into account the combined issues of the mining, recycling, and dumping of the respective elements. That is particularly obvious in the case of rare earth elements (Chapter 12), where elemental resources tend to be scarce, the mining extremely harmful to the environment, and therefore recycling technologies are required. However, it would also cover every management of dumping places (Chapter 8), and, more general, the pollution of air, water, and soil by elemental compounds, including large-scale attempts at capturing carbon dioxide to reduce climate change (Chapter 14).

Polymer chemistry has only recently become under ethical scrutiny, beyond the unspecific public aversion of plastics that tend to neglect the great diversity of environmental properties of polymers, including biodegradability and ultraviolet (UV) instability. On the one hand, the industry seems to be locked in received manufacturing practice that ignore environmental and health standards, due to toxic additives and toxic combustion products as in PVC (Chapters 10 and 11). On the other hand, there are tremendous challenges for polymer chemists to develop new materials that comply with standards of green chemistry, if only these standards would become more widely accepted by chemists.

Because of its relation to medical ethics, *biochemistry* has frequently been a matter of moral disputes. However, there are several distinct ethical issues that chemists should be aware of. One is whether one ought to be allowed to file patents on human (and nonhuman) genes (Chapter 17), or if these genes are considered public good because they are part of (human) nature. Another one, more radically, asks if the potentially artificial creation of life from inanimate matter by way of chemically means would violate any ethical norm. In legal term, at least, researchers are free to explore their own field (Chapter 16).

Most of the ethical issues discussed in this volume are relevant to all of chemistry far beyond the specific case and field. For instance, issues of moral responsibility for unintended adverse consequences go across all fields of science and technologies. Thus, the historical cases are only a starting point to engage in ethical discussions. The two chapters on codes of conduct and chemical regulation (Chapters 18 and 19) as well as the chapter on scientific misconduct (Chapter 2) obviously have no thematic focus but refer to all fields of chemistry.

6. Conclusion: How to Use This Volume in Class

This volume has been composed as a reader to support ethics classes for undergraduate students of chemistry. Each chapter focuses on different ethical issues based on a particular case from chemistry; it can form the basis of a class session. Ideally, the volume covers a one-semester course that discusses the cases in the order of the chapters, thereby introducing students into the full range of both ethical issues and important events of the cultural history of chemistry. For a complementary reading, we point to the 'Recommended Reading' section of each chapter, including the one below.

However, each chapter can also be read independently from the other ones, which allows for selective reading according to different aspects and class needs. Because the field of ethics of chemistry is still in its infancy, we imagine teachers with different backgrounds at the beginning who want to, or are required to, give a mandatory course in ethics of chemistry, including ethicists, historians of chemistry, and chemists. This volume is flexibly composed so as to meet all their needs.

While an ethicist would probably select the cases according to the order of our chapters, a historian of chemistry might want to make a different choice. For that purpose, we recommend the order of reading from Section 4 that arranges the cases according to the cultural history of chemistry. In yet another form of teaching, historical-ethical aspects would directly be integrated into chemistry classes, so as to enrich the traditionally narrow focus of teaching, for example, in physical, inorganic, and organic chemistry. In that way, ethics of chemistry can be part of a regular study-program in chemistry throughout. To that end, we refer to the order sketched in Section 5 that arranges the cases according chemical fields, as well as to the subject index for identifying more specific topics.

This book can also be used by students in their project work. They can be inspired by the case studies when they choose the topic of their projects and when they formulate their problem statement. Moreover, they can draw parallels between their own work and the cases analyzed in the book.

Whatever the selection of chapters and order of reading might be, the lack of teaching material can no longer be an excuse for the lack of courses of ethics of chemistry, nor for inadequate ethics courses that miss what matters in chemistry. The discipline has an extraordinary record of historical events that made it, more than any other one, suspect in moral regard to the broader public. The appropriate response is not distancing oneself ever more from the humanities, but integrating ethics into chemistry curricula. This volume aims at a turnaround by eventually educating future generations of responsible chemists.

Further Reading

Kovac 2018 is a brief introduction to ethics with cases from chemistry. For a list of books, paper collections, case studies, and other materials related to ethics of chemistry, see Børsen & Schummer 2016. Beauchamp & Childress 2009 introduces and discusses moral principles like respect for autonomy, non-maleficence, beneficence, and justice. Good and brief introductions to philosophical ethics are Frankena 1988 and Rachels 2012.

References

Beauchamp, T.L. & Childress, J.F.: 2009: *Principles of Biomedical Ethics*, Oxford & New York: Oxford University Press, seventh edition.

Børsen, T. & Schummer, J.: 2016, 'Editorial Introduction: Ethical Case Studies of Chemistry', *Hyle: International Journal for Philosophy of Chemistry*, **22** (1), 1-8 [online available at: http://www.hyle.org/journal/issues/22-1/editorial.htm, accessed 7 Oct. 2020].

Frankena, W.K.: 1988, *Ethics*, Englewood Cliffs, NJ: Prentice-Hall, second (and later) edition.

Kovac, J.: 2018, *The Ethical Chemist: Professionalism and Ethics in Science*, New York, NY: Oxford University Press, second edition.

Rachels, J.: 2012, *The Elements of Moral Philosophy*, New York, NY: McGraw-Hil, 7th (and later) edition.

Schummer, J.: 2018, 'Why Chemists Need Philosophy, History, and Ethics', *Substantia*, **2**(1), 5-6.

Schummer, J.: 2021, 'Art and Representation in the Twentieth Century: From the 'Mad Scientist' to Poison Gas and Chemical Pollution', in: Peter J.T. Morris (ed.), *A Cul-*

tural History of Chemistry in the Modern Age, 1914-2019, London: Bloomsbury, chap. 8 (forthcoming).

Chapter 2

The Case of the Finicky Reactions: A Case Study of Trust, Accountability, and Misconduct

Janet D. Stemwedel

Abstract: Scientific knowledge-building requires careful engagement with the phenomena being studied and with other researchers. In instances where attempts to replicate research findings fail, it can be challenging to determine whether the source of the problem is an experimental factor or a scientist who is not trustworthy. This chapter presents a case study to illuminate the ethical challenges of responding to a failure to reproduce an experiment in a training environment. Ethical analysis of the case takes account of features of the situation connected to knowledge-building and to competition between scientists for credit, and includes consideration of how obligations to trainees and to the broader scientific community can shift depending on the facts in evidence.

1. Introduction

Chemistry is a field where knowledge-building involves working with a team, making sure the results you find are reproducible, and trying to find the source of the problem when results turn out not to be reproducible. It is also a field whose practitioners use published scientific results and techniques described in the literature as the starting point for their own knowledge-building projects.

Knowledge is not fixed for all time when it is published in a scientific journal. We assume that additional research may reveal new effects, better interpretations of old data, the limitations of particular experimental methods, even mistakes. However, published reports of research cannot play even this role, of tentative knowledge, if they are fabricated or falsified.

Fabrication is making up data or results rather than actually collecting them from observation or experimentation. Falsification involves 'adjusting' real data – changing the values, adding some data points, omitting other data points. Both fabrication and falsification amount to lying about the empirical data, which means that these practices undermine the project of building a body of reliable knowledge about the world. Plagiarism is misrepresenting the words or ideas (or, for that matter, data or computer code, for example) of others as your own. Plagiarism is a sort of dishonesty that robs someone else of their credit in the knowledge-building project. There are other behaviors that can do damage to knowledge-building efforts within the community of science, such as sabotaging someone's equipment or refusing to share important research materials like reagents or cell lines. However, fabrication, falsification, and plagiarism are recognized as 'high crimes' against the scientific endeavor (Zigmond & Fischer 2002) and as such are explicitly included in official definitions of scientific misconduct.

Scientific misconduct provides a special challenge for both knowledge-building and ethical decision-making in science. It is hard enough to understand a phenomenon or to work out how to control an experiment when you and the other scientists with whom you are working are honest about the data observed and the procedures used to obtain the data. If one or more of those scientists is lying about procedures or observations, what conclusions can you draw?

Part of the challenge is that experimental results themselves are not enough to distinguish hard-to-reproduce results from made-up results. If you cannot replicate an experiment, it might mean that you failed to follow all the necessary procedures, that your experimental skill was not sufficient, that your equipment was not precise enough, that your reagents were not pure enough. It might mean that the description of the experiment you are trying to reproduce failed to identify a parameter that matters a lot to producing the desired outcome. But it could also mean that the scientist who described that experiment and reported those results is not trustworthy.

The fact that scientists compete with each other (to be the first to a discovery, or to secure scarce research funding or jobs) complicates things. Researchers are engaged in building a shared body of reliable knowledge about the world, which requires that they cooperate with each other and evaluate each other's results. At the same time, they have individual incentives that pull against cooperation so they can stake their priority claims or win their grants. These dynamics can make it harder to identify misconduct, because an ethical scientist trying to control the flow of information before her findings are published can be hard to distinguish from a scientist who is being secretive because her 'data' are made up. Fierce competition for scarce resources can also make it hard to know whether practices falling short of misconduct might still be ethically problematic, either because they undermine knowledge-building or because they are unfair to other knowledge-builders.

Because doing chemistry means engaging with chemical phenomena and with other people, chemistry students should understand the harm scientific misconduct does to the knowledge-building endeavor. They should also recognize the challenges of identifying and addressing misconduct, and how these challenges can fall heavily on scientific trainees.

In real life, unless you yourself are consciously committing an act of misconduct, it can be hard to tell whether the situation in which you find yourself involves misconduct or something else. But scientists are always placed in the position of drawing the best conclusions they can on the basis of incomplete information. Sometimes these are judgments about the chemical systems they are studying, and sometimes they are judgments about how much they can trust their fellow scientists.

2. The Case of the Finicky Reactions

The case presented here is fictional, although it bears some resemblance to a number of real cases. I have opted for a fictionalized case in part because the central players in real misconduct cases are usually not forthcoming about the details of what they did and why they did it. They often dispute the facts asserted in official findings or by other parties, and they frequently see little benefit in being candid about their motivations.

The case itself is presented in three parts, meant to reflect one way the situation described here could unfold. Remember that an inescapable feature of our ethical decision-making is that we cannot know in advance exactly what consequences will come from our decisions. We make the best decisions we can on the basis of the evidence we have, and we never have the luxury of delaying action until we have complete information.

When thinking through each part of this case, it is worth asking: What do the characters here know? About which of their questions are they uncertain? What is the ethical weight of what they know and what they do not know? It is also worth considering how differing roles (PhD student or professor) bring with them different interests, obligations, and likely consequences for one's actions. Especially important in understanding likely consequences are the power differentials, especially between students and research directors.

Part 1

At the end of the first year of her PhD studies in chemistry, Anna Bijou joins the research group of Dr. Martin Green. The first project Dr. Green assigns her involves extending a set of organometallic reactions that were the centerpiece of the doctoral thesis of Paul Fanning, a recently graduated Ph.D. from the Green group. Fanning's journal articles on these reactions were part of what attracted Bijou to the Green group, so she is excited to carry this research forward.

Bijou's excitement does not last. Her initial attempts to reproduce Fanning's reactions, guided by the 'Materials and Methods' sections of his journal articles, fail to produce the desired products. Reasoning that the reactions might not be working because they are sensitive to experimental conditions that are not fully described in Fanning's journal article, she sets about trying to find Fanning's lab notebooks.

"Paul left eleven notebooks from his six years in the lab," says Tony Pham, a third-year doctoral student in the group. "They're on the bookshelf in Room 203. But there aren't many helpful details in them."

"What do you mean?" asks Anna Bijou.

"If Dr. Green has you trying to reproduce Paul's reactions, the scratchings in those notebooks will probably be of as little use to you as they were to me. But maybe the notebooks I kept will give you something," Pham says, handing them over to her. "You could also look at Carole Lee's notebooks. She actually got a couple of those reactions to work once or twice, back when Paul was still in the lab."

"I don't think I've met Carole yet," Bijou replies. "I should, so I can ask her – and you – what I'm doing wrong."

Shaking his head sadly, Tony Pham answers, "I'll tell you what I can on my way out, and I'll give you my email in case you want to ask more, but Carole's gone and this is my last day. When we couldn't get Paul's reactions to work, the boss decided to cut us loose. I'm not sure how much help I can be. And, if you want to stay on good terms with Dr. Green, you should probably keep your distance from me."

Anna Bijou is startled at this revelation. She thanks Pham for his help, then sits down with a stack of lab notebooks from Paul Fanning, Tony Pham, and Carole Lee, trying to figure out her next approach to her experiments.

By the end of the afternoon, she has discovered that Paul Fanning's notebooks are nearly incomprehensible. In contrast, Tony Pham's and Carole Lee's notebooks contain clear, detailed descriptions of multiple attempts, over many months, to reproduce Fanning's reactions. Lee's notebooks include especially exhaustive details of experimental runs and trouble-shooting as Lee tried to repeat her own limited success and to determine which piece of the experimental set-up had changed beyond the absence of Paul Fanning.

Anna Bijou plans her next experimental attempt, but she fears that Fanning's reactions may not be reproducible. She feels like she ought to communicate her concerns to Dr. Green, but she is scared by his apparent dismissal of Tony Pham and Carole Lee for their inability to get these reactions to work. What if Dr. Green responds by dismissing her, too? As she gathers her things, she takes a deep breath and hopes that her next experimental run will be more successful.

Questions

(1) Should Anna Bijou bring her concerns to Dr. Green? If so, how should she present them? If not, what else should she do instead?

(2) Anna Bijou's decision is difficult because she is uncertain how Dr. Green will react if she shares her concerns. What is the best reaction she might hope for? What is the worst reaction she might fear? How could these reactions translate into consequences for Anna Bijou, in both the short-term and the long-term?
(3) Because Dr. Green has enormous power over her scientific future, Anna Bijou might feel safest just trying to keep him happy and keeping her concerns to herself. If she chooses this course of action, who will benefit? Who will be hurt?
(4) Anna Bijou may feel that Dr. Green's response to the two students he dismissed is unethical. However, telling her supervisor she thinks he is being unethical is risky. As a student, who else could she turn to for help? Who else might have the power to assist her? Who else might have the responsibility to assist her?

Part 2

Dr. Martin Green is a recently tenured Associate Professor of Chemistry at a major research university, running a research group with eight graduate students and two postdocs. Almost two years ago, one of his first graduate students, Paul Fanning, earned his Ph.D. Fanning discovered a series of new organometallic reactions, and the journal articles that Fanning and Green coauthored on these reactions lent significant weight to Green's tenure case.

Since Fanning graduated and left the Green group to pursue a postdoc at another university, things have not been going so smoothly.

Martin Green finds himself frustrated than none of his current personnel seem to have the experimental skills or the patience to get Fanning's reactions to work. Indeed, two of the graduate students he assigned to projects that were to build off of Fanning's reactions wasted more than a year each without success. Rather than recognizing their own experimental shortcomings (or working to remedy them), they tried to excuse their failures by suggesting that Paul Fanning might not have gotten exactly the results reported in the journal articles. One of these students even suggested to Green that the results might have been fabricated (although she did not use the word).

Finally, Dr. Green ran out of patience and dismissed these students, Carole Lee and Tony Pham, from his research group. He had enough to do without having to micromanage students' experimental attempts, and their negativity was hurting morale in the lab.

Green is hopeful about the new doctoral student, Anna Bijou, who has recently joined the research group. Assigned to pick up where Fanning's projects left off, she has been working diligently at the bench, and carefully studying both the journal articles describing Fanning's reactions and the lab notebooks kept by lab members (including Fanning, Lee, and Pham) who have been working on these reactions. Bijou's experimental skills are not yet where Fanning's were, but Green is sure they will improve rapidly, given her work ethic and her determination.

Dr. Green's hope starts to fade when, at their weekly one-on-one meeting, Anna Bijou describes all of the things she has tried in her efforts to reproduce Fanning's reactions. "I have ruled out every possible problem I could think of," she tells him. "Maybe they're very finicky reactions, because I'm setting them up exactly as described and I'm not finding any trace of the products Fanning says he got."

Green expects to feel the same impatience that led him to dismiss Carole Lee and Tony Pham. Instead, he realizes that what he feels is fear. *Maybe something is wrong with the results he and Fanning published.*

Sitting across the desk from him, Anna Bijou asks Dr. Martin Green, "What would you like me to do next?" Green realizes that he is not sure what he should do next.

Questions

(1) What, if anything, should Dr. Martin Green do given his worry that something might be wrong with Fanning's published results?
(2) What responsibilities does Dr. Green have to Anna Bijou? Does the news she is sharing about the finicky reactions change any of his responsibilities to her?
(3) What responsibilities does Dr. Green have to the other personnel in his research group? To Paul Fanning? To the students he dismissed from the group? To his department? To his scientific community?

(4) Dr. Green secured tenure partly on the strength of his scientific research and publications. What negative consequences might he face if the results he published with Paul Fanning turn out to be wrong?
(5) Are there any positive consequences that could come to Dr. Green from identifying and addressing a problem with his publications on the finicky reactions?

Part 3

After many discussions with his graduate student Anna Bijou, some of them in the lab as she was conducting her experiments, Dr. Martin Green is almost certain that something is wrong with the organometallic reactions Paul Fanning said he discovered, reactions that Fanning and Green reported as coauthors. Bijou's experimental skills are very good, her lab records are exhaustive, and her logical reasoning in trying to track down the experimental problem has been impressive.

All the evidence suggests that the reactions described in those journal articles just do not work. The only question that remains for Dr. Green is whether this comes down to an honest mistake on Fanning's part or an intentional deception.

Either way, Dr. Green decides that he will have to retract those articles to correct the scientific record. And, because Paul Fanning is the first author on the articles, Green will probably have to contact Fanning about the planned retractions.

While he contemplates these unpleasant tasks, Dr. Green checks his email and finds a message from the coordinator of the department's NMR facility. When Anna Bijou had been running through possible explanations for the discrepancies between her experimental results and Paul Fanning's experimental results, she decided it was worth checking whether there had been any issues with the NMR facilities (either for her runs or during Fanning's time in the lab) that might have led to misidentification of the products. The NMR facility coordinator's email notes that the equipment was operating without problems for all of Anna Bijou's runs. It also notes that the facility has no record of an NMR account for Paul Fanning.

NMR spectra featured prominently in the journal articles by Fanning and Green on the troublesome reactions, so this news disturbs Dr. Green. He sets it aside to have a quick conversation with his lab manager about why the lab's spending lately seems to have accelerated. "Elemental analyses," she tells him. "Your grad students have been ordering elemental analyses of their products, and those aren't cheap."

"I don't remember them being so expensive a few years ago, when Paul Fanning was running all those reactions and characterizing all of his products," Dr. Green says wistfully.

"Paul Fanning never placed an order for an elemental analysis in the six years he was in the lab," the lab manager replies.

Martin Green adds this to the other pieces of bad news and wonders whether he can get through a phone call with Paul Fanning without screaming. As he imagines what he could say to Fanning, Anna Bijou knocks timidly at his door. "I found something in one of Paul's notebooks," she says. "I think it's pretty bad." She clutches an NMR spectra print-out which, when held up to the light, shows clear signs of being altered with correction fluid.

Questions

(1) Should Dr. Martin Green contact Paul Fanning directly about his plans to retract their papers? If so, should he mention these other issues? If not, what should he do instead?
(2) Have Dr. Green's responsibilities to Paul Fanning changed in light of what he has learned?
(3) What are the potential consequences of trying to handle the situation with Paul Fanning quietly? Who benefits if Dr. Green chooses this course of action? Who will be harmed, and how?
(4) What responsibility does Dr. Green have for conditions and practices in his research group that led to what now looks like fraudulent publications? Which particular practices seem to have enabled sloppy work or wrongdoing? What should Dr. Green do to prevent such problems going forward?
(5) What does Dr. Green owe Tony Pham and Carole Lee? What does the department owe them?

3. Ethical Analysis of the Case

Here, I draw on the strategy for ethical decision-making described by Muriel J. Bebeau (1995). The strategy involves identifying who has an interest in the situation where an ethical decision must be made, considering the consequences (positive or negative, short-term or long-term) that might come to those interested parties depending on what action is taken, and the obligations (responsibilities, duties) of the person trying to make an ethical decision toward each of those interested parties (including to themselves). Once this information is on the table, the task is to choose an ethical course of action. This may be the course of action that maximizes the good consequences and minimizes the bad consequences for all the interested parties in the aggregate (a consequentialist approach). Or it may be the course of action that allows the protagonist in the case to fulfill their duties to all the interested parties insofar as this is possible, recognizing that some duties are stronger than others and that obligations can conflict with each other (a Kantian approach). Or it may be the course of action that a properly functioning scientist would take in the situation, where 'proper functioning' amounts to being able to play their proper role as a knowledge-builder and a trainer of new scientists (a virtue ethics approach). If possible, it is desirable to choose a course of action that is ethical from each of these perspectives.

In this case, we are presented with two protagonists who are trying to make good ethical decisions, Anna Bijou, a graduate student working toward her Ph.D., and Dr. Martin Green, a professor who is supervising Anna Bijou's research and training. These two players have different interests, different obligations, different information about the situation that is unfolding, and vastly different amounts of power.

Anna Bijou has an interest in getting good scientific training in her Ph.D. program. As a chemist in training, she has an interest in building reliable knowledge and cultivating the skills to be a successful knowledge-builder in her scientific field. She also has an interest in cultivating the relationships that will help her further her scientific career. This includes having a good working relationship with Dr. Martin Green, her advisor. As well, Anna Bijou has interests connected to her own sense of herself, including an interest in maintaining her own integrity and in having a clear conscience.

Dr. Martin Green has many similar interests here. As a scientist, he has an interest in building reliable knowledge and in maintaining good relations with other members of the scientific community. He also has an interest in avoiding harm to his reputation and career (including his tenure), and in protecting his own integrity and conscience. But his career stage and professional role bring some additional interests. Because he heads a research group, Dr. Green has an interest in having group members who accept his guidance and respect his expertise. He also has an interest in having group members who do honest work and who communicate important information to him. He has an interest in bringing in grant money to support the work of his research group, in securing credit for his scientific findings, and in providing good training to new members of the scientific community.

Other parties besides Anna Bijou and Dr. Martin Green have relevant interests in this case. Paul Fanning is certainly an interested party. He has an interest in protecting his own reputation and career, as well as his priority claim on his scientific discoveries. He also has an interest in being treated fairly by others in his scientific community. Tony Pham and Carole Lee are also interested parties who, among other things, have an interest in some recognition of their efforts to build reliable scientific knowledge and to develop good experimental skills. The other members of Dr. Green's research group have an interest in a research environment where they can trust each other's work and they can share their concerns without fear of retaliation. Other faculty, students, and staff in Dr. Green's department and university have an interest in maintaining the integrity and the reputation of the organization. Members of the scientific community have an interest in being able to trust that what is reported in the scientific literature reflects honest efforts to build objective knowledge – and an interest in being able to trust other members of the scientific community more generally. Members of the general public also have an interest here; since public money is allocated to support scientific research, the public has an interest in that money going to good use to build reliable knowledge and to train researchers who are skilled and ethical.

In contemplating what to do, the person confronted with an ethical decision ought to consider potential consequences for the course of action they are contemplating. It is not enough to choose a path likely to result in the best outcomes for oneself. Rather, an ethical decision ought to give consideration to producing good outcomes and avoiding bad outcomes for the other interested parties as well.

In Part 1 of the case, Anna Bijou is weighing the consequences of telling Dr. Green that the reactions he has asked her to reproduce are not reproducible against the consequences of doing something else instead. The fact that she is a relatively new member of Dr. Green's research group means that she may not have much information from which to predict how Dr. Green will respond to her. Given what she has learned about his dismissal of Carole Lee and Tony Pham, though, Bijou may suspect that telling Dr. Green that she cannot get the reactions to work could lead to her own dismissal from the research group. If that happens, depending on the policies in their department and university, Bijou may need to find another faculty member with the funding and space in their research group to accept her as a student; otherwise, being dismissed from Dr. Green's research group is *de facto* being dismissed from the graduate program. Even if Bijou finds another research group willing to have her, this may require her to shift her area of research. The time and effort she has already invested in her research for Dr. Green would be lost.

Beyond the consequences for Anna Bijou in terms of her progress to her Ph.D., telling Dr. Green that she cannot reproduce Paul Fanning's experiments could have an impact on her relationship with Dr. Green. He might judge her to be a poor experimentalist, or a quitter. Given the importance of Dr. Green's assessment of her for her future as a chemist in his subfield, if he develops a negative opinion of her it could have lasting repercussions for her career.

But there are also potential consequences for Anna Bijou if she decides to keep her concerns to herself. She may engage in more futile attempts to get the reactions to work, and after that additional time and effort Dr. Green may still judge her a poor experimentalist. Perhaps she will figure out a way to make the reactions work after all. Or, faced with mounting pressure from Dr. Green to get the reactions to work, she may feel pressured to fake experimental success. This response would probably impact Bijou's conscience and, in the case that her fakery is discovered, it could well mean the end of her scientific reputation and career.

Anna Bijou's choice here will have consequences for the other interested parties as well. If she keeps her concerns to herself, Paul Fanning maintains his reputation and his career trajectory. The Green group maintains its stellar publication record. But, if there is actually something wrong with the results reported in Fanning's journal articles – if those reactions really are irreproducible – keeping her concerns to herself means that other scientists might be trying to base their own research projects on these erroneous results. That might mean that they are wasting time and effort on a dead-end – and that scarce research funding is being wasted, too.

On the other hand, if Anna Bijou shares her concerns with Dr. Green, he potentially faces a number of consequences. His relationship with Paul Fanning may suffer. His confidence in the research output of his lab, and in the decisions he has made in managing his lab personnel (including dismissing Carole Lee and Tony Pham) may also be undermined. Because of the important role his research collaboration with Fanning and the publications that resulted from it played in his tenure case, Dr. Green may worry about whether he ought to have been tenured – and whether, if Fanning's results are not reproducible, his employment status at the university might be at risk. On the other hand, if Fanning's reactions really do not work, it might be better for Green's scientific reputation if this news comes from his own research group rather than from a scientific competitor.

What are Anna Bijou's obligations as she decides what to do about her concerns? She has obligations to Dr. Green, as her mentor and boss, to perform the tasks assigned to her, to respect his knowledge, experience, and judgment, but also to share relevant information with him. She has an obligation to her department and university to uphold their standards and not to sully their reputation. She has an obligation to other members of the scientific community to play her part in building a shared body of reliable knowledge, including identifying problems that ought to be corrected with the scientific record. To the other members of the scientific community she also has an obligation to cultivate a good reputation with the public. This includes an obligation not to make baseless accusations against other scientists. Anna Bijou also owes it to the scientific community to cultivate the kind of honest communication between scientists without which the knowledge-building project is impossible. Finally, Anna Bijou has obligations to herself: to look out for her own reputation and career, not to derail her educational progress, not to undermine her integrity or to make a decision that will make it hard for her to live with herself.

Note that the focus on Anna Bijou's obligations here, in the context of her efforts to choose an ethical course of action, does not erase Dr. Martin Green's obligations! Given the details of the case, there is ample reason to believe Green has fallen down on a number of his obligations, especially on those to his trainees. However, the power imbalance means the consequences for Anna Bijou of telling Dr. Green his conduct has been unethical could be extreme. While she has an obligation to be honest, her obligations to herself and her own well-being in the short-term and long-term mean that she must be strategic.

Question

(1) Part 2 of the case shows us Anna Bijou communicating her concerns to Dr. Martin Green. Is there anything else she could have done here that would have been ethical?

Part 2 of the case shifts to the point of view of Dr. Martin Green. Green is trying to decide what, if anything, he should do about the possibility that there is something wrong with the results he published with Paul Fanning.

The interested parties are the same as those Anna Bijou might have been considering in deciding whether to communicate her concerns to Dr. Green. As she did when choosing her ethical course of action, Dr. Martin Green must now consider the potential consequences, for himself and for the other interested parties, of what he does next.

An option that may be tempting is to set Fanning's reactions aside and to shift the group's focus to devising other organometallic reactions. This would let Anna Bijou and other members of the group spend their time more productively, rather than to keep them stuck on what seems to be a dead-end. However, making this shift internal to the group without sharing his concerns with the larger scientific community could result in scientists in other research groups wasting their time on this dead-end. Without Green's intervention, his articles with Fanning may stay in the scientific record asserting that these reactions are possible when maybe they are not.

No matter what Green decides to do here, his opinion of Fanning is likely impacted. As things stand, Green may not know whether Fanning just missed an elusive factor necessary to make a finicky reaction run reliably, or whether he missed a more obvious factor due to carelessness, or whether he was dishonest about the research he conducted. If Green contacts Fanning to discuss his concerns with the reactions, Fanning might help him figure out what the problem is – or, Fanning might tell Green something aimed at protecting his own reputation rather than at getting to the bottom of the problems with the reactions.

If Green decides to investigate what is wrong with the reactions without involving Fanning, it might hurt Fanning's reputation within the research group, which could in turn impact Fanning's reputation in the wider scientific community. Keeping Fanning out of the loop could also mean that Anna Bijou and other members of Dr. Green's research group do not have crucial pieces of information about how to get the experiments to work that they might otherwise be able to get from Fanning. Then again, asking Fanning for help here will only be useful if Fanning was honest about being able to run the reactions in the first place. If the problem here is less a finicky set of reactions than a dishonest chemist, involving Fanning gives him the opportunity to interfere. (In Part 1, Tony Pham told Anna Bijou that Carole Lee was only able to get Paul Fanning's reactions to work herself when Fanning was still in the lab. If Bijou has shared this information with Dr. Green, he needs to consider what kind of conditions for a replication would convince him that the reactions – and Fanning – are reliable.)

If Dr. Green decides to dig deeper into what is going on with these reactions, and with Fanning, there could be significant consequences for Green, members of his research group, his department and university, and his scientific community, depending on what he finds. Green might get a reputation as someone whose group produces sloppy results (or worse), or he might get a reputation as an honest researcher interested in maintaining the quality of the scientific record. He might be looked upon as a scientist committed primarily to the truth, even if it means correcting findings that have come from his own research group, or he might get a reputation as someone who played favorites, to the extent that he would dismiss graduate students making reasonable attempts to correct the record if it meant protecting himself and his favorite former student. The department and university could lose some prestige for their connection to published results that do not hold up (and to Fanning, who produced them). The fact that Lee and Pham were dismissed might also discourage future graduate students from choosing this department or Green's research group. If there is a scandal around Fanning's findings, the public's opinion of these scientists – and of scientists in general – might be harmed. Potentially, this could have an impact on public funding for scientific research in the future.

As a professor, Dr. Martin Green's obligations are somewhat different from Anna Bijou's obligations. He has an obligation to Anna Bijou, and to all the graduate students in his research group, to provide good scientific training, to provide good mentoring, and to set a good scientific example. Arguably, Dr. Green has an obligation to treat his trainees as future colleagues, not simply as cheap labor to generate scientific results. Part of this is a duty to take seriously what they tell him about their experiments and the work conditions and interpersonal interactions in the lab group.

Green has also obligations to his former students. He has an obligation to Paul Fanning to respect him as a member of their scientific community. This includes not assuming wrongdoing on Fanning's part without evidence, as well as looking out for Fanning's reputation and career. Even though Fanning has received his Ph.D. and moved on from the Green group, Green arguably has a duty to continue mentoring Fanning and to set a good example of how to be a responsible scientist for him.

As well, Green has obligations to Carole Lee and Tony Pham, the graduate students he dismissed from his research group. If he dismissed them primarily because they were unable to reproduce Fanning's reactions, Green arguably has an obligation to address the harm this dismissal may have done to their education and their careers.

Dr. Green has obligations to his department and university to do research, to train students, to bring in funding, to uphold standards, and to contribute positively to their reputation. He has obligations to his research group to provide funding, safe lab conditions, guidance, and oversight, and to ensure that the group is functioning well.

Green has obligations to the journal that published his papers with Fanning. These include the obligation to make corrections or retractions if the situation warrants it. He also has obligations to the scientific community to contribute to honest knowledge-building, to train future members of the community to be responsible scientists, to help correct known errors in the scientific record, and to identify wrongdoers in the community lest they deceive other scientists.

Finally, Dr. Martin Green has obligations to himself. Among these are the obligation to protect his own reputation and career, his own integrity and conscience. Green has a duty to be the kind of scientist he knows he ought to be. This includes admitting past mistakes and addressing them.

Whatever particular course of action Dr. Martin Green elects to pursue at the end of Part 2 of the case, it is reasonable to assume he has an obligation to do what he can to get to the truth of what is happening with these chemical reactions – and with Paul Fanning.

In Part 3 of the case, Dr. Green becomes aware of new information that may shift his obligations, especially in terms of the prospects for being able to fulfill them all or having to recognize some conflicts between them. From working closely with Anna Bijou and observing her experimental efforts, he has good empirical evidence that the reactions as described in Fanning's journal articles just do not work. From his lab manager, he has evidence that Fanning did not order elemental analyses of the products of his reactions. From the coordinator of the NMR facility, he has evidence that Fanning did not have an NMR account. Finally, he has the altered NMR spectra Anna Bijou found in Fanning's lab notebook. Taken together, these facts seem to support the hypothesis that the problem is not just that the desired reactions are finicky, but that Paul Fanning may have fabricated or falsified the results he reported.

It is also worth noting that Dr. Martin Green now knows that there are other people who have reason to believe there is a problem here (Anna Bijou, the lab manager, and the NMR facility coordinator). Arguably, whether or not other people know there is a problem should not wholly determine Green's response, since *he* knows there is a problem. However, the fact that other people are aware that all is not well might well lead to consequences that are relevant. For example, students and staff members who know that something is wrong but who see the cognizant faculty member doing nothing to address the problem may become cynical and start to believe that the 'rules' of honest science can be sacrificed to get publications or to protect one's reputation. This is a potential consequence that could be corrosive to the cooperation and trust scientific communities need to build reliable knowledge.

How have Dr. Martin Green's obligations shifted by Part 3 of the case? Given the facts in evidence, the likelihood that there has been a significant problem, and that the problem is the result of misconduct, is too high to ignore. Given his relationships (with Paul Fanning and Anna Bijou, among others) and his interests (to be an honest scientist, to protect his reputation), Dr. Green needs to recognize that he may not be the right person to untangle what really happened. Green has an obligation to recognize his own biases here. On account of these biases, Green should pass on the evidence he has to someone who can be more objective, like the university Research Integrity Officer, so that they can investigate whether Fanning has committed scientific misconduct.

It's worth noting that bringing in the Research Integrity Officer is a best practice. Someone involved in the research (especially someone who has earned tenure on the basis of it) will have biases on the basis of their relationships with the other personnel and their attachments to their own scientific hypotheses and intuitions. The Research Integrity Officer is more likely to have enough distance from the matter to make a more objective determination of the facts, and has the institutional authority to report findings to the appropriate parties, including funding agencies and journals.

In light of what he knows at this point, Dr. Green has an obligation to other scientists who may consult those journal articles he coauthored with Fanning to signal that there is a problem with them. When the exact nature of the problem is known (perhaps through the Research Integrity Officer's investigation), he has a duty to be transparent about it, contacting the journal editors to communicate corrections or retractions as needed.

It may be tempting for Dr. Martin Green to blame all the problems in this case on Paul Fanning, identifying him as a bad apple in an otherwise ethical research group. However, as head of the lab, Green has an obligation to think hard about whether the way he has been running the lab may have contributed to this situation. Has he been providing enough guidance for his graduate student researchers? Has he provided too little oversight, either of experiments and data or of interactions with the NMR facility or requests for elemental analyses? Especially in large research groups, it can be tempting for the lab head to delegate oversight to a lab manager and transmission of best practices (*e.g.*, standards for keeping notebooks and other research records) to advanced graduate students. A case like this illustrates the risks inherent in that approach.

Has Dr. Green conveyed to his lab members, explicitly or implicitly, that experimental failures are unacceptable? Did he come into this research with such a strong expectation that these reactions should work that he failed to exercise sufficient skepticism about positive results, or that he was overly skeptical of negative results? Has he engaged in favoritism? Green has a duty to address any issues in his lab management style that incentivize cheating, that create barriers to honest communication, or that otherwise undercut his ability to train responsible scientists while building reliable knowledge.

How do Dr. Green's duties to Paul Fanning shift by Part 3? While Green still has obligations to Fanning, an obligation to cover up Fanning's misdeeds is not among them. Concealing misconduct would run counter to his duties to the larger scientific community. Handing the investigation over to a more objective party like the Research Integrity Officer would mean that Green understands that his duty to uphold standards of scientific integrity outweighs his duty to protect Fanning's reputation. But Green still has duties to Fanning as a scientific colleague – indeed, as a colleague he mentored and trained as a scientist. One of these may be an obligation to share his assessment of Fanning's potential to contribute to the scientific community, and to share how the lab environment and his own management style may have contributed to Fanning's bad choices, with those investigating Fanning's conduct.

In his role as a mentor to Fanning, Green arguably has a duty to hear Fanning's side of things. As well, he has an obligation to communicate to Fanning just how damaging fabrication and falsification are to knowledge-building within their scientific field – and how personally hurtful Green finds it to have been lied to and to have his reputation on the line with Fanning's. However, it is not clear that Green has an obligation to engage in this direct communication with Fanning *before* a neutral third party has had a chance to investigate the situation. Communicating with Fanning before an investigator can secure the relevant evidence might give him a chance to tamper with it or destroy it.

Finally, Dr. Martin Green needs to consider his obligations to Carole Lee and Tony Pham, the graduate students he dismissed from his lab group when they failed to reproduce Paul Fanning's experiments. Given the facts in evidence by Part 3 of the case, it is reasonable to conclude that these failures had more to do with problems with the purported reactions than with Lee's and Pham's experimental skills or scientific potential. Indeed, it now seems that they were doing exactly what individual scientists must do for science to be 'self-correcting'. They attempted to reproduce a finding, and when they could not, they shared that information with other scientists. Here, they shared the information with their lab head, Dr. Green, rather than taking it to a higher authority in the department or the university. Had they gone to a higher authority with their concerns, as 'whistleblowers' they might have been given some legal protection from retaliation, but they still might have lost their places in the Green lab. At this point, with compelling evidence that Lee and Pham were not wrong to be concerned about these chemical reactions (or about Paul Fanning), Green has an obligation to do what he can to make things right for them to the extent possible. This might mean making room for them in his lab group again, or using his professional network to help find them reasonable educational or employment situations elsewhere. It might also involve publicly acknowledging their role in identifying the problems with reactions he and Fanning reported in the literature. To fulfill his obligation to make things right for Lee and Pham,

Green should probably start by *asking* them what kind of help they need from him.

It is important to note here that any ethical path forward in this case will involve some negative consequences for Dr. Martin Green. There are some options that might spare Green some negative consequences in the short-term, but these are likely to bring him negative consequences in the longer term. Plus, these options have a good chance of bringing more bad consequences for most of the other interested parties. Even Paul Fanning may be better off in the long run if Martin Green intervenes now and argues for the possibility of rehabilitation. If Fanning continues on and racks up a record of more frauds before being caught, the prospects of mercy for him are slim.

4. Some Comparable Cases

Some key details of this case are inspired by the case of Bengü Sezen and Dalibor Sames in the Chemistry Department at Columbia University. The press coverage of the case began, in 2006, with a set of retracted papers and a dispute between Sames, the senior author, and Sezen, his former graduate student, about whether the papers ought to have been retracted, as well as about whether the experiments reported in those papers were reproducible (Chang 2006a/b). By the time the findings of the United States Department of Health and Human Services on the matter were published in the Federal Register in 2010, Sezen had been found guilty of falsification, fabrication, and plagiarism of research data in three published papers and in her doctoral dissertation. As well, the investigation conducted at Columbia University found that Sames asked two graduate students who had devoted significant time and effort to attempts to replicate and extend Sezen's work to leave his group (Schulz 2011).

There are other cases involving disputes about whether a finding is reproducible in which misconduct is suggested but not proven. An interesting example of this sort is the case of Duke University biochemist Homme W. Hellinga and his graduate student, Mary Dwyer, who coauthored, and then retracted, a pair of papers on enzymes designed using computational methods. According to news coverage around the retractions, Dwyer was concerned that her experimental results were too variable to be ready for publication, while Hellinga thought that the amount of variability they were seeing was normal for this type of system. (The published papers, however, failed to note this experimental variability.) When other researchers tried repeating these experiments and found no enzymatic activity from the designed enzymes, Hellinga at first assured them that the experiments worked, and that he knew this because the Hellinga lab had run a number of negative controls. Later, according to Dwyer, Hellinga confronted her and said "I find it really hard to believe that you didn't make this up" (Hayden 2008, p. 277). Hellinga retracted the papers, but other researchers remained skeptical that the results reported in them could have been produced using the assays the papers described (Arnaud 2008, p. 41). An inquiry at Duke University cleared Dwyer of the allegations of falsification and fabrication of results. The Hellinga case raises a question that is also central to *The Case of the Finicky Reactions*, namely, how much responsibility does the senior researcher have for scientific work co-authored with his graduate student?

The differentials in experience and power between graduate students and principal investigators can complicate the kind of communication that is essential for knowledge-building. It is hard enough to share the news that one's experimental efforts have been unsuccessful. It is even harder to confront one's supervisor with concerns about misconduct. A number of real-world cases of scientific misconduct have come to light because graduate students or postdocs decided to be whistleblowers, reporting problems in their research groups so they could be addressed. Among these is the case of Diederik Stapel, a social psychologist forced to retract more than 50 publications because he fabricated the results that they reported. Some of his graduate students, concerned about what seemed to be anomalies in experimental results, asked Stapel if they could examine the raw data. When told he no longer had the raw data, they became suspicious but feared that reporting their suspicions would be damaging to their careers. Eventually, Stapel's students were able to persuade his department chair that something was amiss, which ultimately led to Stapel's dismissal (Bhattacharjee 2013).

Another case illustrates the high costs for students of blowing the whistle on an advisor's misconduct. Students of Elizabeth Goodwin in the genetics department at the University of Wisconsin-Madison brought their concerns about experimental data and manipulated figures in Goodwin's grant proposals to their department chair. This resulted in a university inquiry that found Goodwin had falsified data in proposals, after which Goodwin resigned her post. Of the six students who brought forward these concerns, three left the Ph.D. program without their degrees, and two others were starting over in new graduate programs, essentially losing the years that they had invested as students in Goodwin's lab (Couzin 2006, p. 1222).

5. Conclusion

Determining whether research results are reproducible is a crucial piece of scientific knowledge-building, one that requires experimental skill, good record keeping, and a thoughtful approach to problem solving. Questions of reproducibility also involve the competence, and the honesty, of other researchers in generating and sharing their results. Scientific knowledge-building takes place in the context of competition between scientists for credit and for resources like funding and jobs. The stakes of the competition make it harder for scientists to be critical of their own positive research findings – especially once they have been published – which, in turn, makes it harder to correct the scientific record when errors creep in.

Especially in research environments that are also training environments, researchers with less power may fear significant negative consequences when they share negative findings or research difficulties with researchers with more power. Their results may be dismissed, or they themselves may be labeled as incompetents or troublemakers. Yet part of learning how to be a responsible scientist is working out how to share information, good or bad, that is relevant to the research with one's supervisor and research team.

This case explores how power disparities complicate the task of sharing information within a research team. From the perspective of graduate student Anna Bijou, we see the challenge of addressing irreproducible results with a lab head who seems to have reacted quite negatively to other students in the same situation. From the perspective of lab head Dr. Martin Green, we see a web of competing interests and obligations, a series of past decisions that have gone badly and the question of how to make things right going forward. Good knowledge-building requires awareness of where you might be mistaken, whether about results or people. Practices that make it less dangerous to recognize and address mistakes are vital.

Further Reading

On the Sezen/Sames case see particularly Chang 2006a/b and Schulz 2011. The Hellinga case is covered by Arnaud 2008 and Hayden 2008; the Stapel case by Bhattacharjee 2013; and the Goodwin case by Couzin 2006. On whistleblowing see Gunsalus 1998; on the responsibilities of mentors, see Weil 2001. More details on misconduct and rehabilitation can be found in Stemwedel 2014 and Djerassi 1991.

References

Arnaud, C.H.: 2008, 'Enzyme Design Papers Retracted', *Chemical & Engineering News*, **86**, 40-41.

Bebeau, M.J.: 1995, 'Developing a well-reasoned response to a moral problem in scientific research', in: M.J. Bebeau, K.D. Pimple, K.M.T. Muskavitch, S. L. Borden, D.H. Smith & E. Agnew, *Moral reasoning in scientific research. Cases for teaching and assessment*, Bloomington: Indiana University, pp. 13-18.

Bhattacharjee, Y.: 2013, 'The mind of a con man', *The New York Times Magazine*, 26 April, available at: http://www.nytimes.com/2013/04/28/magazine/diederik-stapels-audacious-academic-fraud.html, accessed 10 November 2016.

Chang, K.: 2006a, 'Ex-Columbia Student Says Disputed Chemistry Research is Sound', *The New York Times,* 18 March, available at: http://www.nytimes.com/2006/03/18/science/18chem.html, accessed 10 November 2016.

Chang, K.: 2006b, 'Columbia Chemistry Professor is Retracting 4 More Papers', *The New York Times,* 15 June, available at: http://www.nytimes.com/2006/06/15/ science/15chem.html, accessed 10 November 2016.

Couzin, J.: 2006, 'Truth and Consequences', *Science*, **313**, 1222-1226.

Djerassi, C.: 1991, *Cantor's Dilemma: A novel*, New York: Penguin.

Hayden, E.C.: 2008, 'Designer Debacle', *Nature*, **451**, 275-278.

Schulz, W.G.: 2011 'Reports detail a massive case of fraud', *Chemical & Engineering News*, **89**, 4.

Stemwedel, J.D.: 2014, 'Life after Misconduct: Promoting Rehabilitation while Minimizing Damage', *Journal of Microbiology & Biology Education*, **15**, 177-180.

Weil, V.: 2001, 'Mentoring: Some ethical considerations', *Science and Engineering Ethics*, **7**, 471-482.

Zigmond, M.J. & Fischer, B.A.: 2002, 'Beyond fabrication and plagiarism: The little murders of everyday science', *Science and engineering ethics*, **8**, 229-234.

Chapter 3

Ethics of Chemical Weapons Research: Poison Gas in World War One

Joachim Schummer

Abstract: This chapter first provides a brief narrative of the research, development, and deployment of poison gases in WWI as well as of the subsequent history of chemical warfare and international conventions to ban it. Because chemical weapons research is still allowed by national and international laws, and indeed widely conducted, it is a primary case for ethical investigation. The analysis shows that chemical weapons research is morally wrong by all major ethical theories, *i.e.* by both utilitarianism and deontology. That conclusion has frequently been blurred by confusions, such as between patriotism and ethics and between weapons research and deployment, which are clarified. The chemical communities who have honored the heroes of poison gas research seem to disregard ethics in their honoring system.

1. Introduction

In WWI thousands of academic and industrial chemists, including leading figures of all major belligerent countries, voluntarily and eagerly engaged in the research, development, large-scale production, or deployment of the first weapons of mass destruction, poison gases. That was the largest ever engagement of scientists in warfare until then, which left an enduring mark on the public image of the chemical profession (Schummer 2019).

It would seem that, after one hundred years, chemists would have drawn ethical lessons from that experience and openly condemn any involvement in weapons research. After all, the Chemical Weapons Convention (CWC) bans both the use and possession of these weapons, such that the issue appears to be only of historical importance. But this is not the case for several reasons.

First, the CWC explicitly allows chemical weapons research with amounts of poisons that could theoretically kill almost a billion people. In fact, at least eighty countries reportedly operate such research units (see below). Military research has long been the largest single area of governmental research funding in many countries. For example in the heyday of the Cold War about two thirds of the US R&D budget for science went into that area.[1] Chemical weapons, all prepared by chemists, still continue to be deployed, such as in Syria and in terrorist attacks or assassinations. Second, the CWC has a very narrow definition of chemical weapons that is focused on specific physiological effects and excludes, for instance, napalm, which was used in firebombs to intentionally kill hundreds of thousands of civilians by suffocation or burning during WWII, the Korean War, and the Vietnam War. The chemical arsenal of weapons is therefore much richer than poisons and includes all chemical preparations that are produced for the purpose of killing, harming, or threatening people.

Third, almost all ethical debates on chemical weapons have focused on deployment and ignored research. However, the one-time deployment of a weapon and the creation of an entirely new weapon to be used in the future and by any party are two fundamentally different actions that require different ethical assessment. The confusion has left an ethical vacuum on chemical weapons research that the professional societies of chemistry have hardly addressed, despite their efforts at writing codes of conduct.

This chapter first provides a brief narrative of the research, development, and deployment of poison gases in WWI (Section 2.2) within the general setting of the war (Section 2.1). Then I take the best researched case of Fritz Haber in Germany to illustrate the complex interaction between academia, industry, military, and government in chemical warfare (Section 2.3). The next two sections survey the aftermath: first I summarizes the research and deployment of chemical weapons up to now, despite numerous international conventions (Section 2.4); then I

[1] For data, see the OECD database (http://stats.oecd.org).

look at how the scientific communities and their honoring systems have dealt with the protagonists of chemical weapon research (Section 2.5). After a brief introduction to ethics (Section 3.1), I assess chemical weapons research both from utilitarian and deontological theories (Section 3.2) and analyze various standard excuses by weapons researchers (Section 3.4), before drawing some general conclusions (Section 4).

2. Historical Narrative

2.1 The 'Great War'

After a 19-year old Bosnian-Serbian separatist assassinated the Archduke of Austria in Sarajevo on 28 June 1914, which triggered the First World War, almost 70 million soldiers worldwide were mobilized. A month after the assassination Austria-Hungary declared war on Serbia; Russia immediately sided with Serbia and asked France for help; Germany allied with Austria-Hungary and invaded on their path to Russia and France both Poland and neutral Belgium, which made the United Kingdom declare war on Germany; and so on. In the course of the war, which lasted from 28 July 1914 to 11 November 1918, about 40 countries, as well as their colonies and dependencies were involved. These included Japan, the Ottoman Empire (Turkey), Italy, Romania, Bulgaria, China, USA, Greece, Siam (Thailand), and Brazil, with major battlefields in Europe, the Middle East, Africa, and East Asia, as well as in the Atlantic, Pacific, and Indian Oceans. About 9 million soldiers died, several millions remained missing, and more than 21 million were injured, while the many millions of killed and wounded civilians worldwide have been left uncounted. In addition, the new colonial structures built during and immediately after the war turned into internal conflicts, from which particularly the Middle East, which was created by a secret British-French agreement from remains of the Ottoman Empire, has never recovered.

Like no prior war, WWI was an experimental battlefield for new weapons that drastically changed warfare, upending the received military knowledge. While war propaganda still upheld the face-to-face combat and particularly the cavalry – the heroic horsemen fighting each other with sabers – soldiers in muddy trenches faced machine gun fire and tanks that overran and buried them alive. Other novel weaponry included submarines attacking battleships and coastal cities unnoticed; airplanes that suddenly appeared and fired with cannons and machine-guns into the crowds from above; shells filled with high explosives that killed or injured anybody in the surroundings of its explosion and which were shot from several kilometers away and noticed too late to protect oneself; and poison gases.

2.2 Chemical warfare in WWI

Before WWI, a series of peace conferences initiated by Tsar Nicholas II of Russia had resulted in the first ever international treaties on the conduct of warfare, called the Hague Conventions of 1899 and 1907, after the city in the Netherlands where the treaties were signed and deposited. The articles defined by the conventions include rules of how to avoid, judge, and end wars; how to deal with civilians in combat zones, with merchant ships, neutral countries, and prisoners of war. They further specify which kind of weaponry and what kind of deployment should be considered a war crime. According to the Hague Convention of 1899 (IV.2) "the use of projectiles the object of which is the diffusion of asphyxiating or deleterious gases" was prohibited.[2] The second Hague Convention of 1907 more generally added the use of "poison or poisoned weapons" (IV, annex, 23).[3] Before the beginning of WWI all major belligerent countries ratified these parts of both Hague Conventions, except the USA that has never signed the first one.

In the course of the war, all major countries violated numerous clauses of the Hague Conventions, famous cases include the German air raids on Belgian, French, and English towns and the sinking of the ocean liner Lusitania. As to chemical weapons, perhaps the first violation of the Hague Conventions was the use of tear 'gas' grenades (ethyl bromoacetate) by the French military in August 1914 against the Germans.

---

2 See http://avalon.law.yale.edu/19th_century/dec99-02.asp.

3 See http://avalon.law.yale.edu/20th_century/hague04.asp#art23.

However, the French could have argued that ethyl bromoacetate is not a gas, as the convention required, but a liquid at room temperature that was dispersed by the explosion of the grenade. The Germans in turn developed grenades filled with the comparably toxic, and non-gaseous, tear 'gases' dianisidine chlorosulphonate, first used in October 1914 against the British in Northern France, and xylyl bromide, first shot on Russian troops in Poland in January 1915. The British likely used cloroacetone hand grenades since early April 1915. All these deployments had little to no of the desired military effects though. They were preliminary experiments in both chemical weapons development and in the transgressions of boundaries of international law, opening the door to an arms race.

Although the distinction between 'asphyxiating or deleterious' substances and 'poisons' is a matter of debate (and concentration, of course), the tear 'gas' grenades violated the first Hague Convention in that they employed projectiles to spread deleterious substances. The next transgression was a move to more toxic gases without the use of projectiles, a clear violation of the second Hague Convention, but not of the first one. On 22 April 1915 near Ypres in Belgian, the German military under the scientific supervision of the physical chemist Fritz Haber released about 168 tons of chlorine gas from a battery of more than 2,000 conventional gas cylinders making use of the wind that transported the toxic cloud to the lines of the unprepared British and French enemy, which caused hundreds of dead and probably thousands of wounded soldiers. Several chlorine gas attacks in the same manner followed during the next weeks in this Second Battle of Ypres, which brought only a small and temporary advantage for the Germans. After these first 'demonstrational experiments', and still in May 1915, Haber moved with now tens of thousands of chlorine gas cylinders to the eastern front at Bolimov, Poland, repeating the gas assaults on Russian soldiers on a much larger scale, causing thousands of deaths.

In response the British and French developed both effective gas masks and their own chemical weapons. In September 1915 the British launched their first gas attack with ethyl iodoacetate grenades and chlorine gas after Haber's model in Loos, France, but under wind conditions that caused many casualties to their own troops. The French, who also had a strong chemical weapons program under the guidance of Charles

Moureu, filled grenades and shells with more toxic substances, introducing, among others, ethylsulfuryl chloride (June 1915), iodoacetone (August 1915), perchlormethylmercaptan (September 1915) benzyl iodide, chloroacetone (November 1915), and the very toxic hydrocyanic acid combined with arsenic trichloride (July 1916) into the chemical weaponry. Haber's team continued to switch between the western and eastern fronts and worked on various methods to deploy poisons. On one hand, they perfected the gas cloud approach by filling cylinders with the much more toxic but less volatile phosgene, propelled by admixtures of chlorine gas, probably first tried out in May 1915 on the eastern and then in December on the western front, which was soon copied by the British. On the other, they developed grenades and shells fired from mortars, filled with poisons, including methylsulfuryl chloride, chlormethylchloroformate, dimethyl sulfate, chloropicrin, diphosgene, thiophosgene and the notorious mustard gas (dichlorethylsulphide, first used in July 1917) as well as various arsenic compounds (phenyldichloro-, ethyldichloro-, phenyldibromo-, diphenylchloro-, and diphenylcyanoarsine). Although gas cloud releases from cylinders continued over the war, gas grenades and shell fillings became the dominant form of deployment, such that in the final years of the war most shells contained a poison. For instance, in the Battle of Messines (7-14 June 1917) the British bombarded the Germans with 75,000 chloropicrin shells; during the Third Battle of Ypres (31 July – 10 November 1917) the Germans fired about 50,000 mustard gas shells against the British, each shell filled with several kilograms of poison (Freemantle 2014, chap. 1).

In addition to Germany, France, and the UK, Austria-Hungary, Russia, Italy, and the US developed, produced, and deployed chemical weapons during WWI, although on a smaller scale (Freemantle 2014, p. 197). They thus all committed war crimes according to the Hague Conventions, which included no exemption for retaliation. In general, successful inventions by one party were soon copied by the others. However, detailed historical research on Austria, Italy, and Russia is still poor. Austria-Hungary employed various tear gases in grenades almost from the beginning of the war, released phosgene from cylinders since June 1916, and introduced new poisons such as the very toxic cyanogen bromide (September 1916) which is easily absorbed by the skin, making gas masks useless (Rauchensteiner 2014, pp. 542-545). Russia seems to have deployed chemical weapons early on too, and massively so since the Brusilov offensive in June 1916, whereas Italy seems to have embarked, despite rumors, on chemical warfare quite late (Zecha 2000). When the US entered the war in April 1917, they were well equipped with and used many of the aforementioned poisons and developed new ones, such as the highly toxic Lewisite (chlorvinyldichloroarsine), which they planned to spray from military aircraft on German cities, but shipped it to Europe only shortly before the armistice (Irwin 1921, pp. 37-42, Freemantle 2014, p. 197).

The chemical war induced a two-fold arms race of developing both more effective poisons and protective devices, of which the gas mask was the most important one. It was not only a defensive tool, as one might think, but also an important offensive one, particularly in trench warfare: while the enemy line was bombarded with poison grenades, shock or assault troops with gas masks advanced to conquer the poisoned territory. Thus the gas mask had to be effective against both the own poisons and those of the enemy, which were regularly analyzed by chemists at the frontline. As the number of chemical weapons increased, soldiers had to be equipped with a battery of filters, each effective only against a certain group of chemicals. The art of chemical weapons research consisted both in finding ways to circumvent the protective devices of the enemy and in being ahead in the development of new devices for the own troops. Initial efforts focused on chemicals that standard filters did not easily absorb, which multiplied the types of filters. A second approach employed some of the previously mentioned arsenic compounds (known as Clark I, Clark II, Clark III, and Adamsite) in the form of fine powders or aerosols that passed through all of the known filters. Because these substances are not lethal but only irritating, they forced soldiers to remove their gas mask and thereby made them vulnerable to phosgene and other lethal gases that were shot at the same time. A third approach, exemplified by mustard agent, consisted in liquids that penetrate cloth and poison the victims through skin contact, which made the gas mask useless.

Overall, about 150,000 tons of some 48 different poisons were produced during the war by chemical factories in Germany, France, UK,

USA, Austria, Italy and Russia, in descending order.[4] It is estimated that about 300,000 soldiers were killed and about one million injured by chemical weapons, but these numbers are questionable for several reasons. First, shells, the most deadly weapons of WWI, likely killed more than 5 million soldiers, and it is difficult to distinguish if they actually died from shrapnel or from the poisons that many shells were filled with. Second, many poisons develop their lethal effects after days, weeks, months, or even years. For instance, phosgene ($COCl_2$), the most lethal chemical weapon of the war, recognized only by its hay odor, reacts with water in the respiratory tract to form hydrochloric acid, which dissolves the lung tissue over several days. The oily liquid mustard 'gas' ($(ClCH_2CH_2)_2S$) causes severe blisters about 24 hours after skin contact and was the most effective non-lethal agent, but could wield its damage many days after the attack via incidental contact with a contaminated surface. In addition, because of its strong carcinogenic effect, mustard gas can cause lethal cancer after years. Third, various endemics, including tuberculosis, typhus, syphilis, shigellosis, cholera, and malaria affected large numbers of troops and particularly killed those whose health conditions were previously weakened by injuries or poisoning. Fourth, it is difficult to imagine that military hospital personnel had sufficient time and know-how to determine if the poisons were the ultimate cause of death. Fifth, casualty statistics poorly cover battlefields in Eastern and Southern Europe and in the Middle East, where chemical weapons were deployed on victims who rarely had protective devices.

2.3 Haber's academic-industrial-military-governmental complex

To understand the role of scientists in chemical warfare, I focus on the best studied case, Fritz Haber's poison gas project in Germany. However, because chemical warfare was rather new, similar contexts for the involvement of chemists might be found in other countries.[5]

At the beginning of the 20th century the kind of state-funded scientific research institutes that exist today were largely unknown. One of the

[4] Freemantle 2014, p. 197, on the industrial production see Johnson 2017.

[5] This section mainly draws on Szöllösi-Janze 1998 to which the following page references refer if not otherwise indicated.

first was a private-public partnership between the wealthy banker and entrepreneur Leopold Koppel and the state of Prussia that led in 1911 to the foundation of the Kaiser-Wilhelm Society in Berlin, which later became the Max-Planck Society that nowadays runs 84 publicly funded research institutes. Initially there were only two in adjacent buildings: the KW Institute for Chemistry and the KW Institute for Physical Chemistry and Electrochemistry, of which Fritz Haber was the founding director.

Haber had already made himself a name by his considerable improvement of catalytic ammonia synthesis from hydrogen and nitrogen, in close collaboration with the chemical company BASF. Another collaboration, with the Berlin firm Auergesellschaft, which produced gas lantern mantles for street lights and which was owned by Koppel, was crucial to his appointment (Szöllösi-Janze 1998, 215f.). When Haber moved from the then still provincial Technical University of Karlsruhe to Berlin, he quickly connected with the political and academic establishment, including physical chemist Walther Nernst and organic chemist Emil Fischer at the University of Berlin.

At the beginning of the war these three chemists offered their scientific service to the government. Fischer, the leading German chemist of the time, had excellent connections to government and industry, and became the primary advisor in all war related chemistry matters. Nernst, together with Carl Duisberg (the CEO of the chemical company Bayer), was commissioned to research and develop tear gas grenades, which were first deployed in October 1914 (see above). The little military success of the tear gas grenades made the factual commander-in-chief of the German army, Erich von Falkenhayn, change his mind in favor of lethal poisons. Haber, who became scientific advisor in the war department, suggested the use of chlorine, a side-product from his ammonia synthesis, and elaborated a plan for its deployment to which Falkenhayn agreed. After the new weapon was, under Haber's supervision, successfully tested in the Battle of Ypres on 22 April 1915, his KW institute became a well-financed center for chemical warfare research (pp. 337 ff.).

From a small research institute with less than 20 employees in 1914, the KW institute turned into a 'Big Science' center that employed close to 2000 people in 1918 (pp. 263, 348 ff.). About 150 scientists – including many who would later became famous, such as Otto Hahn, Heinrich Wieland, James Franck, and Gustav Hertz – worked on the research, development, training, and scientific supervision of the deployment of chemical weapons on the battlefields, as well as on the further development of gas masks, for which Koppel's Auergesellschaft produced the filters. The largest part of the technical staff was responsible for the control of the gas mask production. Also the adjacent KW Institute for Chemistry, under the directorship of Richard Willstätter, was soon incorporated into the gas mask program. In addition, various other institutions were attached to Haber's institute, including soldiers training centers, weapons testing grounds, and shell filling stations with thousands of workers. Fischer and Nernst, who were both on the scientific board of the institute, served as advisors. In total, about 1,000 German scientists eventually worked for chemical warfare, most of them in industry.

New as it was, chemical warfare required building up new convictions, networks, and responsibilities within the established social structure. Haber's role was to fill that gap in any regard (pp. 332 ff.). First, in order to establish a link to the military, the professor of chemistry voluntarily enrolled in the army as a captain to supervise chemical attacks in the battlefields. Second, Haber could rely on his pre-established industrial connections and established many new ones with the chemical industry that produced the poisons. Third, he was appointed to a post in the war department that controlled the industrial support of all war relevant chemicals. Together with the directorship of his research institute, Haber was thus the crucial figure connecting academia, the military, industry, and government, and therefore, almost single-handedly, established for the first time what was only much later called the academic-industrial-military-governmental complex.

In the final years of the war Haber undertook various efforts to save his poison gas project, and his complex social network, for peacetime activities (pp. 419 ff.). He succeeded in transforming a portion of the project into a research facility for pest control (pp. 452 ff.). A particular irony of the history is that this facility in 1922 developed under the supervision of Haber, who was of Jewish descent, Zykon B, the main poison used in the gas chambers of the Holocaust two decades later (pp. 462 ff.).

After the war, Haber never denied that he was responsible for the German chemical weapons R&D program – although, of course, Erich von Falkenhayn and Kaiser Wilhelm II, commander-in-chief of the army by German constitution, were responsible for the weapon deployments. However, Haber's notion of ethical responsibility was limited in several regards. First, he argued that he had never cared about the Hague Convention and its interpretation because that had been Falkenhayn's responsibility (p. 326). Second, he was convinced that in times of war ethical standards are to be replaced by patriotism, such that warfare engagement becomes a moral duty for scientists (p. 428). Third, he was fully aware that his weapons program initiated an arms race among the enemies, a systemic force that, once put into action, all sides had to follow if they did not want to lose the war (p. 332). We will come back to these strategies of diminishing the responsibility of scientists in Section 3.3.

2.4 Aftermath I: Chemical weapons deployments and international treaties

Chemical weapons deployment did not stop after the official end of WWI. Many of the major hostile countries moved their forces to Russia to interfere in the Russian Civil War, where at the least the British (1919) and the Red Army (1921) reportedly used poison gas. Furthermore, several European countries used chemical weapons of mass destruction, dropped from air planes, in their efforts to control and extend their recently acquired colonies in the Middle East and North Africa. These likely include the British use of arsenic compounds against Arab and Kurdish rebellions (1920); Spanish massive air bombardment of Northern Morocco with mustard agent and other lethal chemicals (1923-26); and large scale deployment of mustard agent and other toxins in Libya (1928) and Ethiopia (1935-39) by the Italians. Moreover, during the Second Sino-Japanese War (1937-1945), which turned into WWII, the Japanese employed a battery of chemical weapons, including mustard agent and lewisite, as well as biological weapons (fleas infected with Bubonic plague) to kill Chinese civilians and troops.

In the late 1930s industrial chemists at IG Farben in Germany, while allegedly seeking insecticides, discovered the phosphate esters tabun and sarin, a new generations of extremely toxic nerve agents. Nobel Laureate Richard Kuhn soon started a chemical weapons research program and in 1944 discovered the even more toxic nerve agent soman. Nerve agents interact with the metabolism of neurotransmitters in the synaptic cleft, particularly by blocking the enzyme acetylcholinesterase that breaks down the neurotransmitter acetylcholine after successful neural signal transfer, resulting in permanent neural signals, muscle contraction, and a quick death. Although Germany produced large stockpiles of sarin and tabun for military purpose from 1940 onward, they did not deploy any chemical weapon in WWII, nor did the Allied Forces. However, after the Russian and US armies discovered the nerve agents in occupied Germany, a new chemical arms race began and numerous countries mass-produced and stockpiled nerve agents. In 1952 British scientists at ICI, again presumably searching for pesticides, discovered another new class of nerve agents among thiophosphonates, the so-called V-agents, of which particularly VX was soon manufactured for military purpose in the US and elsewhere. From approximately 1960 onwards the Russian chemical weapons program developed the Novichok nerve agents, which included a large number of organophosphate compounds, some of which are probably more toxic than VX.

Many countries replicated the chemical armament of the Cold War enemies, albeit on a much smaller scale, including those that had suffered chemical attacks by colonial powers during or after WWI. And some would use it. For instance, when Egypt interfered in the Yemeni Civil War (1962-67), they reportedly dropped phosgene and mustard aerial bombs and likely nerve agents (sarin). During much of the Iran-Iraq war (1980-1988) Iraq extensively used chemical weapons, particularly mustard agent and tabun, against Iranian and Kurdish soldiers and civilians, with tens of thousands of casualties. The latest confirmed deployments of chemical weapons have occurred during the still ongoing Syrian Civil War (since 2011) and included chlorine, mustard agent, and sarin.

Shortly after WWI international negotiations began about a more precise and efficient ban on chemical weapons, eventually resulting in the

Geneva Protocol of 1925 that prohibited the "use in war of asphyxiating, poisonous or other gases, and of all analogous liquids, materials or devices".[6] However, it is difficult to assess the actual effect of this treaty. By 1939 it was ratified by 40 parties, including almost all European countries, who indeed abided by the treaty during WWII. On the other hand, Italy and Egypt (who both had ratified it by 1928) have never been prosecuted for their war crimes. UN sanctions against Iraq and Syria, who ratified the treaty in 1931 and 1968, respectively, were either inefficient or blocked by members of the Security Council. Japan ratified the treaty only in 1970; and the US, who had still deployed tear gases during the Vietnam War, as late as 1975.

In 1992, following the Treaty on the Non-Proliferation of Nuclear Weapons from 1968 and the Biological Weapons Convention of 1972, the more ambitious Chemical Weapons Convention (CWC) was signed. Effective since 1997 this treaty prohibits, with unprecedented scientific and legal clarity, the use of chemical weapons, including tear gases and temporarily incapacitating chemicals. Moreover, it also prohibits the development, stockpiling, and transfer of chemical weapons, and requires that all member states declare and destroy by April 2012 their existing stocks and production facilities specific poisons. The CWC allows for the supervision and unannounced inspections by the Organization for the Prohibition of Chemical Weapons (OPCW) that was established for that purpose.[7] By 2017 all members of the UN have ratified the treaty, except Egypt, Israel, North Korea, Palestine, and South Sudan. Nine other countries have declared chemical weapons stocks and production facilities. Of those countries the US and Iraq were, in October 2017, still behind the schedule in the destruction of their stockpiles, and at least Syria appears to have originally made wrong declarations.[8]

[6] For the text and the dates of signatures and ratifications, see http://disarmament.un.org/treaties/t/1925.

[7] See https://www.opcw.org/chemical-weapons-convention/.

[8] The nine countries are Albania, India, Iraq, Japan, Libya, Russia, (presumably) South Korea, Syria, United States. Japan, which declared abandoned chemical weapons from WWII located in China, is also behind schedule. Syria declared fulfillment of the destruction in August 2014 after which numerous chemical weapons deployment have been confirmed.

Table 1. Selected toxicological data for some warfare poisons, based on rat experiments if not otherwise indicated; in approximate order of toxicity, but note the different values for different routes. Toxicological information is still very poor, not always comparable, or not available for many WWI poisons. Source: https://chem.nlm.nih.gov/chemidplus.

	boiling point (°C)	**LD_{50}, skin (mg/kg)**	**LD_{50}, subcutaneous (mg/kg)**	**LC_{50}, inhalation (ppm)**	**LC_{Lo}, inhalation (ppm)**
Chloroacetone	119	141^{b}		262 (1 h)	
Chlorine	-34.0			137^{3} (1 h)	800^{d} (30 min)
Arsenic trichloride	130	80			200^{e} (20 min)
Phosgene	8.2				190^{e} (15 min)
Hydrocyanic acid	26		3.0^{c}	160 (30 min)	
Chloropicrin	112			$111^{a,b}$ (20 min)	
Phenyldibromo arsine	265	15			
Mustard agent	216	5	1.5	$40.0^{a,b}$ (10 min)	
Lewisite	197	15	1.0		6.0^{f} (30 min)
Sarin	147	2.5	0.103	$0.81^{a,c}$ (30 min)	
Soman	198	7.8	0.071	$0.13^{a,c}$ (30 min)	
VX	300	0.25^{b}	0.012		

LD_{50} and LC_{50} are the lethal dose per kg body weight and the lethal air concentration, respectively, which kill 50% of a sample of test animals, differentiated by the species of the animal, the uptake route (skin absorption, subcutaneous, oral, inhalation, *etc.*), and exposure time in case of LC_{50}; LC_{Lo} is the lowest concentration reported to have caused the death of certain animals after a certain exposure time.

^{a} converted from mg/m^{3}; ^{b} rabbit; ^{c} mouse; ^{d} dog; ^{e} cat; ^{f} human.

Of course, the CWC had to consider numerous exceptions to their requirements for declaration and destruction, such as chlorine gas that is widely used in the manufacture of various civil chemicals. Therefore, the treaty includes both criteria and an explicit list of toxic chemicals divided up into three classes, according to their toxicity and the use in other areas, for which different declaration requirements apply.

In the present context, the most important exemption is that the CWC does not prohibit chemical weapons research, but only requires annual reports on small scale production and research facilities. In 2015 eighty countries submitted such declarations, of which 23 referred to the highest toxicity class (OPCW 2016, p. 8). For research purposes ("medical, pharmaceutical, or protective") poisons, such as sarin and VX, can be produced and stockpiled up to an amount of one metric ton (Verification Annex, Part VI, A.1). Note that, based on the LD_{50} of Table 1, it is theoretically possible to kill almost one billion people with one ton of VX.

2.5 Aftermath II: Post-war honors for chemical warfare scientists

In August 1919, shortly after the formal peace Treaty of Versailles was signed, Fritz Haber escaped to neutral Switzerland, fearing the persecution by the Allies for war crimes. Two months later he received a message from Sweden that he would be awarded the Nobel Prize for chemistry, retrospectively for 1918, for his achievements in ammonia synthesis. However, by then Haber's invention had not much been used for the manufacture of fertilizers, which only became feasible and cheap by massive catalytic improvements that earned the former CEO of BASF, Carl Bosch, the Chemistry Nobel Prize as late as 1931. Instead, by 1918, ammonia by the Haber process was mostly used (via nitrogen dioxide and nitric acid to react with various aromatics) for the large scale production of high explosives for shells, such as trinitrotoluene (TNT), trinitroglycerin (TNG), and nitrocellulose. Moreover, in order to obtain hydrogen for ammonia synthesis (by electrolysis of aqueous NaCl solutions, the chloralkali process) equal amounts of chlorine were produced to be used as poison gas. Thus, immediately after the war, the Nobel committee honored the crucial chemical reaction that enabled the mass production of both high explosives and poison gas – a cynical prize for chemical warfare as contemporary critics called it.

By many other Nobel Prizes, which were already then considered the highest international awards in science, the Swedish Academy honored major figures in German chemical weapons research and development (Van Der Kloot 2004). Richard Willstätter, head of the national gas mask research unit, received the Nobel Prize for chemistry already in 1915. Walther Nernst, who like Haber escaped after the war out of fear first to Sweden and then to Switzerland after he had sold his estate in Germany, was awarded the same prize in 1920. Haber's most talented recruitments of his poison gas team, Gustav Hertz and James Franck, were the physics Nobel Laureates of 1925. Otto Wieland, the German co-father of mustard agent and Adamsite, won the chemistry prize in 1927. Otto Hahn did so only in 1947 after his co-discovery of nuclear fission had been developed into the 'atomic bomb'.

Despite notable exceptions, starting with Nobel Laureate Hermann Staudinger during and after WWI, the scientific community has never seriously questioned the reputation of these scientists because of their engagement in chemical weapons research. Instead, their names have been upheld as models for future generations. For instance, the KW institute that Haber once turned into the biggest weapons research unit worldwide, is now named the 'Fritz Haber Institute of the Max Planck Society'. The German Physical Chemical Society (Bunsen-Gesellschaft) calls its highest award the 'Walther Nernst Medallion' and its young scholars award the 'Nernst-Haber-Bodenstein Prize'. The top award of the German Chemical Society for organic chemistry is named after Emil Fischer.

The honoring of former 'heroes' of chemical weapons research is not confined to Germany (Freemantle 2014, pp. 44 ff., 219 ff.). For instance, the medical chemist Fritz Pregl, a leading figure in the Austrian chemical warfare project, received the chemistry Nobel Prize in 1923. When the International Union of Pure and Applied Chemistry (IUPAC) was founded in 1919, they elected as their first President Charles Moureu who had been head of the French offensive chemical warfare department during the war. Both Moreau and Nobel Laureate Victor Grignard, who was the major scientific innovator of French chemical warfare, are still honored by numerous monuments in France. In 1922 the British chemist Sir William Jackson Pope, who was knighted for his chemical warfare achievements, followed Moureu as President of IUPAC. Such as Pope had made for himself a name in the synthesis of mustard agent in Britain, so had James B. Conant in the US as a young scholar, whom historians of science mainly know from his mentoring of Thomas S. Kuhn. After the war, Conant rapidly advanced from chemistry professor to Harvard University President to one of the most influential science policy advisor in the US during and after WWII, particularly on nuclear weapons research and deployment. Before the rise of nuclear weapons physics, chemical weapons research appears to have been one of the most promising fields to make a career in science and science administration.

3. Ethical Analysis

3.1 A brief introduction to ethics

Ethics or moral philosophy, one of the oldest philosophical disciplines, serves various purposes. One is to justify or criticize new and existing national and international laws on the basis of accepted ethical principles. For instance, before a new law comes into force in democratic societies, ethical deliberations and debates are usually conducted to ensure that it is in accordance with the prevailing ethical standards. Equally important is the role of ethics in providing moral guidance in areas not covered by law. Because the law cannot – and for various reasons should better not – control all human behavior, this leaves ample room for ethics. For instance, chemical weapons research is not forbidden by any law. If you research a potential substance for chemical warfare, you could always argue that you are just studying the compound to understand its interesting chemical structure, or to find a new agent for pest control. No judge would be able to read your mind, although your colleagues might guess what you are after.

Ethical theories aim to make impartial judgments about what is morally right or wrong, regardless of any personal, corporate, or national interest, *i.e.* their first principle is impartiality. Like any theory, they do so by providing general principles and methods to derive judgments for particular cases. All ethical theories for the moral assessment of human actions fall into two main groups, utilitarianism and deontology, which both provide respectable moral positions. They differ to some extent in their moral judgments, but not, as we will see, about weapons research.

Utilitarian theories are based on a single principle, a normative rule for actions: act so that the consequences of your actions maximize the benefit of all people. Theories greatly differ in what they understand by 'benefit', how to calculate and balance it with harmful consequences, how to distribute it best, and to what extent 'all people' include future generations and nonhuman living beings. In the present context, these differences are unimportant. What matters is that utilitarian (and more generally consequentialist) theories judge actions in retrospect only according to their actually beneficial and harmful consequences, including the unintended adverse consequences, *i.e.* the naive good will that brings about harm is a major cause of moral failure.

Deontological theories (from Greek *deon*: duty, obligation) are based on two or more principles that are all general normative rules or duties. These duties are frequently organized by values (commandments of what one should strive for) and evils (prohibitions of what should be avoided). They all incorporate the utilitarian norm in the form of the commandment of benevolence and the prohibition of doing harm, which is typically of higher rank such that the prospective benefit rarely justifies doing harm. Unlike in utilitarianism, however, benefit and harm cannot simply balance each other out. On the one hand, there are additional absolute prohibitions, *e.g.* of harming human integrity or dignity. On the other, benefit and harm should each be fairly distributed according to values of justice. For specific contexts, such as biomedical ethics, core lists of further principles have been developed, albeit with vague priority rankings, which is one of the weaknesses of deontology.

The most famous deontological approach, by Immanuel Kant, provides a meta-rule for deriving normative rules in each kind of context, which is a sophisticated version of the Golden Rule: chose only those general rules of which you can reasonably want that it becomes a universal rule applicable to anyone. In deontology, actions are not judged on the basis of the actual consequences but according to whether one acts out of ethically justified duties or not, which of course includes the duties to foresee, based on the available knowledge, any possible harm and to avoid it. Here again, as in utilitarianism, naivety is a significant moral failure, not an excuse.

3.2 Ethical assessment of weapons research

Compared to ordinary life activities, science and engineering are special in that they potentially create entirely new entities that did not exist before, say a new chemical substance or a new weapon. Moreover they discover, and usually make public, the ways in which these new entities can be made. Research, development, and publication are the actions for which scientists and engineers are to be hold responsible and which are to be ethically assessed.

3.3.1. Utilitarianism

Let us begin with utilitarianism and ask what the likely consequences of successful weapons research are. They are of two kinds. First, those who get access to your knowledge will try to build the weapon and use it to threaten other people, some of whom will deploy it in order to kill or harm other people, if only to illustrate their power. That has been true throughout history, and includes the 'hydrogen bomb' that a growing number of countries have rebuilt. In general, despite all efforts to classify it as secret, scientific and engineering knowledge about powerful weapons quickly leaks, by espionage or the analysis of weapon tests, to the rest of the world, including your enemies, 'rogue states', and terrorists. It is difficult to find any exception to that historical law.

Therefore, according to utilitarianism every deployment of your weapon by anyone in the future belongs to the consequences of your research, according to which your action is morally judged. You might only have wanted your weapon used only as a means of deterrence, or for a certain one-time deployment by 'good guys' in a special situation, but that naivety is no excuse in ethics. Even if there are such special situations in which the possession or use of a weapon by one party has beneficial consequences, the overall consequences in the future, which includes any deployment by any party, are by all reasonable foresight harmful and outweigh any possible positive consequence. Thus from an utilitarian point of view weapons research is clearly morally wrong.

Moreover, as scientific research strictly builds on itself, which is an almost unique feature of science (and technology) among all cultural activities, so does weapons research. The second type of consequences includes further research by others who modify and improve your weapon, making it more effective. If that is done by your enemy, it becomes a step in an arms race that develops ever more devastating weapons, of which poison gas research during WWI is a particularly instructive example because it quickly escalated once the international ban had been broken. In this case your original research does not literally cause the follow-up research, but it enables it, such that subsequent step(s) in an arms race are the consequences of your research activity for which you are co-responsible. You might desire to reach an immediate advantage for the good, but you actually contribute to an ever worsening development of weaponry. It is difficult to imagine another research situation where the utilitarian verdict, no matter what specific theory, is so clear as in weapons research.

3.3.2 Deontology

Much of what has been discussed above also applies to deontological ethics because the prohibition of doing or causing predictable harm is a major duty in all systems. Because all weapons research causes easily predictable harm in the future, it is forbidden.

And yet, the appeal to duty has frequently been abused for justifying weapons research and other crimes. For instance, Haber in the interwar period argued that he had performed his projects out of duty to his home country. Even Heinrich Himmler, leader of the Nazi SS, claimed in his notorious Posen speech (1944) that the extermination of Jews would be a "moral right", a "duty to our people". However, the alleged duties to one's nation, corporation, or gang are not moral duties but only self-imposed rules by a group. They all violate the principle of impartiality that defines the scope of moral rules. In contrast a moral duty is a duty to anybody regardless of membership of a group, or to humanity as a whole. Hence, patriotism is not to be confounded with morality.

Kant's meta-rule, which implements the principle of impartiality, is a useful test instrument for moral rules: chose only those general rules of which you can reasonably want that they become universal rules applicable to anyone. The rule to be examined is thus not 'My chemical weapons research is permitted', but instead 'Chemical weapons research is generally permitted'. Can you reasonably want that anyone else follows this, now and forever? If you think that there are irresponsible people who should not be allowed to do that, then you consider the rule morally wrong. Moreover, if you think that such unrestricted weapons research leads to an arms race to make ever more poisonous substances that threaten the existence of humanity and all living beings, then you would even more strongly oppose the rule. Only a suicide candidate might want that, but that does not count as a 'reasonable' volition.

In sum, chemical weapons research and development is morally wrong according to all major ethical theories, all of which were well-known during WWI. This does not only include the synthesis of new poisons, but also the research and development of effective deployment methods in the form of actual weapons. All chemists who contributed to that during WWI and thereafter morally failed.

Note that you do not need to be a pacifist to accept the conclusion as some have argued (Kovac 2013). Even if you are willing to support the use of weapons under certain circumstances, you can strictly disagree with weapons research in general for ethical reasons.

3.3 Standard excuses by weapons researchers

After WWI chemical weapons researchers have expressed various excuses that, strangely enough, became popular vindications for moral wrongdoing in science. Because their ethical refutations are less known, it is worthwhile to point out the underlying misconceptions of the thirteen most common excuses.

My weapons research was a moral duty to my country

Surely most weapons researchers felt some obligation during their work, including patriotism and commitment to their research unit, and thus considered it only right to fulfill their duties. However, as has been shown above, 'moral duty to my nation' (or to any other group) is a contradiction in terms because morality implies impartiality, whereas patriotism includes a nationalistic bias. This position thus confounds patriotism with ethics. You cannot take holidays from ethics, not even in wartime.

I did only the research, others are responsible for deployment

That most common misunderstanding in science takes weapons research to be ethically neutral, only the deployment is to be blamed. However, this assumption is wrong by any ethical theory. Everybody is co-responsible for the consequences of one's action. Hence, also the creators of new weapons are co-responsible for any future deployments of their creations because those are the consequences of their action. There is no ethical theory that would allow for an exemption or excuse.

I did it to prevent greater harm

Even though there might be situations where the deployment of a weapon prevents greater harm than it causes, the arguments confuses research with deployment. It focuses on a specific deployment as the consequence of one's research, but neglects all future uses and misuses of the weapon that are to be considered in an ethical assessments too. This is the standard form of moral naivety that neglects the unintended but easy to foresee consequences.

Chemical weapons research is morally justified by Just War theories

This sophisticated form of the previous excuse refers to 'Just War' theories, according to which a particular war can be morally justified under certain conditions, which includes several elaborations of the Hague Conventions and the prohibition of weapons of mass destruction (*i.e.* nuclear, biological and chemical weapons), such that the argument is pointless. In general, 'Just War' theories are irrelevant for the moral assessment of weapons research, unless all possible hostile parties, including terrorists, will always comply to those rules, which is more than unlikely. 'Just War' theories can only be applied to very particular war situations, whereas weapons researchers increase the arsenal of weaponry for any party in any future war.

Chemical weapons are more humane than other weapons

After WWI many chemists, including Haber, argued that chemical weapons are more humane because of their lower death toll compared to other weapons.[9] On the one hand, it is impossible to calculate and

---

9 Note that during the war, Haber argued that chemical weapons are more humane because they would save lives by ending the war faster, meaning that they would bring a soon German victory by their devastating effect (Szöllösi-Janze 1998, p. 327).

compare the 'degree of humanity' of different weapons. For instance, is killing slowly over years more humane than killing fast? Moreover, during WWI and thereafter chemical weapons employed ever more toxic substances that were dropped or sprayed from airplanes to kill anyone living beneath. This indiscriminate or uncontrolled effect made them weapons of mass destruction, like biological and nuclear weapons. On the other hand, the argument again confuses weapons research with deployment. Research and development of any new weapons adds another tool for killing and harming people and is morally wrong as such, regardless of what other weapons already exist.

I did it only to have my people be prepared for retaliation

If chemists produce chemical weapons that existed before, they do no original research and development, but only production work. Then the arguments above do not apply. However as creators they are co-responsible for the use of the stockpile they produce, even if they have no control over the particular deployment. And they help their country to commit war crimes for which they are co-responsible. If, on the other hand, they develop new chemical weapons for 'retaliation', then they engage in morally wrong research and, even worse, contribute to an arms race that develops ever more devastating weapons.

We had to do it because the enemy forced us to do it

In times of war the enemy is usually made guilty by downplaying one's own activity as purely defensive and exaggerating that of the enemy as aggressive. This creates the dangerous constellation of an arms race in which each step further is justified as an allegedly defensive or responsive measure. Systemic forces seem to take over the responsibility of the individuals. However, systemic forces cannot assume ethical responsibility. Only individuals can be held ethically responsible, either alone or as members of a group who share the responsibility, such as a weapons research team or all chemists involved on either side. Pointing to systemic forces is therefore no moral excuse, but a way to shift responsibility to an abstract entity. Moreover, using this argument to justify ad hoc research in an arms race once again confuses the different responsibilities of weapons research and deployment.

If I had not done it, somebody else would have done it

A frequent excuse tries to downplay one's own role by arguing that one was a replaceable actor: my refusal would have made no difference, others would have worked in my position such that the consequences were unavoidable. What at first glance looks like a moral argument is actually not. Imagine a man is lying unconscious on the street with money in his hand. You steal his money thinking, 'If I don't do it, somebody else will do it'. If you are tried for your crime, your excuse would make no impression because it is morally irrelevant for the judgment of your culpability in committing a robbery. Pointing to other possible criminals does not diminish one's responsibility, nor does it excuse wrong-doing, be it robbery or weapons research.

I was ordered/forced to do it

Many weapons researchers have tried to diminish their responsibility in retrospect, arguing that they had no other choice, were forced or ordered to do so. However, there is no single reported modern case of forced research; it is even questionable if creative research is possible under force. The social pressure on weapons researchers is usually not different from that of any other employee. Leaving a weapons program might bring some disadvantage, for example, in one's personal research career or earnings. But that does not count as a moral excuse, such as the need of money does not excuse a thief. The question is: why does somebody get involved in such a research program in the first place?

My research is only for protective purposes

The Chemical Weapons Convention permits the small-scale production of highly toxic substances for medical, pharmaceutical and protective research (see above). However, that permission can easily be abused. First, as we have seen above, protective devices such as special gas mask filters are part of offensive equipment. They allow offenders to use poisons while being themselves protected. Second, any new highly toxic substance that might be researched for some pharmaceutical effect is at the same time, by its toxicity, a new potential chemical weapon. Thus, if you first synthesized it, you are co-responsible for its possible military or terrorist abuse by anyone in the future.

I did not intend terrorists to use my weapon

As a weapons researcher you are supposed to know at least what anybody else knows, that knowledge about powerful weapons readily leaks out to the rest of the world, including terrorists. You might not have intended terrorists to use your weapon, but that is an unintended consequence of your research that is easy to foresee and for which you are co-responsible.

My research serves the purpose of keeping peace by way of mutual deterrence

According to Cold War standard rhetoric, a war between two enemies becomes unlikely if both are equally equipped with weapons of mass destruction that are ready to be deployed as retaliation for any possible first strike by the other. Apart from the general deficits of the argument, it can hardly be applied to weapons research. The argument presupposes a weapons balance. However, research on either side to create more sophisticated or disastrous weapons is an attempt to destroy exactly that balance, which triggers an arms race rather than enabling stable conditions for peace.

I didn't know what my research was used for

The development of complex weapons systems requires a division of labor. Various individuals or research groups each work on a small element of the entire system. If the project is secretly coordinated, it might be possible that some researchers, particular young scientists, are not aware of the overall goal of their individual work. However, such conditions hardly apply to chemical weapons research aimed at poisons or explosives, where military purposes are likely intended. If, nonetheless, senior researchers trick young scientists into weapon research projects without their knowledge and consent, they commit a major ethical offence.

3.4 Ethical analysis of honoring warfare chemists

As we have seen in Section 2.5, many leading figures of chemical warfare research in WWI had excellent careers afterwards. Moreover, they received numerous Nobel Prizes and have been honored by the scientific community in the names of educational buildings, scientific institutes, and awards till the present day. No doubt they all made other important contributions to science that are worth commemorating. However, they also morally failed according to all major ethical theories. And many were honored, not despite, but because of their warfare engagement.

Using someone's name for a scientific institution or an award honors the person's integrity as a whole rather than a particular achievement. For instance, Germany's biggest research institute for physical chemistry is called Fritz Haber Institute, not Nitrogen Fixation Institute, because it honors and commemorates Haber entire lifetime work beyond his contribution to nitrogen fixation. It singles out the person as an outstanding role model for a younger generation, to be admired and copied. How can this still be justified in today's world?

One could argue that Haber's scientific achievements and his unquestionable personal engagement for his employees outweigh his moral failure. But how does one balance these factors? Does not such a compromise deliver a dangerous message: your scientific achievements can outweigh your moral failure? In the same vein one could honor the Nazi physiologists who, by brutal or lethal experiments on concentration camps prisoners, produced valuable physiological knowledge.

It seems more likely that many chemists take moral failures to be marginal and ignore them. For instance, biographies of WWI warfare chemists, except for Haber, written by fellow chemists typically omit their war engagement or mention it only in passing, as if nobody should not know about that. They thereby miss the chance of engaging young chemists in historical and ethical issues of their discipline and the chance to draw valuable lessons. Moreover, they further isolate chemistry from a civil society that learn about such topics from public media. Keeping moral failure as an open secret leaves the impression as if the chemical community has not come to terms with ethics since WWI.

4. Conclusion

The story of poison gas in WWI is an instructive example of the academic-industrial-military-governmental complex. Actors from different fields collaborated in a network with the aim of committing a war crime. Such complexes invite confusion about who is responsible for what. Thus the first task of an ethical analysis is to disentangle the network and define the primary responsibilities according to the different kinds of actions and decisions that occurred: The scientists and engineers research and develop the new weapon; industry produces it; the government has the ultimate decision on its deployment; and the military decides when and how it will exactly be deployed. In the case of Haber, parsing the responsibilities is particularly difficult because he had positions in virtually all fields of the academic-industrial-military-governmental complex. The next step consists in eliminating all non-ethical duties and commitments, such as by patriotism, public pressure, business contracts, and local law. Based on an ethical theory you can then perform an ethical assessment of the individual contributions and, on a more advanced level, of the interactions of the actors.

In the case of weapons research conceptual confusion abounds, particularly between ethics and patriotism and between research and deployment. Based on patriotism weapons researchers have constructed a pseudo-moral legitimation for their work. And moral debates on weapons research have either made researchers responsible only for certain deployments or rejected any responsibility for deployment. However, ethically the creators of a new weapon are co-responsible for all subsequent uses and misuses of their creation, which they enable and of which they are supposed to know the overall harmful consequences. There is no excuse of not-knowing or not-intending.

Since WWI governments have employed or contracted scientists on a large scale for researching new weapons that would soon spread worldwide. While politicians might feel responsible only for their own use of these weapons, they tempted scientists into becoming ethically responsible for all possible uses and misuses of their creations in the future. By upholding conceptual confusions about responsibilities, or by having a blind spot towards weapons research, scientific societies have never adequately responded to that large-scale abuse of science (*e.g.* by condemning it in their codes of conduct).[10] This has made science, and chemistry in particular, suspicious to the public, and rightly so according all major ethical theories. Obviously there are still important lessons to learn from WWI.

Further Reading

There are numerous books on poison gas in WWI. Freemantle 2014 provides a broad and up-to-date view; Friedrich *et al.* 2017 is the latest anthology. Still worth reading is the classic Haber 1986, written by the son of Fritz Haber. For a comprehensive history of chemical warfare, see Tucker 2006. The best researched Haber biography, with numerous valuable insights, is Szöllösi-Janze 1998, for a very short English essay, see Szöllösi-Janze 2017. Papers on ethics of chemical weapons research typically confuse research with deployment. For an introduction see Kovacs 2016 and Schummer 2001. Some of the standard excuses are dealt with in Ryberg 2003.

References

Freemantle, M.: 2014, *The Chemists' War: 1914-1918*, Cambridge: Royal Society of Chemistry.

Friedrich, B.; Hoffmann, D.; Renn, J.; Schmaltz, F. & Wolf, M. (eds.): 2017, *100 Years of Chemical Warfare: Research, Deployment, Consequences*, Heidelberg: Springer.

Haber, L.F.: 1986, The Poisonous Cloud: Chemical Warfare in the First World War, Oxford: Clarendon.

Irwin, W.: 1921, *The Next War*, New York: Dutton, 1921.

Johnson, J.A.: 2017, 'Military-Industrial Interactions in the Development of Chemical Warfare, 1914-1918: Comparing National Cases Within the Technological System

---

[10] The only code that mentions chemical weapons at all is the one by the German Chemical Society, but it condemns only their production and not their research.

of the Great War', in: B. Friedrich et al. (eds.), *100 Years of Chemical Warfare: Research, Deployment, Consequences*, Heidelberg: Springer 2017, pp. 135-149.

Kovac, J. 2013, 'Science, Ethics and War: A Pacifist's Perspective', *Science and Engineering Ethics*, **19**, 449-460.

Kovac, J. 2016, 'Ethics of Chemical Weapons Research', *Bulletin for the History of Chemistry*, **41**, 56-63.

OPCW: 2016, 'Report of the OPCW on the Implementation of the Convention on the Prohibition of the Development, Production, Stockpiling and Use of Chemical Weapons and on Their Destruction in 2015' [available online at https://www.opcw.org/documents-reports/annual-reports/, accessed 5 October 2017].

Rauchensteiner, M.: 2014, The First World War and the End of the Habsburg Monarchy, 1914-1918, Wien: Böhlau.

Ryberg, J.: 2003, 'Ethics and Military Research: On the Moral Responsibility of Scientists', in: B. Booss-Bavnbek, & J. Høyrup (eds.), *Mathematics and War*: Basel: Birkhäuser/Springer, pp. 352-364.

Schummer, J.: 2001, 'Ethics of Chemical Synthesis', *Hyle: International Journal for Philosophy of Chemistry*, **7** (2001), 103-124.

Schummer, J.: 2019, 'Art and Representation', in P. Morris (ed.), *A Cultural History of Chemistry*, vol. 6, London: Bloomsbury, forthcoming.

Szöllösi-Janze, M.: 1998, *Fritz Haber (1868-1934): Eine Biographie*, München: Beck.

Szöllösi-Janze, M.: 2017, 'The Scientist as Expert: Fritz Haber and German Chemical Warfare during the First World War and Beyond', in: B. Friedrich et al. (eds.), *100 Years of Chemical Warfare: Research, Deployment, Consequences*, Heidelberg: Springer, pp. 11-23.

Tucker, J.B.: 2006, War of Nerves: Chemical Warfare From World War I to Al-Qaeda, New York: Pantheon.

Van Der Kloot, W.: 2004, 'April 1915: Five future Nobel Prize-winners inaugurate weapons of mass destruction and the academic-industrial-military complex', *Notes and Records of the Royal Society*, **58**, 149-160.

Zecha, W.: 2000, 'Unter die Masken!': Giftgas auf den Kriegsschauplätzen Österreich-Ungarns im Ersten Weltkrieg, Wien: öbv, 2000.

Chapter 4

Ethical Responsibilities in Military-Related Work: The Case of Napalm

Stephen M. Contakes and Taylor Jashinsky

Abstract: Two case studies are presented illustrating how leaders of chemical enterprises addressed ethical questions posed by the incendiary napalm. The first one examines how the chemist Louis Fieser grappled with the ethical questions posed by his development of the napalm incendiaries used against military and civilian targets in the Second World War. The second involves the Dow Napalm Controversy, in which Dow Chemical engaged protests over its role as a supplier of napalm to the American military in Vietnam. Dow weighted the protesters' charges that napalm was being used indiscriminately on civilians against what it saw as an obligation to support the American government and soldiers in their aims of defending South Vietnam against a communist insurrection. Both cases are examined from a Just War viewpoint to illustrate chemists' responsibilities in the weapons development pipeline and the dilemmas that can arise in weapons development and over foreseeable misuses of chemical products.

1. Introduction

The 20th century provides numerous examples of ethical questions associated with chemists' involvement in military work. Perhaps the most notable one is chemists' leadership in developing chemical weapons during the First World War. Even though poisoned projectiles were outlawed under the 1907 Hague convention, eminent chemists like Fritz Haber, Walter Nernst, Emil Fischer, and Carl Duisberg of Germany and William Ramsay of Great Britain promoted poison gas weapons (Haber 1986, Moy 1989, Stoltzenberg 2000, 2004). Of these Haber's efforts had the largest impact on the war and chemists' subsequent involvement in military research more generally. In addition to leading Germany's poison gas warfare program and advocating his ammonia synthesis process as a means of sustaining munitions production amidst the privations of blockade, Haber transformed his Kaiser Wilhelm Institute for Physical Chemistry and Electrochemistry into the world's first large scale military research lab (Stoltzenberg 2004, Charles 2005).

Since then chemists have been involved in the development of numerous military technologies, including some which found wide use (Chemical Corps Association 1948, Cornwell 2003, Jeffreys 2008, Remers 2000). Of these the present chapter examines jeilled gasoline incendiaries, which arguably had the greatest impact on 20th-century warfare. During the Second World War a number of chemists and engineers in the United States and Great Britian pursued improved incendiaries for use in flamethrowers and bombs (Brophy *et al.* 1959, pp. 167-9; Neer 2013). Their discoveries let to one of chemistry's greatest military successs stories, napalm, which was produced in large quantities by major chemical concerns and employed with devastating effectiveness on the battlefield. However, due to its potential to induce massive firestorms and casuse particularly cruel wounds, napalm occasionally became the subject of public ethical scrutiny and reevaluation, in which chemists and chemical corporations had to consider what it means to develop and manufacture weapons responsibly. Thus the story of napalm provides an opportunity for chemists to consider the ethical issues associated with military work. This will be done using two case studies.

Our first case examines the organic chemist Louis Fieser's development of the incendiary napalm, particularly his decision to participate in incendiary research and his responsibility for the large-scale use of napalm against civilians during the closing stages of the war. Unlike the political and military leaders who have been the subject of intense scholarly and popular study,[1] Fieser neither ordered nor participated in the attacks. Any responsibility he bears is as napalm's inventor. The second case considers the responsibilities of chemical corporations engaged in the large-scale industrial manufacture of munitions. This issue rose to the fore when Dow Chemical faced protests over its manufacture of napalm

1 See Schaeffer 1980, Crane 1993, Bess 2006, and Grayling 2006.

for the American military during the Vietnam War. Unlike Fieser, Dow could not claim ignorance over how napalm *might* be used but instead wrestled with the relative harms and benefits of its napalm operations in the midst of questions about the merits of the war and uncertainty over whether napalm was *actually* being employed indiscriminately against civilians.

2. Ethical Responsibility in Military Research

Arguments that scientists are not accountable for the outcomes of their work since they should be free to pursue any truth (Hoffmann 1975) or are mere functionaries[2] have been sharply criticized (Weeramantry 1987, pp. 157-166; Ryberg 2003; Wolpe 2006). More helpful then are accounts that recognize scientists' ethical responsibilities (Brown 1971, Coulson 1966, Douglas 2003, Forge 2008, Hildebrand 1955, Hoffmann 1995, Jacob & Walter 2005, Kovac 2007/2013/2015, Lonsdale 1955, Mitcham & Siekevitz 1989, Fischhoff 2014, National Research Council 2014, Weeramantry 1987),[3] including those which address the synthesis of new compounds (Schummer 2001), the dissemination of results (Hoffmann 2008), noting or failing to note hazards (Jacob & Walters 2005), and other consumer safety and environmental impacts of chemicals.[4]

Chemists and engineers are responsible for the outcomes of any work to which they in soundness of mind freely contribute, either as individuals or members of a larger organization. Nevertheless an individual's degree of responsibility might depend on how their actions or failure to act contributed to those outcomes. In general they would only be judged praiseworthy or blameworthy for expected or anticipatable outcomes of their actions or omissions and, in deontological ethical systems, whether they fulfilled or neglected their duties. This is illustrated by the post-World War II war crimes trials of chemist-managers at I.G. Farben and other concerns (DuBois 1952; Cornwell 2003, pp. 367-376; Hayes 2005;

---

2 For example, the biochemist Harold Urey argued that scientists are first responsible to their governments (Chalk 1989).

3 Herein we limit ourselves primarily to English language sources.

4 See, for example, green chemistry as "an expression of environmental ethics" (George 2010).

Jeffreys 2008). Consider the example of Bruno Tesch (Hayes 2005), who in the 1910s helped develop the Zyklon B delousing agent that the Nazis used to murder civilians in the gas chambers of the holocaust over twenty years later. Tesch is not considered morally blameworthy as Zyklon B's developer since its use in genocide was not foreseeable at the time. However, when Tesch later managed a firm that supplied Zyklon B to the death camps and learned that it was used in gassings, he had an opportunity to avoid further complicity. He neglected to do so and was later found responsible for the killings and sentenced to death by a postwar tribunal.

Because moral blame and praise is conditioned on individuals' choices and actions, our analysis of Dow's and Fieser's responsibilities will focus exclusively on whether they were or should have been aware of ethical issues associated with their work and whether they did or should have acted in their professional capacities to address those issues.[5] The issues themselves will be identified and examined using a Just War approach.[6]

3. Just War Thinking

Just War thinking is the dominant framework currently used to consider the morality of military research. It forms the backdrop to international laws governing war; combines consequentialist and deontological criteria; and yet can be appropriated by deontological, consequentialist, and virtue ethics frameworks. Here we outline its main features and illustrate how it might apply to chemists engaged in military work.[7]

---

5 We do not deny the existence of other responsibilities, but only limit our treatment to professional ones associated with anticipatable outcomes.

6 Other perspectives on the morality of war are pacifism and realism. Realism opposes moral restraint in war and by extension in military research. For pacifism and a pacifist perspective see Kovac 2013. Examples of chemists who refused to conduct military research are Frederic Soddy (Anonymous 1921) and Kathleen Lonsdale (1964, p. 54).

7 For more comprehensive overviews of Just War Theory see Lazar 2016, Johnson 1999/2011, and Moseley (2004). On weapons and military research from Just War and related perspectives see Dinegar 1989, Fichtelberg 2006, Forge 2004, Malsch 2013, and O'Donovan 2003, pp. 78-94. For the relevant international law, its relation-

Just War Theory envisions war as an unfortunate means of preventing harms, protecting social order, or appropriately chastising aggressors. It calls for right conduct in war and delimits when wars may be waged.[8] Specifically, its *Jus ad Bellum* provisions allow a *sovereign authority* to wage war in a *just cause* with *right intent* (*i.e.* it seeks a just resolution). Additional subordinate *prudential criteria* serve as criteria for overcoming a presumption against the use of force and hold that war should be a last resort, with a reasonable likelihood of success, and a favorable calculus of risks, benefits, and harms (*proportionality criterion*).

The *Jus ad Bellum* criteria typically apply to state actors but can also be interpreted to prohibit chemists' engagement in military work that supports war-related efforts of an entity that fails to meet them.[9] This judgement is reflected in the post-Second World War trials of I.G. Farben and Krupp managers (May 2008, pp. 185-206; Jeffreys 2008). However, the *Jus ad Bellum* criteria have also been employed to argue for military research on the grounds that new technology might enhance states' capabilities to protect the oppressed, promote global security, or achieve some overall noble aim (Dinegar 1989). For example, in the Cold War Western nuclear weapons research was justified as a means of forestalling the expansion of Soviet oppression. However, *Jus ad Bellum* arguments for and against military work require scientists to accurately assess their country's present and future war aims without succumbing to uncritical patriotism or overly optimistic ideas about technology's potential to deter wars or render them more humane. The fragility of scientists' judgments is illustrated by World War I era chemists' diverse and in hindsight dubious justifications for poison gas work. Fritz Haber justified it by appealing to the justice of Germany's cause and gas' potential to save lives by promoting a speedy end to the war.[10] American chemists

ship to values, and some difficulties in applying it to specific weapons and situations see Greenwood 1998, Schmitt 2005, and Lietzau 2004.

8 Additional *Jus post Bellum* provisions address right postwar conduct and outcomes. These might restrain the use of chemical weapons and defoliants with persistent environmental or human health effects.

9 Or an industry that supplies such entities; see Fichtelberg 2014.

10 The colleagues Haber sought to convince included future Nobel laureates Hermann Staudinger (Weber & Deußing 2013), James Franck, and Otto Hahn (1970), the latter two of whom later publicly opposed atomic weapons. Other arguments Haber employed either addressed technicalities of international law or invoked Germany's

talked about America's national security and chemical weapons' potential to deter conventional warfare (Whittemore 1975, p. 157).

Just war theory addresses right conduct in war through its *Jus in Bello* criteria of *discrimination* and *proportionality. Proportionality* holds that one's conduct should be commensurate with the justice of one's aims. Specifically, the means employed should not create greater evils than those the war is intended to avert and even then any harms inflicted should be necessary and kept to a reasonable level. *Proportionality* considerations feature in arguments against chemical weapons as unnecessarily cruel or as an unjustifiable existential threat to Western ideals of civilization[11] and in arguments against nuclear weapons as presenting an unjustifiable threat of nuclear annihilation.

The debates over napalm incendiaries involved the *discrimination* criterion and the closely-related concept of noncombatant immunity. The former holds that only the killing of enemy combatants (*i.e.* those actually contributing to the fighting) is morally justified while the latter stipulates that noncombatants may not be intentionally targeted. For absolutists noncombatant immunity it means that combatants must seek to avoid foreseeable harms to noncombatants.[12] Other Just War thinkers only hold that noncombatants may not be the primary target of attacks. For example, arguments for civilian bombing typically rely on the doctrine of double effect, which permits the killing of noncombatants as long as their deaths are a side effect of efforts to achieve an overall morally good end.[13]

"need and helplessness" (Cornwell 2003, p. 535) – an argument similar to the 'supreme emergency' used to justify the British Second World War civilian bombing campaign. Later, Haber and the biochemist J.B.S. Haldane would argue that Chemical weapons were preferred as being nonlethal relative to conventional weapons.

11 For example, the 'taboo' against chemical weapons is reflected in the 1907 Hague Convention's judgment that chemical weapons' barbarity constitutes an existential threat to civilized warfare (Price 1997).

12 For a critical review of arguments for absolute noncombatant immunity see Lazar 2014.

13 Other nonabsolutist understandings of noncombatant immunity tend to justify killing civilians from the standpoint of the morality of individual acts and cases rather than of the general conduct of states in war (Arneson 2006, McMahon 2011, Frowe 2014), a viewpoint difficult to reconcile with the way war is actually conducted (Lazar 2010, Strawser 2013).

Exactly what constitutes reasonable care and who is a noncombatant are matters of judgment. One complicating factor is the reliance of modern armies on technological expertise and civilian-manned industry located in or near cities – factors which tend to blur the combatant-noncombatant distinction and render the selective targeting of military enterprises more difficult. Thus the Second World War Allied civilian bombing campaigns avoided this difficulty by claiming civilians as legitimate targets by virtue of their supposed support for the Axis war effort and their complicity in upholding an immoral regime.[14] However, not all enemy civilians worked in arms factories or supported aggressive warfare and it was impossible for bombardiers to determine which civilians had been sufficiently militarized to constitute a legitimate target anyway. Thus proportionality and the doctrine of double effect were also employed to rationalize the bombings as a way of redressing the 'supreme emergency' which the Axis war effort posed to civilization;[15] and in the case of Japan, to avoid the casualties expected on ground invasion of the Japanese mainland.[16]

In principle, scientists can use the principle of discrimination to determine whether their work might result in unjustifiable foreseeable harms. However, this assumes they are able to accurately assess the relative justice of their nations' war aims and the likelihood their discoveries may be inappropriately employed against civilians. That such judgments are far from simple and require constant reassessment is well illustrated by the contrast between the widespread approval of indiscriminate Allied civilian bombing during World War II and the storm of protest provoked by America's relatively more discriminate use of napalm in Vietnam.

---

[14] For an example see Russell (2006, pp. 131-32).

[15] By disrupting infrastructure and de-housing, displacing, and inconveniencing workers. See Walzer (2006, pp. 251-268).

[16] Combatants' decision to pursue civilian bombing depended on whether they considered it necessity and effective. Americans objected to civilian bombing in Europe as ineffective and immoral (Crane 1993, Schaffer 1980). In the Pacific they considered the projected immense potential harms of a ground invasion to outweigh the harms of a devastatingly effective civilian bombing campaign. Nevertheless, even then they attempted to preserve noncombatant immunity by using warning leaflets (Crane 1993, pp. 133-35; Neer 2013, pp. 84-85).

Scientists should also consider whether the weapons they are developing might be inherently discriminate. For example, nuclear weapons are regarded as indiscriminate due to their destructiveness and the lingering radiation they produce. In contrast smart munitions designed to deliver a measured explosive payload to within an accuracy of a few feet generally are not. Whether a weapon or strategy is regarded as discriminate depends on its social and technological context. For instance, America's Second World War precision bombing campaign was considered discriminate given the inaccurate bombing technology of the time but would be regarded as indiscriminate today.

4. Louis Fieser and the Firebombing of Cities

4.1 Historical description

By July 1940 Nazi Germany had overrun Poland and much of Western Europe.[17] Concerned that America itself might soon face an aggressive and capable foe, its leaders initiated a variety of military research projects under the auspices of a National Defense Research Committee (NDRC). Among the scientists it recruited was the Harvard organic chemist Louis Fieser. Originally assigned to work on explosives, Fieser's work on incendiaries began opportunistically. After learning of industrial explosions involving sticky divinylacetylene at a 1941 NDRC conference, Fieser judged his postdoctoral researcher Emanuel B. Hershberg "ideally qualified" to develop them as incendiaries (Fieser 1964, pp. 9-11).

Fieser and Hershberg soon found peroxidized divinylacetylene gels were not very shock sensitive, although they "burned with an impressive sputter" and retained a "viscous, sticky consistency" – ideal characteristics for "a bomb that would scatter large burning globs of sticky gel" (*ibid.*, p. 12). With support from the NDRC and British Air ministry they soon investigated related materials as incendiaries, including particularly effective rubber-benzene and rubber-gasoline gels. By November they

[17] For a comprehensive account of napalm and its development see Neer 2013. For an English-language account of the European bombing war see Overy 2013.

were even testing prototype bombs at Harvard's stadium and demonstrating their effectiveness to military officials.[18] However, when the Japanese overran the Malaysian rubber plantations that December, Fieser's work on rubber-benzene bombs was suspended.

Undeterred, Fieser and his team investigated several alternatives, including metal salts of fatty acids which were used to thicken lubricating oil into greases. One such material, aluminum naphthenate, had earlier been investigated by another NDRC incendiary group and found to give suitable gels, but only after a heating step that was impractical for battlefield use. Fieser formulated mixtures of aluminum naphthenate and 'aluminum palmitate' which gelled when mixed with gasoline at room temperature, and named the resulting material 'napalm' after its components.[19]

Fieser and his group soon developed napalm bombs and designed a burster which scattered "large, burning, globs over a 50 yard radius" (*ibid.*, p. 36). At various proving grounds these incinerated wooden buildings and proved stable to rough transport, demonstrating such effectiveness that Fieser proclaimed Japan's conquest of Malaysia "a blessing to the allied nations" (*ibid.*, p. 53).

Napalm-filled bombs and flamethrowers soon demonstrated their military utility; the latter proving especially useful for burning Japanese soldiers out of caves and other emplacements (*ibid.*, pp. 44, 52-53). Its most fateful use, however, was in the M-69 cluster bomb, an "aimable cluster" of 38 bomblets, each containing several pounds of napalm (Bess 2006, p. 98.) These were designed to ignite after they had penetrated the top floor of German and Japanese houses, whereupon they would eject a flaming stream of napalm with the intent of rapidly turning the structure into a raging inferno (Davis 1999a/b).

Napalm bombs proved extremely effective against German and Japanese cities, particularly when used in combination with high explosives that destroyed water mains and killed any firefighters who might extinguish blazes before they burned out of control. By the war's end, napalm

---

18 Hydrocarbon-based incendiaries like napalm still produce the most heat per gram (International Committee of the Red Cross 1973).

19 Fieser later discovered that his 'palmitate' was actually lauric acid and improved naplam's effectiveness by adding unsaturated oleic and linolenic acids to the mix.

and related incendiaries had burned away large swaths of many Axis cities. Particularly devastating were the firebombings of Dresden and Tokyo in 1945, in which massive incendiary-fueled firestorms generated tens of thousands of dead civilians (Crane 1993, pp. 113-119, 132-136). One incendiary raid on Tokyo was perhaps the single deadliest night in the history of warfare, in which the American General Curtis Le May claimed more people were "scorched and boiled and baked to death [...] than went up in vapor in Hiroshima and Nagasaki combined" (quoted in Grayling 2006, p. 147).

4.2. Fieser's responsibility

Our consideration of Fieser's responsibility for the use of napalm against civilians examines whether he was or should have been aware napalm might be unjustifiably employed against civilians and, if so, could have acted to limit unjustifiable harms.

Fieser's own position thirty years later was that he did not foresee napalm's use against civilians and in any event was not responsible for how napalm was used. Specifically, he claimed to "certainly [have had] no thought about the use of napalm against non-military personnel" (Neilands 1971, p. 82) and disclaimed responsibility for such uses.

> I discovered that a jelled fuel burns more efficiently than a free fuel [...]. I don't think I have to be ashamed of having made that discovery. And I would be the first to suggest that antipersonnel use be outlawed. But how in the world do you make the distinction? Why should the investigator be called on to rule on the uses? [Fieser quoted in Lemann 1973]

Fieser's repudiation of responsibility may be rejected as inconsistent with his role as developer of a weapon that might cause anticipatable harms. However, his claim of ignorance about the antipersonnel use of napalm is worthy of scrutiny.

Fieser could likely not have foreseen indiscriminate uses of napalm in postwar contexts[20] or even America's Second World War use of napalm

---

[20] Although this raises the issue of military researchers' responsibility for foreseeable impacts associated with proliferation.

against Axis civilians, at least not during most of 1941 when America had not yet entered the war and Britain was unable to prosecute a serious bombing campaign. The prospect of Allied civilian bombing also seemed precluded by American and British outrage over German bombing attacks on civilians early in the war and, later, by America's commitment to precision bombing. However, the use of napalm against noncombatants was anticipatable when Fieser began his work and grew increasingly likely as the war progressed. In fact, Germany had used incendiaries against Warsaw in 1939 and against residential areas of London from September 1940 onward while the British had been attacking German cities since the end of 1940. Perhaps because these employed conventional explosives, Fieser justified his work with the British Air ministry on the grounds he thought it unlikely the British would use napalm against civilians. If so, this might factor into whether his napalm work might be justified, but it does not absolve him of blame on the grounds that napalm attacks against civilians were unforeseeable. Such excuses would also certainly not apply to Fieser's mid-late war work, especially after a 1943 Hamburg raid in which British-dropped magnesium incendiaries fueled a firestorm that killed over 40,000 people (Neer 2013, p. 62).

Furthermore, whatever thoughts Fieser initially harbored about the potential uses of napalm, he did not long remain ignorant of its utility against civilian structures. While early tests on napalm were carried out on nondescript buildings, in his memoir *The Scientific Method* (1964) Fieser makes it clear that his napalm research envisioned the destruction of German and Japanese houses from the start. Perhaps he imagined these would be empty when the bombs fell, although this seems hard to reconcile with his knowledge of tests conducted at Dugway proving ground in Utah. There villages designed to model German and Japanese 'working-class housing'[21] were repeatedly bombed and rebuilt in an effort to determine optimal conditions for burning out Axis cities (Davis 1999a/b). Moreover, even if Fieser was unaware of the Dugway tests' civilian-oriented nature, he could hardly claim ignorance of America's abortive 'bat bomb' project, in which he personally served as a major driving force in developing incendiary-carrying bats designed to

[21] They were complete with authentic furnishings including children's toys.

incinerate Japanese urban areas more effectively than ordinary napalm bombs (Couffer 1992).

The following spectrum of possible projects may help clarify Fieser's responsibility for the use of napalm against civilians:

(1) Incendiaries for the safe burning of pathogen-contaminated dead livestock;[22]
(2) Incendiaries discovered by accident or with no concern for their utility;
(3) Incendiaries for a variety of military applications;
(4) Incendiaries designed to make flamethrowers more effective against military emplacements (bunkers, machine gun nests, *etc.*);
(5) Incendiary bombs for use against factory workers' housing;
(6) Incendiary weapons designed to efficiently and indiscriminately destroy cities;
(7) Incendiary bombs designed to spray incendiary at child height for maximal effectiveness against schools, playgrounds, and daycare centers.

Of these, case 1 is likely to be perceived as a positive benefit to mankind, whereas cases 5-7 target noncombatants, as either connected to the enemy's war effort (5) or indiscriminately (6), or selectively (7). No Second World War combatant deliberately aimed for the latter. However, throughout the war Fieser's work moved from predominately considering cases 3 and 4 (with cases 5 and 6 being foreseeable but unlikely) towards a preoccupation with realizing cases 5 and 6. Given this, we suggest that Fieser first failed to consider anticipatable Second World War anti-civilian uses of napalm and then later even promoted them.

4.3 Ethical analysis

Fieser's later strong disapproval of napalm's use against civilians[23] seems hard to reconcile with his wartime enthusiastic avowal of the bat

[22] This is a use of napalm considered in anthrax and foot-and-mouth disease epidemics (Neer 2013, pp. 200-1)
[23] That included urging Nixon to "promote an international agreement to outlaw further use of napalm or napalm-type munitions" (Lemann 1973).

bomb project[24] and the positive portrayal of napalm in his wartime memoirs. The latter depicts Fieser as a man involved in military research through a desire to help his country resist a potential aggressor, pursuing napalm as a morally preferable weapon compared to the "inhumane" vesicants which he had initially been assigned to develop (Fieser 1964, p. 14). Fieser's postwar disclosure in *Industrial and Engineering Chemistry* even presented napalm as a laudable contribution to America's war effort and, by implication, the welfare of mankind (Fieser *et al.* 1947).[25]

Nevertheless, because Fieser's moral judgments are somewhat opaque our analysis engages a speculative reconstruction of his moral reasoning, specifically one that merges Fieser's enthusiastic participation in incendiary work with popular American wartime sentiments that approved of the Allied civilian bombing campaign. In this respect it perhaps assesses the general merits of Second World War research on incendiaries designed for use against cities more than the actual views of Fieser himself.

First, it may be helpful to address some interpretations of the *Jus in Bello* principles which would forbid the development of incendiary weapons for use against civilian structures:[26]

(1) Absolutist interpretations of the principle of discrimination which forbid any targeting of noncombatants;
(2) Fieser's even more stringent 1973 interpretation of the principle of proportionality, which condemned all use of napalm against persons;
(3) Ethical judgments reflected in the UN Convention on Certain Conventional Weapons, which prohibits the use of incendiaries against noncombatants in any form (United Nations 1980) and consequently has the potential to render chemists so engaged liable for war crimes.[27]

---

[24] As explained in Greenwood (1998, p. 226), the US Naval Commander's handbook on the laws of war includes Fieser's "bat bombs" in its list of weapons (along with German V-1 & V-2 Rockets, and Japanese "balloon bombs") that are forbidden as inherently indiscriminate (US Department of the Navy, 1997).

[25] Though it is suggestive that Fieser emphasizes napalm's battlefield role over its use in the civilian bombing campaign.

[26] That might reasonably be expected to contain noncombatants.

[27] However, we are not sure American chemists could be charged, because the US did not sign the relevant portions of the convention (Neer 2013, pg. 192).

Each of these three interpretations depends on accepting particular judgments about the principle of discrimination. The first relies on an absolutist understanding of noncombatant immunity and the latter two judge the use of napalm in specific antipersonnel applications a great enough evil to *a priori* outweigh any good that might result. Moreover, interpretations two and three are historically-grounded in a retrospective awareness of the suffering caused by napalm's 'stickiness' when used as an antipersonnel weapon, something that was foreseeable but not as readily apparent at the time of Fieser's work as it would become later.

Thus it may be more interesting to consider whether Fieser's napalm work might be justifiable using interpretations of the *Jus in Bello* criteria that in principle allow for attacks against civilians. As explained in Section 3, the Allied civilian bombing was justified using a web of arguments based on proportionality, the doctrine of double effect, and the militarization of civilians. These rest on questionable assumptions and appear less compelling in view of the questions about their effectiveness and necessity which have arisen in hindsight. However, they were widely accepted by wartime Americans (Hopkins 1966) and remain a topic of considerable recent debate (Grayling 2006; Bess 2006, pp. 88-110). In short, Fieser's actions may be justifiable in principle, but involve abandoning previously-accepted norms concerning noncombatant immunity. At minimum the rejection of established norms represents poor practice, at least when done outside the scope of a sustained, professionally-informed, and appropriately public dialogue about the implications of such a move.[28]

---

[28] A recent NRC report on military research (2014) cites the public ethical debate over the human genome project as an example of fruitful dialogue. Review programs to assess weapons' legality like that advocated by Lawand (2006, International Committee of the Red Cross 2006) provide convenient opportunity for ethical discourse in military research, although in these cases care should be taken not to conflate legal and ethical considerations.

5. The Dow Napalm Controversy

5.1 Historical description

In the 1960s and early 1970s the American and South Vietnamese effort to resist North Vietnamese aggression was beset by thorny problems. Unable to risk the expansion of Cold War hostilities that might attend invasion of the communist north, they faced an enemy who employed terror and anticolonial appeals to supplement their invasion of the South with a guerilla war. Moreover, the latter involved attacks on American forces by ununiformed Vietcong fighters who, after attacking from within or near a village, quickly blended back in with the local populace, who through a combination of terror, anticolonial patriotism, or simple lack of trust would often do nothing to help the offended Americans track down the offending Vietcong.

The specific strategies American military commanders employed to address this problem, their rationale, and precedents are described by Walzer (2006, pp. 186-196). The bottom line is that they adopted rules of engagement that militarized Vietnamese civilians who did not evacuate to designated strategic hamlets or actively seek to expel Vietcong fighters. The British had successfully employed a similar strategy to squelch an earlier communist insurgency in Malaya. Whatever its strategic value, the strategy effectively shifted responsibility for upholding the principle of noncombatant immunity from the combatants onto the Vietnamese civilians themselves.

As a result the strategy ended up rationalizing incidents like the attack on a "communist rest center" witnessed by Bernard Fall in 1965:

> As we flew over the village it looked to me very much as any normal village would look [...] a peaceful scene. [...] I could see the napalm bombs dropping from the wings [...] an incredibly bright flash of fire. [...] The napalm was supposed to force the people – fearing the heat and the burning – out into the open. Then the second plane was to move in with heavy fragmentation bombs to hit whatever – or whomever – had rushed out into the open [Fall 1965, p. 25].

Fall's account was reported in the New Leftist *Ramparts* magazine, along with his skepticism about whether any communists were actually in the village at the time (*ibid.*, p. 26). Additional accounts followed, some of which included graphic images of horrible napalm-produced wounds.[29] When coupled with popular unease over America's involvement in Vietnam, these provoked a storm of protest that spread to napalm's suppliers, including the United Aircraft Corporation, Witco Chemical, and Dow.

These protests soon centered on Dow and included calls for a boycott of its most popular product, the plastic 'Saran Wrap'; vandalism; picketing; and over two hundred campus protests, some of which involved harassment of Dow recruiters (Wells 1996, pp. 84-88). Featuring slogans like 'Napalm Burns Babies, Dow Makes Money' and 'Nazi Ovens in 44, Napalm in 66' the protests evoked images of industrial collusion in the holocaust and elicited much publicity (Whitehead 1968, p. 264; Brandt 1997, p. 353).

Dow was caught by surprise. When the protests started no member of Dow's governing troika was even aware of its napalm operations, which involved only ten employees and accounted for a mere ~0.25% of its annual revenue.[30]

Despite the relative unimportance of its napalm operations, Dow refused to give in to the protester's demands that it cease production. Through its chairman Carl Gerstacker it argued:

> […] we are a supplier of goods to the defense department and not a policy maker. We do not and should not try to decide military strategy or policy.
>
> Simple good citizenship requires that we supply our government and our military with those goods which they feel they need whenever we have the technology and capability and have been chosen by the government as a supplier.
>
> We will do our best, as we always have, to try to produce what our defense department and our soldiers need in any war situation. Purely aside from our duty to do this, we will feel deeply gratified if what we are able to provide helps to protect our fighting men or to speed the day when fighting will end. [Brandt 1997, p. 353; also in Brandt 2003, p. 95; Whitehead 1968, pp. 264-265]

---

[29] For examples see Pepper 1967 and Colaianni 1966.

[30] Whitehead 1968, pg. 264; Brandt 1997, pg. 352. These figures perhaps understate napalm's contribution to Dow's profits (Friedman 1973, pg. 130).

This seemed to imply Dow was deflecting moral responsibility for the use of napalm wholly onto the US government, adopting a 'Nuremberg defense' like those deemed legally inadequate in post-Second World War trials of German industrial leaders. However, Dow's President Ted Doan later clarified that Dow's continued napalm operations reflected its judgment that America was fighting a just war:

> All of the debate in the world about how we got [into the war] or how we get out […] doesn't change the fact that we are there nor the fact that our men are there and need weapons to defend themselves.
>
> […] We reject the validity of comparing our present form of government with Hitler's Germany. In our mind our government is still representative of and responsive to the will of the people.
>
> Further, we as a company have made a moral judgment on the long-range goals of our government and we support these. We may not agree as individuals with every decision of every military or government leader but we regard these leaders as men trying honestly and relentlessly to find the best possible solution to very, very complex international problems. As long as we so regard them, we would find it impossible not to support them. This is not saying as the critics imply that we will follow blindly and without fail no matter where our government leads. […] Should despotic leaders attempt to lead our nation away from its historic national purposes, we would cease to support the government.
>
> Our critics ask if we are willing to stand judgment for our choice to support our government if history should prove us wrong. Our answer is yes. [Doan 1967]

In short, Dow argued that the US effort in Vietnam met the *Jus ad Bellum* Just War criteria of sovereign authority and right intent and that consequently Dow had a duty to provide the American military with napalm needed to defend itself and achieve its aims.

However, Dow had not publicly addressed whether napalm was used indiscriminately, an issue with which it was in private deeply concerned. Its leaders spent two days of a 1967 board of directors meeting discussing the "moral and ethical considerations involved" (Brandt 1997, p. 360). Ultimately Dow decided to continue its napalm operations. They judged it had an obligation to support American soldiers and emphasized governmental assurances that reasonable precautions were taken to avoid hurting civilians as well as the findings of doctors who had reported few or no cases of napalm burn injuries in Vietnamese hospitals.[31]

Nevertheless, Dow president Ted Doan remained open to the possibility napalm was being used indiscriminately. Two years later he told protesters that "if they could prove to him that napalm was being used, intentionally or not, primarily on a civilian population, he would do all he could to get the company out of the contract" (Wells 1996, p. 295). Unfortunately the ensuing discussion produced an impasse, in which at least one antiwar demonstrator privately wondered exactly what evidence might convince Doan (*ibid.*). In turn the company's own accounts betray that Dow's otherwise sympathetic leaders had difficulty seeing past protesters' rudeness and occasional incoherence (Brandt 1997/2003). When combined with the protesters' all-or-nothing attitude towards the use of napalm in Vietnam, this effectively directed attention away from the rules of engagement which allowed attacks on questionable targets, and from the issue of whether napalm satisfied the principle of proportionality, both of which later became foci of postwar discussions about whether and how incendiaries might legitimately be employed (Lumsden 1975, International Committee of the Red Cross 1973, Björnerstedt 1973).

However, by then the issue had passed. Doan's apparent skepticism notwithstanding, later that year Dow's contract to supply napalm to the US military was not renewed. Rumors circulated that Dow had chosen not to bid competitively, although its official history provides no evidence that was the case (Brandt 1997).

5.2. Dow's responsibilities

To assess Dow's moral responsibilities it is helpful to first clarify what was at stake. The protesters argued:

---

[31] Friedman 1973, pp. 127-29; Brandt 1997, pp. 357-8; Brandt 2003, p. 9. Defense secretary Robert MacNamara assured Doan that napalm was "a military necessity" used with precautions "as painstaking as we can make them without hamstringing our military operations" (Whitehead, 1968, p. 268), leading Doan to later claim "napalm is a good discriminate, strategic weapon" (Friedman 1973, p. 115).

(1) America's involvement in Vietnam at minimum did not satisfy several *Jus ad Bellum* criteria.
(2) Napalm was being used in ways that violated the *Jus in Bello* criteria of discrimination.
(3) Napalm itself violated the *Jus in Bello* criteria of proportionality when used as an antipersonnel weapon.

If true, argument 1 meant that industrial concerns like Dow ought not to provide war-related goods at all, and arguments 2 and 3 that in supplying napalm Dow contributed to disproportionate warfare and unjustified attacks on noncombatants.

Dow understood its responsibility to address these charges. It also recognized that if they were true Dow would face a responsibility dilemma between its desire to avoid complicity in unjust warfare, on one hand, and both its contractual obligations to the government and its civic[32] and personal[33] duty to support American soldiers, who greatly valued napalm's effectiveness against attacking enemy troops, on the other. Thus, after an initial misstep in which Dow seemingly deflected responsibility onto the government, it addressed the *Jus ad Bellum* criteria directly. It took steps to assess whether it was indeed supplying napalm to parties who were misusing it to unjustly harm civilians. In this respect, it was unfair to compare Dow with holocaust-associated industrialists like the Zyklon B supplier Bruno Tesch, who deflected responsibility for genocide onto the state and offered that the killings would have been achieved by other means had he refused complicity.[34]

---

[32] For an argument that military research is a civic duty see Kemp 1994.

[33] Here Dow leaders explicitly cited employees' family members engaged in the fighting.

[34] Tesch's lawyer argued, "if Tesch did know the use to which the gas was being put, and had consented to [supplying Zyklon B for use in murder], this happened only under enormous pressure from the S.S. [and], had Tesch not co-operated, the S.S. would certainly have achieved their aims by other means." In fairness the comparison should not be pressed too far because Tesch argued after the fact and likely realized that the harms of genocide far outweigh the benefits of pest control. For details see the UN Commission on War Crimes Report (1997).

5.3 Ethical analysis

Dow ultimately decided the American war effort in Vietnam met the *Jus ad Bellum* criteria of legitimacy and right intent. Thus it did not view itself as supporting an unjust war of aggression or expansionist conquest but rather a defensive one aimed at preventing the spread of Soviet-style repression in Southeast Asia.[35] Dow refused to commit to a position on the more difficult issue to evaluate *Jus ad Bellum* proportionality and necessity criteria, however. It resolved any associated moral dilemmas in favor of acknowledging the reality of the fighting and consequently decided in favor of its contractual, civic, and personal obligations to support the American military engaged therein.

In continuing as a napalm supplier, Dow also accepted a responsibility to avoid contributing to unjustified attacks on noncombatants. However, it again weighted the potential misuse of napalm against the impact of its loss on the ability of the largely conscript US forces to defend themselves in genuine combat situations. By all accounts napalm was very effective and fulfilled a vital role in such situations. In doing so Dow rejected claims that napalm is inherently disproportionate with respect to the *Jus ad Bellum* criteria. Instead it concluded napalm may be used to both advance just war aims and support combatants' in asserting a right of self-defense. Dow also shared the protesters' judgment that targeting noncombatants in napalm attacks was unjustified and sought assurances that napalm was being used discriminately in Vietnam. In this respect Dow was not so much adjudicating benefits and harms associated with napalm's intended use. Instead it weighted benefits associated with napalm's intended use against harms from its expected misuse, ultimately deciding in favor of the former.

However, in doing so Dow understood that it could still take steps to mitigate against misuse, although the strategy of warning, education, and restricting access that are commonly employed to limit the misuse of chemical products[36] might need to be modified in the case of a governmental user. Thus Dow sought assurance that its napalm would not be

---

[35] Albeit in the form of a Soviet-supported movement with its own anticolonial aims.

[36] Examples include warning labels on environmentally-harmful pesticides and the criminalization of harmful drugs and their precursors.

misused through a strategy that combined governmental assurances with eyewitness testimony and medical reports. When these were set against the easily-discounted assertions of the often incoherent protesters, Dow understandably concluded that napalm was largely used discriminately but remained open to reevaluating its conclusions in case of later changes.

However, because of its distrust of the protesters and its policy of non-interference in military affairs, Dow did not push the crucial questions about the US rules of engagement which governed how napalm could be used. These rules represented a case of moral slippage by redistributing the burden of discrimination onto the noncombatants themselves. This practice effectively served to minimize US casualties while giving the "appearance of attending to the combatant/noncombatant distinction" (Walzer 2006, p. 193).[37]

6. Conclusions and Possible Lessons

Taken together the cases of Louis Fieser and Dow Chemical illustrate how Just War Theory may be employed by chemists engaged in military work. They also suggest that it is important for chemists and managers engaged in the chemical enterprise to:

(1) Accept responsibility for undertaking ethical reflection about both expected outcomes and foreseeable unintended harms and misuses.
(2) Consider social context. Louis Fieser was not just developing napalm; he was developing napalm in a world experimenting with civilian bombing. Dow was not just supplying napalm for self-defense purposes; it was supplying napalm to a military employing dubious rules of engagement.
(3) Commit to a sustained and charitable ethical discussion in which no question is off-limits and where opposing arguments are strengthened before being addressed. Dow's engagement with the protesters helped it avoid several ethical pitfalls and produced a coherent justification for its napalm operations. However, had it undertaken a more charitable reconstruction of protesters' arguments, its execu-

[37] General Telford Taylor, US chief counsel at Nuremburg, judged the types of attacks these rules permitted violations of the laws of war (1970, pp. 144-45).

tives might have better understood the systemic issues that at least allowed for napalm's misuse. So equipped, Dow might likely have been better able to exercise moral leadership in promoting the responsible use of its products.

(4) Anticipate shifts in responsibility as projects evolve. Louis Fieser's early napalm work did not clearly violate Just War principles and we suspect that in 1941 he would have balked at the prospect of designing bombs for use against Japanese civilian housing. Yet later in the war he did just that.[38] It is not enough to consider ethical issues at the start of an enterprise. Ethical discussion should be made part of that enterprise and kept up to date.

References

Anonymous: 1921, 'Christian Conscience and Poison Gas', *The Literary Digest*, January 8, p 38.

Arneson, R.J.: 2006, 'Just Warfare and Noncombatant Immunity', *Cornell International Law Journal*, **39** (3), 663-688.

Bess, M.: 2006, Choices Under Fire: Moral Dimensions of World War II, New York: Knopf.

Björnerstedt, R.: 1973, Napalm and Other Incendiary Weapons and All Aspects of Their Possible Use, New York: United Nations.

Brandt, E.N.: 1997, *Growth Company*, East Lansing: Michigan State University Press.

Brandt, E.N. & Gerstacker, C.A.: 2003, *Chairman of the Board*, East Lansing: Michigan State University Press.

Brown, M.: 1971, The Social responsibility of the scientist, New York: Free Press.

Chalk, R.: 1989, 'Drawing the Line: An Examination of Conscientious Objection in Science', *Annals of the New York Academy of Sciences*, **577** (1), 61-74.

Charles, D.: 2005, Master Mind: The Rise and Fall of Fritz Haber, the Nobel Laureate who Launched the Age of Chemical Warfare, New York: Ecco.

Chemical Corps Association: 1948, The Chemical Warfare Service in World War II: A Report of Accomplishments, New York: Reinhold.

Colaianni, J.F.: 1966, 'Napalm: A Small Town Diary', *Ramparts*, August, 46-50.

Cornwell, J., 2003, Hitler's Scientists: Science, War, and the Devil's Pact, New York: Viking.

Couffer, J.: 1992, *Bat Bomb: World War II's Other Secret Weapon*, Austin: University of Texas Press.

Coulson, C.A.: 1966, *Responsibility*, London: Christian Socialist Movement.

---

[38] For similar examples in computer science and ethics research see Singer 2010, pp. 304-306.

Crane, C.C.: 1993, Bombs, Cities, and Civilians: American Airpower Strategy in World War II, Lawrence: University Press of Kansas.

Davis, M.: 1999a, 'Angriff auf "German Village"', *Der Spiegel*, October 11 (online available at http://www.spiegel.de/spiegel/a-50307.html, accessed 10 November 2016). English translation in: *Sealand Letter*, no. 19, March 2005 (online available at http://www.mudarchitecture.com/assault-on-german-village/, accessed 10 November 2016).

Davis, M.: 1999b, 'Berlin's Skeleton in Utah's Closet', *Grand Street*, **69**, 92-100.

Dinegar, R.H.: 1989, 'The Moral Arguments for Military Research', *Annals of the New York Academy of Sciences*, **577** (1), 10-20.

Doan, H.H.: 1967, 'Why Does Dow Chemical Make Napalm?', *Wall Street Journal*, December 8.

Douglas, H.E.: 2003, 'The Moral Responsibilities of Scientists (Tensions Between Autonomy and Responsibility)', *American Philosophical Quarterly*, **40** (1), 59-68.

DuBois, J.E.: 1952, *The Devil's Chemists*, Boston: Beacon.

Fall, B.B.: 1965, 'This Isn't Munich, It's Spain', *Ramparts*, December, 23-29.

Fichtelberg, A.: 2006, 'Applying the Rules of Just War Theory to Engineers in the Arms Industry', *Science and Engineering Ethics*, **12** (4), 685-700.

Fieser, L.F.: 1964, *The Scientific Method*, New York: Reinhold.

Fieser, L.F.; Harris, G.C.; Hershberg, E.B.; Morgana, M.; Novello, F.C. & Putnam, S.T.: 1947, 'Napalm', *Industrial and Engineering Chemistry*, **38**, 768-73.

Fischhoff, B.: 2014, 'Ethical and social issues in military research and development', *Telos*, **169**, 150-154.

Forge, J.: 2004, 'The Morality of Weapons Research', *Science and Engineering Ethics*, **10** (3), 531-42.

Forge, J.: 2008, *The Responsible Scientist: A Philosophical Inquiry*, Pittsburgh: University of Pittsburgh Press.

Friedman, S.: 1973, 'This Napalm Business', in: D. Obst (ed.), *In the Name of Profit*, New York: Doubleday, pp. 115-36.

Frowe, H.: 2014, *Defensive Killing*, Oxford: Oxford University Press.

George, D.B.: 2010, 'Green Chemistry as an Expression of Environmental Ethics', in: S.K. Sharma & A. Mudhoo (eds.), *Green Chemistry for Environmental Sustainability*, Boca Raton: CRC Press, pp. 105-25.

Grayling, A.C.: 2006, *Among the Dead Cities*, New York: Walker.

Greenwood, C.: 1998, 'The Law of Armed Conflict: Into the Next Millennium', in: M.N. Schmitt & L.C. Green (eds.), *International Law Studies Volume 71*, US Naval War College, pp. 183-232.

Haber, L.F.: 1986, The Poisonous Cloud: Chemical Warfare in the First World War, Oxford: Oxford University Press.

Hahn, O.: 1970, *My Life*, New York: Herder.

Hayes, P.: 2005, From Cooperation to Complicity: Degussa in the Third Reich, Cambridge: Cambridge University Press.

Hildebrand, J.H.: 1955, 'The Social Responsibility of Scientists', *Proceedings of the American Philosophical Society*, **99** (2), 46-50.

Hoffmann, R.: 1975, 'Scientific Research and Moral Rectitude', *Philosophy*, **50** (194), 475-77.

Hoffmann, R.: 1995, *The Same and Not the Same*, New York: Columbia University Press.

Hoffmann, R.: 2008, 'Should've', online available at: http://www.roaldhoffmann.com/sites/all/files/Should've%209%20with%20cover%20page.pdf, accessed 30 March 2016.

Hopkins, G.E.: 1966, 'Bombing and the American Conscience During World War II', *Historian*, **28**, 451-73.

International Committee of the Red Cross: 1973, Weapons That May Cause Unnecessary Suffering or Have Indiscriminate Effects, Geneva.

International Committee of the Red Cross: 2006, A Guide to the Legal Review of New Weapons, Means and Methods of Warfare, Geneva.

Jacob, C. & Walters, A.: 2005, 'Risk and Responsibility in Chemical Research: The Case of Agent Orange', *Hyle: International Journal for Philosophy of Chemistry*, **11**, 147-66 [also in this volume, chapter 7].

Jeffreys, D.: 2008, Hell's Cartel: IG Farben and the Making of Hitler's War Machine, New York: Metropolitan.

Johnson, J.T.: 1999, *Morality and Contemporary Warfare*, New Haven: Yale University Press.

Johnson, J.T.: 2011, Ethics and the Use of Force: Just War in Historical Perspective, Burlington: Ashgate.

Kemp, K.W.: 2007, 'Conducting Scientific Research for the Military as a Civic Duty', *Annals of the New York Academy of Sciences*, **577** (1), 115-21.

Kovac, J.: 2007, 'Moral Rules, Moral Ideals, and Use-Inspired Research', *Science and Engineering Ethics*, **13** (2), 159-69.

Kovac, J.: 2013, 'Science, Ethics and War: A Pacifist's Perspective', *Science and Engineering Ethics,* **19** (2), 449-60.

Kovac, J.: 2015, 'Ethics in Science: The Unique Consequences of Chemistry', *Accountability in Research*, **22** (6), 312-29.

Lawand, K.: 2006, 'Reviewing the Legality of New Weapons, Means and Methods of Warfare', *International Review of the Red Cross*, **88** (864), 925-30.

Lazar, S.: 2010, 'The Responsibility Dilemma for *Killing in War*: A Review Essay', *Philosophy & Public Affairs*, **38**, 180-213.

Lazar, S.: 2014, 'Necessity and Noncombatant Immunity', *Review of International Studies*, **40** (1), 53-76.

Lazar, S.: 2016, 'War', in: E.N. Zalta (ed.), *Stanford Encyclopedia of Philosophy,* (Winter 2016 Edition), forthcoming URL: http://plato.stanford.edu/archives/win2016/entries/war/).

Lemann, N.: 1973, 'Napalm's Daddy – 31 Years Later', *The Harvard Crimson,* October 12.

Lenoir, D. & Tidwell, T.T.: 2009, 'Louis Fieser: An Organic Chemist in Peace and War', *European Journal of Organic Chemistry*, no. 4, 481-91.

Liberman, P.: 1996, 'Does conquest pay? The exploitation of occupied industrial societies', Princeton: Princeton University Press.

Lietzau, W.K.: 2004, 'Old Laws, New Wars: Jus Ad Bellum in an Age of Terrorism', *Max Planck Yearbook of United Nations Law*, **8** (8), 383-455.

Lonsdale, K.: 1951: 'The Ethical Problems of Scientists', *Bulletin of the Atomic Scientists*, **7** (7), 201-204.

Lonsdale, K.: 1964, I Believe: The Eighteenth Arthur Stanley Eddington Memorial Lecture, Cambridge: Cambridge University Press.

Lumsden, M.: 1975, *Incendiary Weapons*, Cambridge MA: MIT Press.

Malsch, I.: 2013, 'The Just War Theory and the Ethical Governance of Research', *Science and Engineering Ethics*, **19** (2), 461-86.

May, L.: 2008, *Aggression and Crimes against Peace*, Cambridge: Cambridge University Press.

McMahon, J.: 2011, *Killing in War*, Oxford: Oxford University Press.

Mitcham, C. & Siekevitz, P. (eds.): 1989, *Ethical Issues Associated with Scientific and Technological Research for the Military*, Annals of the New York Academy of Sciences, vol. 577.

Moseley, A.: 2004, 'Just War Theory', in: *The Internet Encyclopedia of Philosophy* (online available at: http://www.iep.utm.edu/justwar/, accessed 10 November 2016).

Moy, T.D.: 1989, 'Emil Fischer as 'Chemical Mediator': Science, Industry, and Government in World War One', *Ambix,* **36** (3), 109-20.

National Research Council and National Academy of Engineering (USA): 2014, Emerging and Readily Available Technologies and National Security: A Framework for Addressing Ethical, Legal, and Societal Issues, Washington: The National Academies Press.

Neer, R.M.: 2013, *Napalm: An American Biography*, Cambridge MA: Harvard University Press.

Neilands, J.B.: 1971, 'Chemical Warfare', in: M. Brown (ed.), *The Social Responsibility of the Scientist,* New York: Free Press.

O'Donovan, O.: 2003, *The Just War Revisited*, Cambridge: Cambridge University Press.

Overy, R.J.: 2013, *The Bombing War : Europe 1939-1945*, London: Allan Lane.

Pepper, W.F.: 1967, 'The Children of Vietnam', *Ramparts Magazine*, January, 45-68.

Price, R.M.: 1997, *The Chemical Weapons Taboo*, Ithaca: Cornell University Press.

Remers, W.A.: 2000, Chemists at War: Accounts of Chemical Research in the United States During World War II, Tucson: Clarice.

Russell, E.: 2006, War and Nature: Fighting Humans and Insects with Chemicals from World War I to "Silent Spring", Cambridge: Cambridge University Press.

Ryberg, J.: 2003, 'Ethics and Military Research', in: B. Booss & J. Høyrup (eds.), *Mathematics and War*, Basel: Birkhäuser, pp. 352-64.

Schaffer, R.: 1980, 'American Military Ethics in World War II: The Bombing of German Civilians', *The Journal of American History*, **67**, 318-34.

Schmitt, M.N.: 2005, 'Precision Attack and International Humanitarian Law', *International Review of the Red Cross*, **87** (859), 445-66.

Schummer, J.: 2001, 'Ethics of Chemical Synthesis', *Hyle: International Journal for Philosophy of Chemistry,* **7**, 103-24.

Singer, P.W., 2010, 'The Ethics of Killer Applications: Why Is It So Hard to Talk About Morality When It Comes to New Military Technology?', *Journal of Military Ethics*, **9** (4), 299-312.

Strawser, B.J.: 2013, 'Revisionist Just War Theory and The Real World', in: F. Allhoff, N.G. Evans & A. Henschke (eds.), *Routledge Handbook of Ethics and War,* Abingdon: Routledge, pp. 76-89.

Stoltzenberg, D.: 2004, *Fritz Haber: Chemist, Nobel Laureate, German, Jew*, Philadelphia: Chemical Heritage Press.

Stoltzenberg, D.: 2000, 'Scientist and Industrial Manager: Emil Fischer and Carl Duisberg', in: J.E. Lesch, (ed.), *The German Chemical Industry in the Twentieth Century*, Dordrecht: Kluwer, pp. 57-90.

Taylor, T.: 1971, Nuremberg and Vietnam: An American Tragedy, Chicago: Bantam.

United Nations: 1980, Convention on Prohibitions or Restrictions on the Use of Certain Conventional Weapons Which May be Deemed to be Excessively Injurious or to Have Indiscriminate Effects (and Protocols) (As Amended on 21 December 2001) 10 October 1980, 1342 UNTS 137 (online available at: http://www.refworld.org/docid/3ae6b3ac4.html, accessed 10 November 2016).

United Nations Commission on War Crimes: 1997, The Zyklon B Case: Trial of Bruno Tesch and Two Others, British Military Court, Hamburg, 1-8th March 1946, in Law Reports of the Trials of War Criminals, vol. 1, Buffalo: Hein.

United States Department of the Navy: 1997, *Annotated Supplement to the Commander's Handbook on the Law of Naval Operations* (NWP 1-14M/MCWP 5-2.1/COMDTPUB P5800.1), para. 9.1.2, n. 12 (online available at: http://www.jag.navy.mil/distrib/instructions/AnnotatedHandbkLONO.pdf, accessed 18 November 2016].

Walzer, M.: 2006, Just and Unjust Wars: A Moral Argument with Historical Illustrations, 4th ed., New York: Basic Books.

Weber, M., Deußing, G.: 2013, 'Courageous Questioning of Established Thinking: The Life and Work of Hermann Staudinger', in: V. Percec (ed.), *Hierarchical Macromolecular Structures: 60 Years after the Staudinger Nobel Prize*, vol. 1, Heidelberg: Springer, pp. 81-138. Springer.

Weeramantry, C.G.: 1987, *Nuclear Weapons and Scientific Responsibility*, Wolfeboro: Longwood.

Wells, T.: 1996, The War Within: America's Battle over Vietnam, New York: Holt.

Whitehead, D.: 1968, The Dow Story: the History of the Dow Chemical Company, New York: McGraw-Hill.

Whittemore, G.F.: 1975, 'World War I, Poison Gas Research, and the Ideals of American Chemists', *Social Studies of Science*, **5** (2), 135-63.

Wolpe, P. R.: 2006, 'Reasons Scientists Avoid Thinking About Ethics', *Cell*, **125** (6), 1023-25.

Chapter 5

Corporate and Governmental Responsibilities for Preventing Chemical Disasters: Lessons from Bhopal

Ingrid Eckerman and Tom Børsen

Abstract: In the evening of December 2nd, 1984, a series of unfortunate events led to 43 tons of methyl isocyanate being vaporized and spread over the city of Bhopal in central India. The accident, which continued throughout the early hours of December 3rd, resulted in the deaths of several thousand people and left hundreds of thousands more with permanent injuries. The Bhopal toxic leakage is widely regarded as the largest chemical industrial disaster in the world. It is also one of the most publicly scrutinized disasters, leading to vivid evidence about the threat to public health posed by the chemical production industry. This chapter argues that Union Carbide Corporation, through its local subsidiary Union Carbide India Limited, as well as the governments of India and Madhya Pradesh, are those who were most accountable for the leakage. Furthermore, the chapter shows that several ethical rules such as: the Golden Rule, the human rights to safety, the precautionary principle, and the principles for a sustainable society were violated. In the final sections, the chapter discusses the measures that should be taken to prevent future exposure to toxic chemical substances which may occur due to accidents at chemical production facilities.

1. Introduction

The gas leakage from the Union Carbide plant in Bhopal, India, in 1984 is the largest industrial chemical accident to have ever occurred in the world. Over 500,000 people were exposed to poisonous gases, resulting in between 3,000 and 10,000 people dying within the first weeks of the accident and between 100,000 and 200,000 being left with possible permanent injuries. Due to its magnitude, the catastrophe remains one of chemistry's most important ethical cases – the lessons of which should never be forgotten by practitioners of chemistry or by any other individuals involved in the production of potentially dangerous compounds.

In the aftermath of the disaster, volunteers and journalists quickly invaded Bhopal's Old Town resulting in a large quantity of documented information about the event. Interviews with employees, health care staff, and inhabitants were published in newspapers and collected by non-governmental organizations. A mine of detailed material, intended solely for internal use, was made public. This depth of investigation resulted in specific details regarding the procedures which led to the leakage (such as the effects of the gases and the actions of the company, the government, and the medical and scientific establishments) being recorded within thousands of articles, reports, books, and films. For example, the reference list for the book *The Bhopal Saga* alone contains some 200 items (Eckerman 2005b). Activists, volunteers, and survivors continue their investigations and battle for justice even after more than 30 years since the accident.

The investigations illustrate the threat to public health posed by the chemical industry. They highlight how decision-makers, citizens, and sometimes even the experts themselves are unaware of the dangers of the chemical compounds that they might be exposed to due to accidents at chemical plants. The responsible parties do not know how, or to what degree, these compounds might be harmful to themselves either in the short or long term. The case also illustrates the roles played by transnational companies, and how they often succeed in influencing decisions made at the regulative level. These mechanisms are not only confined to less developed countries, but are also found in the so called 'richer' countries.

In relation to minimizing the risk of industrial hazards, three main actors can be identified: companies, governments, and the civil society (which in this case is represented by trade unions and other NGOs). The most powerful entities are the companies which have expertise and money. The governments, especially in low income countries, rely on these companies for economic development. The trade unions are mostly concerned about jobs, but also about the health of their workers. These, along with other non-governmental organizations such as those protecting human rights and the environment, typically hold less power.

In the Bhopal case, the major actors identified are Union Carbide Corporation (UCC) and its subsidiary Union Carbide India Limited (UCIL); the Governments of India (GoI) and Madhya Pradesh (MP Gov); and, finally, the survivors' organizations and plant workers' trade unions. Each of these actors had several ethical aspects to consider which will be analyzed and discussed in this chapter. The chapter begins by providing a review of the catastrophe, based on the previously mentioned book (Eckerman 2005b).

2. Background

2.1 The green revolution

During the 1950's, it was widely held that the expanding global population, particularly rapid in lower income countries, necessitated an increase in global food production. Therefore, increased global efforts to develop new agricultural techniques were made. Countries most regularly threatened by famine eagerly turned to solutions developed in the chemical industry to increase their own food production. In India, the introduction of synthetic fertilizers and pesticides were major developments in a wider reform of farming practices known as the 'green revolution'. A decade after colonial rule had come to an end, the Indian government also targeted foreign investment as being a key route to industrial development. Large corporations were encouraged to establish subsidiary companies within the country. The US owned Union Carbide Corporation (UCC) formed the subsidiary Union Carbide India Limited (UCIL), 50.9 % of which was controlled by UCC with the remaining 49.1 % being held by Indian investors, including the Government of India and government-controlled banks.

The company's pesticide plant was constructed in 1969 in Bhopal – a city located on a key railroad junction in the middle of the country with a nearby lake that acted as a source of fresh water. Being one of the first major foreign investment projects, the plant was regarded as the crown of Indian industrialism. The upper social strata of Bhopal were closely involved in the plant's activities while working for UCIL became a prestigious affair and offered an excellent salary, even for those on the lower social tiers.

Methyle isocyanate (MIC) was used as an intermediate in the production of carbamate pesticides (brandmark Sevin). To start with, MIC was imported from the US, but later a new plant for producing MIC was planned. The MP Government designated a zone southeast of the town that, due to the prevailing wind patterns, would specifically accommodate industries involved with dangerous substances. The MIC plant was at first directed to this area. However, an exception was made, and UCIL received the permission to build the MIC plant within the existing plant area close to the railway station in the Old Town. In 1980, the Bhopal plant started to manufacture MIC, applying knowledge and designs supplied from the parental company based in the USA.

2.2 MIC and its properties

Between 1958 and 1973, Union Carbide used a method of producing the carbamate pesticide without using MIC (Figure 1b). For economic reasons, this was later changed to a more hazardous process (Chouhan 1994).

A run-away-reaction was described in the UCC manuals (UCC 1976, UCIL 1978):

> Water reacts exothermically to produce heat and carbon dioxide. As a result, the pressure in the tank will rise rapidly if MIC is contaminated with water. This reaction may begin slowly, especially if there is no agitation, but it will become violent. [...] MIC reacts vigorously with contaminants such as water, acids, alkalis and amines and can polymerize rapidly if in contact with iron, steel, zinc, tin, galvanized iron, copper and its alloys.

Exposed to high temperatures, MIC breaks down to hydrogen cyanide (HCN) (Ramachandran 1994). Even small concentrations of HCN are known to kill people rapidly.

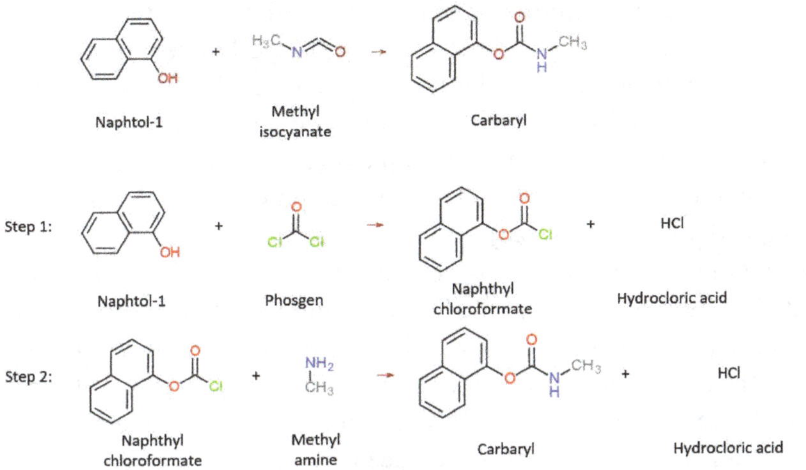

(a)

(b)

Figure 1. The production of Sevin (carbaryl): (a) direct pathway with MIC, (b) original pathway without MIC.

3. The Accident and Its Causes

3.1 The direct cause

The specifics of the accident and its effects are described and remain contested in several books and reports including those released by Union Carbide (UCC 1985, Kalelkar 1988), Indian authorities (Varadarajan 1985), trade unions (ICFTU-ICEF 1985), non-governmental organizations (Delhi Science Forum 1985), foreign experts on disaster medicine (Kulling & Lorin 1987) and former employees (Chouhan 1994). However, what is not disputed today is that during the evening of December 2nd, the supervisor on duty had been transferred from a different unit to the MIC-plant – a plant which he was not familiar with.

According to the account, the supervisor ordered a group of untrained workers to connect a water hose to the pipeline system and to let it run for several hours, but had forgotten to inform them to add a slip bind. Approaching midnight, a group of workers noticed the presence of MIC in the air and informed members of staff who deemed that no action was necessary as leakages were common. Shortly afterwards, the production tank began to rumble, the pressure reading started to rise, and the levels of MIC in the air increased. The staff gradually realized that something was very wrong and attempted to use the safety equipment to solve the issue, but failed. It was not until approximately 2 a.m. that the main alarm, directed towards the surrounding community, sounded. By that stage injured people had already started to enter the hospital.

While the account above has been mostly agreed upon, what remains disputed is the direct cause of the leakage. Union Carbide, now owned by DowDuPont, maintains that it was impossible that the water from the hose could have reached the tank. Instead, it was a disgruntled worker who had directed the water directly into the tank. The worker, who has since identified himself, has never been sued. Activists, on the other hand, argue that the worker would have died immediately. They cling to the idea that it was an extra jumper line that made it possible for the water to enter the production tank (Chouhan 1994).

3.2 Deficient factory design

A number of investigators identified deficiencies in the design of the MIC plant in Bhopal as having contributed to the accident. Researchers include Chouhan (1994), Delhi Science Forum (1985), Council of Scientific and Industrial Research (CSIR) (Varadarajan 1985) and the Disaster Management Institute (Ramachandran 1994).

At the Bhopal plant, a dangerous but cost-effective method of manufacturing of Sevin was chosen. Instead of storing MIC in many small barrels, or producing only small batches of MIC, as was the practice in the UCC plant in Virginia, it was stored long term in two large tanks. For economic reasons the factory in Bhopal was not provided with the required safety equipment or security systems that the engineers had envisaged (Lapierre & Moro 2001). The safety features that were installed were under-dimensioned. Instead of using a computerized alarm system like that in the Virginia plant, the plant in Bhopal relied on manual operations. The poor staffing policy cut the training of unskilled employees and reduced the employment of people in skilled roles – leading to inadequate safety management. Other areas of criticism focused on the poor maintenance of the plant and that no emergency response plans were in force (Eckerman 2005b).

After the leakage, UCC's first line of defense was that the equipment installed in Bhopal was made in the USA to US specifications, and that it included safety equipment and standards virtually identical to the Virginia plant (Jones 1988). However, due to a backlash by the teams in Virginia, UCC was forced to admit that this was not the case.

The MIC unit in Bhopal was over-dimensioned from the start and always run at a loss. The US manager was replaced by an Indian manager who was ordered to achieve cost savings. And so he did, by cutting down on safety. It is this report's opinion that as it was UCC that chose the design and was well represented on the UCIL board, the company can hardly avoid accepting responsibility for the safety status of the plant.

3.3 Neglect of warning signals

The UCIL hierarchy regarded the plant as "one of the safest ships in the modern fleet". The chief medical officer at Bhopal claimed that "the safety precautions we took were the best possible" and stated "We did everything the Americans advised. In fact, we used to think that we were overdoing the safety" (Jones 1988). The reality, however, is that leakages occurred regularly. The trade union distributed 6,000 posters warning about the dangers. A union leader went on hunger strike at the factory's entrance to highlight the risks. Meetings and processions were held in the city and a journalist wrote several articles in the local press to further highlight the dangers. However, neither UCIL nor the MP Gov took any notice. Instead of acting on the warning signals, the officials dismissed the union leaders. It appears that the management's views on safety differed from those of the workers.

Safety audits were completed every year in US and European UCC plants, but only every second year in other parts of the world (Jones 1988). As stated in a 'Business Confidential' safety audit by UCC in May 1982, the senior officials of the corporation were well aware of "a total of 61 hazards, 30 of them major and 11 minor in the dangerous phosgene/methyl isocyanate units" (UCC 1985). This audit expressed severe concerns about the poor state and inappropriate placement of safety equipment; the lack of periodic checks to see that the instruments and alarm systems were functioning correctly; an alarming turnover of inadequately trained staff; unsatisfactory instruction methods; and a lack of rigor in maintenance reports. However, the report ended by stating that "No situations involving imminent danger or requiring immediate correction were noted during the survey". Nevertheless, UCC admitted in their own investigation report (UCC 1985) that most of the safety systems were not functioning on the night of the accident. The report revealed that:

- Tank temperatures were not logged;
- The vent gas scrubber (VGS) was not in use;
- The cooling system was not in use;
- A slip bind disc, that would prevent the water to take the wrong way, was not inserted when the pipes were washed;
- The concentration of chloroform in the production tank was too high;
- The tank was not pressurized;
- Iron was present because of corrosion;
- The tank's high-temperature alarm was not functioning;
- The evacuation tank was not empty.

Both the plant design and safety management were of a lower quality than that at UCC's other plants in US and Europe, meaning that UCC operated with a double set of safety standards. As a result, all the chemical conditions to cause the disaster were made present.

3.4 Contents of the gas cloud

Based on several reports (Varadarajan 1985, UCIL 1978, Subramaniam 1985, Kumar & Mukerjee 1985), it is likely that at areas nearest the factory the cloud would have been mainly composed of MIC and trimers of MIC, but would also have contained hydrogen cyanide (HCN), nitrogen oxides (NOX), carbon dioxide (CO2), and carbon monoxide (CO), all of which replaced the air (Figure 2). The released MIC, once in contact with the moisture in the air, would have been converted in the atmosphere into monomethyl amine (MMA) and carbon dioxide. It is improbable that cyanide (CN) would have been widely dispersed.

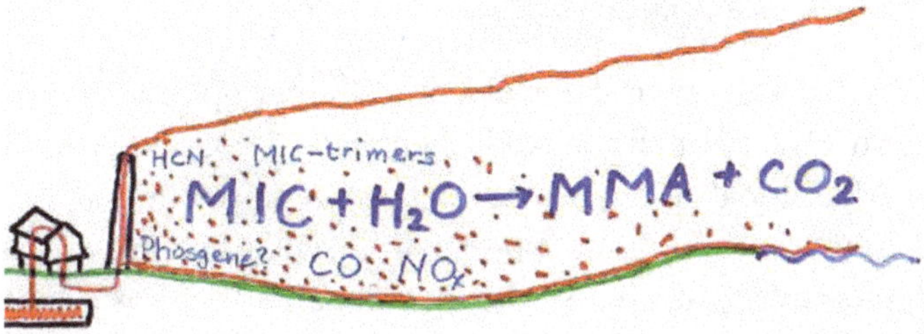

Figure 2. The suggested contents of the gas cloud (from Eckerman 2001).

4. Effects of the Bhopal Accident

4.1 Short term effects

The inhabitants of the surrounding neighborhood were awoken due to coughing, vomiting, a burning feeling in eyes and respiratory organs, and a feeling of suffocation. They attempted to flee from the plant – and thus ending up walking, running, and cycling in the same direction as the wind. Many of those living close to the plant collapsed dead in the streets. Those who reached the hospital arrived breathless with lungs filled with water and suffering from convulsions. The doctors present in the hospital were ill-informed and inadequately prepared, while the UCC doctors claimed, 'it's just like tear gas'. The death toll soon rose so much that the assistants were told to stop counting.

The following morning, the narrow streets in the Old Town were filled with dead bodies. At the railway station, an entire group of gypsies were found dead. Many bodies were collected and taken to the Muslim churchyard or the Hindu crematorium, others were simply dumped into the Narmada River by the police without registration. The town was completely silent, people were too sick to work or to even move. Trading stopped and farmers from outside districts stayed away. 4000 dead cattle had to be disposed of.

More than 500,000 people were exposed to the gases. The still-birth rate increased by 300 %; peri- and neo-natal mortality rates increased by 200 %. While the official death figure in 1991 was stated as 3,928, it is more probable that at least 8,000 died within the first days (Eckerman 2005b).

4.2 Long term health effects

For several years following the accident, different opinions existed about the long-term effects of the leaked chemicals. UCC consistently maintained that MIC could not cause permanent impairment – which was an opinion shared with various sections within scientific and official establishments. However, survivors' organizations and activists have fought to have their opinion acknowledged. It is their view that thousands of people have been left permanently disabled due to the gases. In the medical sphere, the term 'syndrome' refers to a combination of several symptoms and findings. Thus the 'Bhopal syndrome' has been used to refer to the combination of symptoms and damage to human organs that the gas leak caused.

In 1986, Dr. C.R. Krishna Murti, president of the commission that investigated the accident, stated that between 30,000 and 40,000 people had persistent disabilities (Kulling & Lorin 1987) and identified them within the following categories:

- Those who were so seriously disabled that they could not work and who often experienced difficulties in walking or cycling due to bad co-ordination.
- Those who had some persistent dysfunction in the airways and suffered from chronic pulmonary insufficiency, but still managed to work.
- Those who seemed well but who had a strongly decreased resistance to infections, especially in the lungs and airways.

In 1992, the Permanent People's Tribunal on Human Rights recommended that an international medical commission should assess the long-term effects. In 1994, the International Medical Commission on Bhopal (IMCB) met. According to the material and testimonies presented to IMCB, the experiences varied at different distances from the plant. Those living closest to the factory suffered from very severe, acute, as well as long-term issues. For some, the symptoms became more aggravated over time. Those living a little further away, or who had spent the night on the second or third floor of a building, recovered better with only a slight degree of breathlessness remaining. Those living on the slopes of the southern part of the Old Town were left with no remaining symptoms, but continue to suffer from the traumatic memories and worry about the effects on their future generations. These findings were verified to some extent by investigations undertaken by IMCB (Dhara 1992, Cullinan *et al*. 1997, Eckerman 1996).

The impression gained is that the information provided by UCC as well from the Indian Government (GoI) and ICMR was meant to portray a picture of injuries being of a mild degree only. The long-term effects were not illustrated in the legal process that followed the accident. There seems to be several reasons for this fact (Eckerman 2005b): UCC wanted to reduce their liability, and many of the leading doctors of ICMR were closely related to UCIL; careless handling of cyanide might be regarded as murder in the US; multinationals did not want mother companies to be responsible for daughter companies – if this were the case, it would affect the global market; the GoI did not wish to alienate multinational capital and they themselves had invited the chemical industries to establish plants in India; there was pressure on the GoI from the local chemical industry which were opposed to any restrictions; and finally, the GoI did not want to spend large amounts of money on health care.

4.3 Socio-economic effects

The economic situation of the hardest affected part of the Bhopal population in 1984 can be categorized as a dependence on physical and casual work or cattle for income, and male offspring for support in old age. A survey conducted in late December 1984, found that 75 % of the workforce was incapable of work mainly due to issues of breathlessness (National Institute for Working Life 2001). It was also recorded that many families had lost their stock of cattle. In a later survey conducted in 1985, it was reported that 25,000 families suffered from a total or substantial loss of income (Chauhan 1996). Parents did no longer have sons to support them in old age. Young girls did not get married and would remain dependent on their parents for a living.

4.4 Compensation from Union Carbide

In the days following the disaster, American lawyers entered Bhopal and persuaded the victims to file for damages. A short while later, back in the US and equipped with the signatures, the legal experts then began proceedings to claim billions of dollars in compensation (ICFTU 1998). However, in March 1985, the GoI passed the Bhopal Gas Leak Disaster Act which handed the government the statutory right to represent all victims in or outside India. This act alone almost completely disempowered the victims from participating in the adjudication process. For UCC, the act was to their great advantage (Shristi 2002).

In 1985, UCC made an offer of US$ 350 million as a settlement. The Union of India emphatically refused to accept the offer and instead demanded compensation of US$ 3.3 billion. In February 1989, there were few signs of progress, as UCC denied all liability. However, suddenly it was announced that a settlement had been reached between UCC and GoI: UCC agreed to pay US$ 470 million in a final settlement of its civil and criminal liability – corresponding to the insurance sum of US$ 350 million plus interest (Shristi 2002).

4.5 Dismantling the plant

Between 1985 and 1986 the factory was dismantled. Pipes, drums, and tanks from the production unit were cleaned with water combined with a chemical decontaminant, and then sold off to local entrepreneurs. However, the MIC and Sevin plants remain standing and continue to house the control room, the tanks, and storages of various residues. The plant is in decrepit condition with isolation material continually breaking off and spreading over the ground. 350 tons toxic waste remain dumped in bags.

The surrounding area had been used as a dumping site for hazardous chemicals. UCC's own laboratory tests conducted in 1989 revealed that soil and water samples collected from within the factory vicinity were toxic to fish (UCC 1989). The report identified 21 sites within the plant as being highly polluted. Other investigations have revealed chemical contamination in soil, ground water, and vegetables in the larger area

(UCC 1985, BGIA 2000, Down to Earth 2003, UCC 1989, Hindu University 2003).

Under the original land lease agreement, UCC was required to clean the factory site before returning it to the government (MacSheoin 2003). While the company did initiate the waste collection process when their stake was sold in 1994, the work was halted. Activists continued to demand for the cleaning of the site along with the solar ponds, where hazardous chemicals were regularly dumped. On several occasions, incineration at different locations was decided, but activists have argued that the incineration process was not safe enough. They are insistent that Dow Chemical (now DowDuPont) should pay.

5. Ethical Analysis

The following section identifies those with whom the responsibility should lie and discusses the ethical rules that ought to have guided those involved. A particular focus is put on highlighting the ethical values that the responsible actors violated.

5.1 Whose responsibility?

From an earlier study, the conclusion was drawn that to create the mega-gas leak, it was not enough that water merely entered the tank (Eckerman 2001, 2005a, 2005b, 2011, 2013). The analysis argued that while the water washing theory seems most convincing, the direct cause of the leakage should not be regarded as the most pressing issue. The magnitude of the disaster was dependent on other factors such as: the design and the location of the plant; the lack of emergency plans and information to the population; the low competence of the staff; the lack of information from UCIL and UCC about the gases; and the absence of a possible antidote. There were also conflicts between NGOs, police, health authorities, and scientists which hindered the response efforts. The rehabilitation offers including adequate economic compensation were insufficient. The parties identified as being most responsible for the magnitude of the disaster were the two main owners, UCC and the GoI, and to some lesser extent, the MP Gov.

Even if the direct cause was sabotage, the leakage could have been prevented. If the personnel management policy had been better, no 'disgruntled worker' or 'negligent employees' would have existed. The impact on health could also have been reduced if the residents had been given reliable information on how to behave in the case of a leakage and if they had been warned by the siren earlier in the night. As it is probable that in the area closest to the factory, where the most serious health effects occurred, hydrogen cyanide was present in considerable concentrations, early treatment with sodium thiosulphate would have mitigated the effects on the health of those living in that sector. The effects on health caused by the leakage could also have been mitigated if the medical, social, and economic rehabilitation had been adequate.

5.2 Ethical principles

While often viewed as a necessary foundation for any society, ethics can be interpreted and defined in several ways. One such definition characterizes ethics as a particular set of rules offering guidance on how to behave towards each other. A disaster such as the Bhopal gas leak provides points to violations of ethical principles. The set of four principles outlined below incorporate the moral view of these authors, and are applied in the analysis of the actions undertaken by the main actors involved in the Bhopal disaster.

5.2.1 The Golden Rule

The Golden Rule, or the ethical principle of reciprocity, is a rule of altruism that seems to have existed since the emergence of early cultures. Confucius (551-479 BC) wrote "that which you do not desire, do not do to others" (Eno 2015). The *New Testament* (Matthew 7:12) says: "Therefore whatever you desire for men to do to you, you shall also do to them". Hippocrates (460-370 BC) formed the first ethical rules for physicians, nowadays often complemented with a rule stating, "above all, do no harm" (Smith 2005).

5.2.2 Human rights to life, health, and healthy environments

The substantial list of declarations on human rights contains rules for nations and individuals regarding behavior, in an effort to achieve a smooth and peaceful co-existence. One might expect that the listed human rights would include the right to life and health as well as the right to a healthy environment. However, in the International Bill of Human Rights, nothing is mentioned directly about those particular themes (UN 1993).

The International Covenant on Economic, Social and Cultural Rights is clearer on the topics, and along with addressing the right to form and join trade unions includes the need for the improvement of environmental and industrial hygiene (OHCHR 1966).

The OECD Guidelines for Multinational Corporations have existed since 1976 and are signed by 35 governments (Saltas 2002). The principles were revised in 2000 to include declarations regarding the environment and corruption, in addition to human rights, the rights of the employees, consumers' interest, *etc.* The OECD rules are voluntary, and there is no possibility of punishment.

Meanwhile, the international non-governmental community has developed its own declarations. In 1996, the Charter on Industrial Hazards and Human Rights was published by the Permanent Peoples' Tribunal following a series of international conventions and guided by a number of similar declarations (Permanent Peoples' Tribunal 1996). The Declaration on Peoples' Right to Safety was produced at the 5th and 6th World Conferences on Injury Prevention and Control in New Delhi 2000 and Montreal 2002 (6th World Conference on Injury Prevention and Control 2002). The participants of the convention argued that the right to safety should be endorsed by the United Nations.

It took more time for the UN society to finally realize the need of guidelines for companies. The Guiding Principles on Business and Human Rights: Implementing the United Nations 'Protect, Respect and Remedy' Framework (UN 2011) were developed addressing the issue of human rights relating to transnational corporations and other business enterprises.

5.2.3 The Precautionary Principle

Pharmacists or doctors would rarely if ever conclude that 'since no evidence of negative side-effects is known, as no research has been undertaken, the medication can be prescribed'. Instead they would more likely state 'since we are not yet fully aware of potential negative effects, we cannot prescribe the medication'. When referencing the potential leakage of chemicals into the environment, this is called the principle of precaution (UNEP 1992). If any negative side-effects of a substance are suspected, precautionary measures should be taken even if some cause and effect relationships have not been fully established scientifically.

5.2.4 Sustainability

Including future generations among Confucius' 'others', leads us to the subject of 'sustainable development'. The most common definition comes from the Brundtland Report (Brundtland 1987) which states that: "Sustainable development is development that meets the needs of the present without compromising the ability of future generations to meet their own needs".

The four principles of the Natural Step Foundation (2016) delve deeper into this topic and state:

> In a sustainable society, nature is not subject to systematically increasing
> 1. concentrations of substances from the earth's crust [...]
> 2. concentrations of substances produced by society [...]
> 3. degradation by physical means [...]
> 4. And in that society there are no structural obstacles to people's health, influence, competence, impartiality and meaning.

These principles constitute a framework for all kinds of organizations and companies, as well as authorities, to use when planning for future sustainability.

5.3 Ethical responsibilities of the corporation

It is widely held that UCC alone was responsible for the accident to happen and was negligent in its handling of the aftermath. The surviving victims regard the company as a capitalistic monster that did not take the lives of human beings into consideration (Figure 3). Morehouse (1993) concludes that the behavior of UCC was immoral.

Although UCIL has admitted that the sabotage theory was false (Morehouse 1993), and the management shortcomings are well known to the public, UCC still maintains the theory that it was a disgruntled worker who caused the leakage.

Following the accident, it seems likely that UCC did their best to hide the truth and disclaim responsibility. Supporting treatment with sodium thiosulphate would have confirmed that other compounds with serious and well-known effects on health had been released. The company firmly denied that MIC could have any long-term health effects. Prior to the accident, maintaining the Bhopal plant in a safe condition would not have involved a major investment for a large corporation such as UCC. Sufficiently compensating the survivors would likewise not have been too costly. It is clear that they regarded shares and profit as more valuable than human beings.

It was in the interests of the company not to recognize the serious effects of the gases as the amount of compensation the company would ultimately be liable to pay would depend on the extent of the injuries. In addition, the GoI had – and still has – an interest of keeping the compensation claim low as a high claim would discourage other chemical industries from establishing plants in the country.

Dow Chemical (now DowDuPont), which took over UCC in 2001, denies all responsibility for the victims and tries to distance itself from current proceedings. This differs significantly to a case described on the website of The Mesothelioma Center (www.asbestos.com). The multinational corporation ABB acquired in 1991 the US based Combustion Engineering, which had used asbestos as insulation material until the 1970s. Having taken control of the company, ABB assumed responsibility for the compensation claims from those who became ill as a result of exposure to the asbestos. In 2001, it became possible for people who had not yet any symptoms to claim compensation. In 2006, a trust of 1.43 billion USD was established to settle pending and future asbestos claims.

This chapter argues that UCC and UCIL violated the four ethical rules chosen by the authors. Adhering to the Golden Rule would lead one to ask themselves: What risk would be acceptable for my family and for my grandchildren? This would have led to a proper risk–benefit analysis, and more than likely have resulted in a different design and location for the plant. The different declarations on human rights are important statements about what can be accepted in a civilized society. In the Bhopal case, one can raise doubts as to whether the corporations were concerned with such declarations.

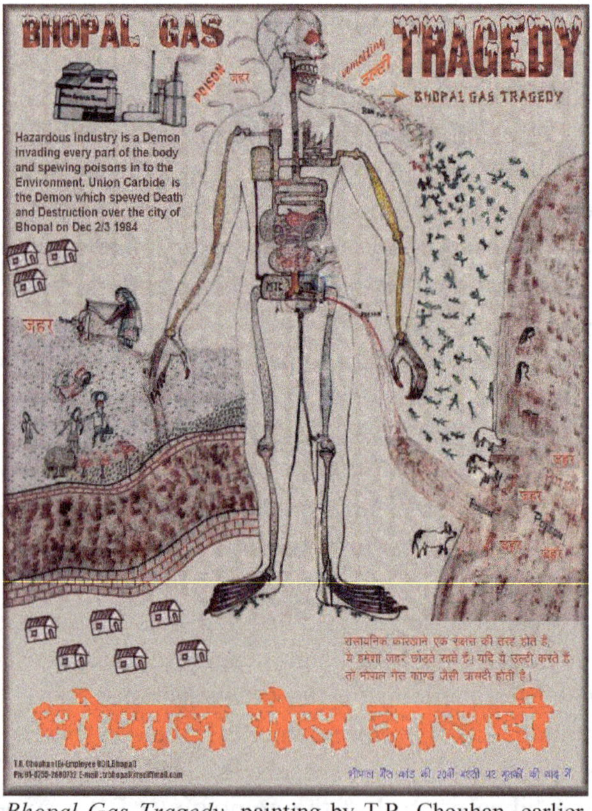

Figure 3. *Bhopal Gas Tragedy*, painting by T.R. Chouhan, earlier employed operator at the UCIL plant in Bhopal. The hazardous industry is depicted as a demon invading the body and spreading poisons into the environment. Reproduced with permission by T.R. Chouhan.

The precautionary principle would have led one to prevent Sevin being produced under such circumstances. It would certainly have cancelled any plans for opening a MIC plant in the first place as knowledge about the adverse effects of carbaryl had already existed in the 1980s. The 2nd principle for sustainability was also violated. The principle states that "in a sustainable society, nature is not subject to systematically increasing concentrations of substances produced by society" (see above) and demands that chemical industry should have complete control of every molecule of the substance it uses.

5.4. Ethical responsibilities of the government

The negligence of both the GoI and MP Governments has been highlighted as an important factor contributing to the leakage. In the aftermath, the GoI itself assumed the role of the victims' lawyer and therefore cannot deny its share of responsibility. Furthermore, Indian society as a whole can be seen to have violated the 4th principle for sustainability, as stated above. The affected population was the poorest of the poor and a great proportion was Muslim. Little interest in the welfare of the affected population has been shown by the authorities (Eckerman 2005b). The caste system and prevalent religious values reinforcing Muslims' lower status may also have contributed to the neglect.

It is noticeable that the attitude of the elite and the ruling class in India towards those on the poorer rungs of society share many similarities with other countries experiencing equally large economic gulfs between different classes. This class issue is relevant on a global as well as a national level. Citizens of poorer countries are more exposed to environmental hazards than those of rich countries, while in each respective country the poorer sections of the population are typically more exposed than the richer.

With a focus on poorer countries in particular, Cassels (1993) investigated the relationship between transnational corporations and governments. He describes its complexity and highlights the dependency that the national governments have on the corporations. This results in a compromising situation in which environmental and safety concerns will inevitably be balanced against the danger of the flight of capital. "What limited power the state does have to respond to the demands of workers and social activists is just that – limited" (Cassels 1993).

In 1992 The Permanent Peoples' Tribunal in Bhopal concluded that fundamental human rights had been grossly violated by the Indian government in relation to several articles in the various international declarations concerning human rights (UN 1993). A large number of national governments have ratified those declarations and many make serious attempts in following them. Therefore, there exists the possibility to prosecute at the International Tribunal for Human Rights in The Hague should a country continue to violate such terms.

5.5 Ethics of non-governmental organizations

While the local trade unions were the first to voice concerns regarding the hazards at the Bhopal plant, we do not know whether they were aware of plans for the closure of the plant or not. It is likely that there would have been forceful protests against the risks of unemployment due to the closure, although from an environmental perspective, closure would have absolutely been the best solution.

In the aftermath of the leakage, several thousand people in total must have been involved in helping the victims of Bhopal. As many of these were experts in different fields, the total knowledge contributed by the NGOs was likely to have been extensive. Many of the NGOs also had contacts with the WHO and other organizations within the United Nations. Their efforts have led to clinics being opened which are chiefly financed through contributions from the civil society. For example, in 1996, the Sambhavna Trust Clinic for gas victims and those exposed to polluted ground water opened, and in 2005, the Chingari Trust founded a rehabilitation center for handicapped children.

The NGOs goals are usually in accordance with the Golden Rule and corresponding with basic human rights, whereas the environmental organizations generally align themselves with the precautionary principle and sustainability. However, the power and influence of each group is much less significant than that of the corporations or governments, resulting in their efforts falling short.

6. Preventing 'New Bhopals'

6.1 Lessons from the Bhopal catastrophe

A government survey in the aftermath of the catastrophe identified over 7,000 potentially hazardous plants in the country (Asia '92 1992). In 1995, it was reported that, of some 9,000 industrial units screened by the Delhi Pollution Control Committee, only 283 had installed effluent treatment plants. Furthermore, 70 % of India's water resources were found to be polluted. These findings along with the numerous investigations into the disaster led to the approval of improved laws and regulations in India. However, few have been implemented.

Due to the complex nature of the political dynamics within India, both the state and central governments have been able to shift the blame to one another for the continued failure of the regulations being implemented. The Indian government remains keen on continuing the establishment of chemical industries within the country. Several incidents have demonstrated how decisions regarding further industrial development have been expedited, without due consideration for the opposition or any anticipated environmental effects (Asia '92 1992). Shristi (2002) also points out that while the laws and regulations proposed address how to manage the toxicity and the waste, the processes that lead to the hazards are not examined. He argues that the laws have provided the government with a sense of security, and, thereby, have legitimized the continuance of a toxic legacy.

The companies who choose to establish themselves within the country have learned from the Bhopal case, and have acted to protect themselves from repeat incidents. An agreement between industrial giant Du Pont and the Indian government included the following statement relating to the company's representatives, who will be harmless from any claims made in the Republic of India against representatives of Du Pont or its assignees alleging bodily harm or death sustained as a direct result of, or in direct connection with, the performance of this Agreement.

On the other side, the International Labour Organisation (ILO) and other inter-state organizations have also acted, by means of conventions and declarations, to ensure that companies adhere to sufficient standards when operating in foreign countries (Sambhavna 1998). The Bhopal Resolution of the European Parliament, for example, calls upon European firms to maintain levels of safety abroad that are comparable to those in place in their home operations. Another movement in the form of Corporate Social Responsibility (CSR) has strived to integrate self-regulation within a company's business model. The aim is to increase long-term profits through positive public relations, high ethical standards to reduce business and legal risk, and shareholder trust by taking responsibility for corporate actions.

It is not yet known whether these laws, regulations, conventions, and declarations will be sufficient in preventing another Bhopal disaster. As long as governments fail to implement the appropriate laws and regulations, no change can be expected. After 33 years the Bhopal catastrophe is not forgotten, and new material is continually being produced. Documentary films support the NGOs in their work for human rights and the environment, while scientific papers, like this one, and new books (Bloch 2016) help us to learn valuable lessons.

6.2 Small 'Bhopals' happen all the time

The Bhopal Gas Leakage has become a symbol of transnational corporate negligence towards human beings and has thus served as a wake-up call. Despite this, industrial disasters continue to occur in India as well as in other industrialized parts of the world – each with mechanisms similar to the Bhopal catastrophe. The accidents most often occur in and affect countries outside those in which the international companies are headquartered, and happen most typically in countries where production regulations are less stringent than in their 'home' country. The affected countries generally have weak trade unions and poorly developed occupational healthcare, which have little influence on the work environment.

Such similarities make it appear as though the catastrophes could have been predicted and prevented. The companies involved have typically reacted in similar patterns – by disputing their own role in the accident, rejecting any health effect claims, and being reluctant to compensate the victims financially. Although they are far from the severity of Bhopal, the sheer number of chemical accidents which have occurred since 1984 could well make a case for them to be considered a public health problem.

6.3 The roads towards prevention

To prevent industrial hazards, regulations and declarations are necessary but are not effective alone. The driving force for companies is financial, rather than ethical or moral. The economic pressure that can actually result in change comes from two directions: from the legal system, by means of fines, taxes, *etc.*, and from the consumer side, *i.e.* the civil society.

A strong legal system requires a state that is democratic and transparent, with little corruption and it is the task of transnational organizations like the UN, EU, OECD *etc.* to support and demand development in this direction. Today, the development in some countries is in fact progressing in the opposite direction. A conflict often emerges between a workers' demand for employment and a society's desire to reduce pollution. This conflict seems to have found a solution where trade unions start to co-operate with employers and governments to make the plants more environmentally friendly.

Further pressure from the civil society might be effective as long as there is real competition in the market – which is not always the case at present. If a transnational company maintains ownership of the fully chain of food productions, from seeds, fertilizers, pesticides and processing to packaging and selling, and if its market is worldwide, there might be very little choice for the consumer – limiting their influencing power. In that sense, consumers' organizations have to cooperate worldwide to effectively make changes. Having said that, sometimes the companies themselves understand that a society's positive opinion of their organization is a valuable asset, leading them to change their corporate image. Many industry leaders also share concerns about possible hazards and understand that a change must take place. There are several methods for planning and avoiding hazards for the companies that are prepared to change their focus from short to long term planning.

7. Conclusion

The two most important factors leading to the catastrophic gas leak in Bhopal were identified as the design of the plant and the company's policy on cutting back on expenses. The negligence of both the company and governmental authorities in the aftermath of the accident were seen to have worsened the impact of the leakage on people's lives.

The case shows that an ethical analysis of industrial accidents should not focus on individuals, but on the responsibilities of companies as well as national and local authorities. A society that is continuously and increasingly breaking the ethical rules can be regarded as an un-ethical society, while a company that has accepted these rules and strives to follow them, is on the right path.

The analysis shows that ethical rules, such as the Golden Rule on not harming others, human rights, the precautionary principle, and the principles for a sustainable society were violated by both UCC and UNCIAL, while the Indian authorities violated, in particular, the 4th principle for sustainability. The chapter has also shown that the prevention of industrial hazards is possible but that it requires a strong legal system, which in turn demands democratic and transparent state governance. Several planning tools exist for companies to utilize in order to reduce the risks of future hazards.

Regarding industrial hazards to public health two antipodes exist: the industries on one side, and the trade unions and non-governmental organizations, working on behalf of human rights and environmental concerns, on the other. While it is common for NGOs and trade unions to fight for what is believed best for human beings, it must be realized that this is not the primary goal for a company. On behalf of the workers and the environment, the NGOs and trade unions typically demand low levels of income disparity, strong employees' rights legislation, manpower-rich companies, effective water and soil protection, and the ability to voice concerns regarding the work environment and the environment as a whole. The company on the other hand, in their pursuit of increased profits, finds it more advantageous to have fewer employees, an ease in hiring and firing laborers, an uncertain labor market (which incentivizes those employed to work harder and refrain from protest out of fear of becoming redundant), and low demands on the work environment and natural environment.

While it is beneficial for a country to strive for increased economic development to improve its citizen's access to education, health care, and public health, it is important to ensure that the methods of achieving the increased prosperity do not adversely affect its citizens. Foreign investments might be a quick way to reach development but often brings with it harmful consequences. It is the task of the government to strike a balance between both needs. Politicians should put the interests of the people, both those of current and future generations, first. NGOs, like trade unions, environmental organizations, and consumers' organizations, should be regarded as resources, not as troublemakers.

Finally, along with the consequences to both the public and corporations discussed above, manmade catastrophes can also result in personal consequences. The UCC chairman and CEO Warren Anderson, together with a technical team, immediately travelled to India having been made aware of the accident. Upon arrival, Anderson was placed under house arrest by the Indian police and was threatened by the government to leave the country within 24 hours. For the rest of his life he was forced to live at a secret address, and never appeared in the Indian courts. The last lawsuit in USA was dismissed in 2012. Anderson did not have much time to enjoy life free from the legal scrutiny before he died in 2014.

Further Reading

Further information can be found in the many books published on the Bhopal tragedy, such as Chouhan 1994, D'Silva 2005, Eckerman 2005b, Hanna et al. 2005, Lapierre & Moro 2001, and Bloch 2016, while a documentary *Bhopal: A Prayer for Rain*, directed by R. Kumar in 2014, is available on DVD. More information can also be found at the following websites:

- Bhopal Gas Tragedy Information. Union Carbide Corporation: www.bhopal.com
- Bhopal Medical Appeal: www.bhopal.org
- International Campaign for Justice in Bhopal: www.bhopal.net

References

6th World Conference on Injury Prevention and Control: 2002, *Declaration on People's Right to Safety,* Montreal, Canada: 6th World Conference on Injury Prevention and Control.

Asbestos.com: 2016, 'Combustion engineering', [availabe online at: http://www.asbestos.com/companies/combustion-engineering.php, accessed 3 August 2017].

Asia '92: 1992, 'Third Session on Industrial and Environmental Hazards and Human Rights', *Permanent People's Tribunal. Findings and Judgements Bhopal,* 19-24 October, Bhopal to Bombay (India).

BGIA: 2000, *Union Carbide Disaster in Bhopal. Fact sheet. 16th Anniversary of the December 2-3, 1984*, Bhopal: Bhopal Gas Pidit Mahila Udyog Sangathan; Bhopal Group for Information and Action.

Bloch, K.: 2016, Rethinking Bhopal. A Definitive Guide to Investigating, Preventing, and Learning from Industrial Disasters, Amsterdam: Elsevier.

Brundtland, G.: 1987, *Our Common Future*, Oxford: Oxford University Press.

Cassels, J.: 1993, *The Uncertain Promise of Law: Lessons from Bhopal*, Toronto: University of Toronto Press.

Chauhan, P.: 1996, *Bhopal Tragedy. Socio-legal Implications*, Jaipur, India: Rawat.

Chouhan, T.R.: 1994, Bhopal: The Inside Story. Carbide workers speak out on the world's worst industrial disaster, New York: Apex Press.

Cullinan, P.; Acquilla, S. & Dara, V.R.: 1997, 'Respiratory morbidity 10 years after the Union Carbide gas leak at Bhopal: a cross sectional survey', *BMJ*, **314** (7077), 314-338.

Delhi Science Forum: 1985, *Bhopal Gas Tragedy*, New Delhi: Delhi Science Forum.

Dhara, V.: 1992, 'Health effects of the Bhopal gas leak: A Review', *Epidemiol Prev*, **14**, 22-31.

Down to Earth: 2003, 'Foul Debris. The UCIL plant is still a health hazard' [availabe online at: http://www.downtoearth.org.in/coverage/foul-debris-13794, accessed 22 September 2016].

D'Silva, T.: 2006, The Black Box of Bhopal: A Closer Look at the World's Deadliest Industrial Disaster, Victoria, BC, Canada: Trafford.

Eckerman, I.: 1996, 'The Health Situation of Women and Children in Bhopal. Report for The International Medical Commission on Bhopal', *International Perspectives in Public Health,* (no. 11-12), 29-36.

Eckerman, I.: 2001, *Chemical Industry and Public Health. Bhopal as an example*, Gothenburg: Nordic School of Public Health.

Eckerman, I.: 2005a, 'The Bhopal Gas Leak: Analysis of causes and consequences by three different models', *Journal of Loss Prevention in the Process Industries*, **18**, 213-217.

Eckerman, I.: 2005b, The Bhopal Saga. Causes and Consequences of the world's largest industrial disaster, Hyderabad: Universities Press (India).

Eckerman, I.: 2011, 'Bhopal Gas Catastrophe 1984: Causes and Consequences', in: J. O. Nriago (ed.), *Encyclopedia of Environmental Health*, Burlington: Elsevier, pp. 302-316 [updated version 2013 available online at: http://www.science-direct.com/science/article/pii/B9780124095489019035, accessed 23 Sept 2016].

Eno, R.: 2015, *The Analects of Confucius*, XV:24 [available online at: http://www.indiana.edu/~p374/Analects_of_Confucius_(Eno-2015).pdf, accessed 21 Sept 2016].

Hanna, B.; Morehouse, W. & Sarangi, S: 2005, *The Bhopal Reader. Twenty Years of the World's Worst Industrial Disaster*. New York: Apex Press [availabe online at: http://bhopal.org/, accessed 21 Sept 2016].

Hindu University: 2003, *Hinterland. A Special Issue on The Bhopal Gas Tragedy*, New Delhi: Department of English, Hindu University.

ICFTU: 1998, Commitments for Sustainable Development. Trade Unions at the Commission for Sustainable Development. Special 'Business and Industry' Segment, Brussels: International Confederation of Free Trade Unions.

ICFTU-ICEF: 1985, *The Trade Union Report on Bhopal*, Geneva: International Confederation of Free Trade Unions-ICEF.

Jones, T.: 1988, *Corporate killing. Bhopals will happen*, London: Free Association Books.

Kalelkar, A.S.: 1988, Investigation of Large-magnitude Incidents: Bhopal as a Case Study, Cambridge, MA: A.D. Little.

Kulling, P. & Lorin, H.: 1987, *The Toxic Gas Disaster in Bhopal December 2-3, 1984*, Stockholm: Försvarets Forskningsanstalt (National Defence Research Institute) [In Swedish].

Kumar, C. & Mukerjee, S.: 1985, 'Methyl Isocyanate: Profile of a killer gas', in: *Bhopal: Industrial genocide?*, Hong Kong: Arena Press.

Lapierre, D. & Moro, J.: 2001, *It Was Five Past Midnight in Bhopal*, New Delhi: Full Circle.

MacSheoin, T.: 2003, *Asphyxiating Asia*. Mapusa, Goa, India: Other India Press.

Morehouse, W.: 1993, 'The Ethics of Industrial Disaster in a Transnational World: The elusive quest for justice and accountability in Bhopal', *Alternatives: Global, Local, Political*, **18** (4), 475-504.

National Institute for Working Life: 2001, *Scientific Basis for Threshold Limit Values. Mehylisocyanate and Isocyanacid*, Stockholm, Sweden: National Institute for Working Life.

Natural Step Foundation: 2016, 'The Natural Step Framework' [availabe online at: http://www.thenaturalstep.org/approach/, accessed 3 Aug 2017].

OHCHR (Office of the United Nations High Commissioner for Human Rights): 1966, *International Covenant on Economic, Social and Cultural Rights* [available online at: http://www.ohchr.org/EN/ProfessionalInterest/Pages/ CESCR.aspx, accessed 20 Sept 2016].

Permanent People's Tribunal: 1996, *Charter on Industrial Hazards and Human Rights* [available online at: https://www.globalpolicy.org/component/content/ article/212/45285.html, accessed 23 Sept 2016].

Ramachandran, C.: 1994, 'Immediate Post Industrial Disaster Management', in: *Refresher course for Top Executives. Management of Chemical Accidents*, Bhopal: Disaster Management Institute.

Saltas, A.: 2002, 'Can UN regulate the transnational companies? Let the UN-organ UNCTAD check the companies' direct investments in developing countries', Stockholm: Sodertorns Hogskola.

Sambhavna: 1998, The Bhopal Gas Tragedy 1984 to ? A report from the Sambhavna Trust, Bhopal, India, Bhopal: Bhopal People's Health and Documentation Clinic.

Shristi: 2002, Surviving Bhopal. Toxic present – toxic future. A report on Human and Environmental Chemical Contamination around the Bhopal disaster site, New Delhi: The Other Media.

Smith, C.M.: 2005, 'Origin and uses of primum non nocere – above all, do no harm!', *Journal of Clinical Pharmacology*, **45**(4), 371-7.

Subramaniam, A.: 1985, 'Bhopal – The dangers of diagnostic delay', *Business India*, (August), 12-25.

UCC: 1976, 'Methyl Isocyanate', Union Carbide Document No. F-41443A – 7/76, New York: s.n.

UCC: 1985, Bhopal Methyl Isocyanate Incident. Investigation Team Report, Danbury, CT: Union Carbide Corporation.

UCC: 1989, Presence of Toxic Ingredients In Soil/Water Samples Inside Plant Premises, Union Carbide Corporation.

UCIL: 1978, Carbide Monoxide, Phosgene and Methyl Isocyanate. Unit Safety Procedures Manual, Bhopal: Union Carbide India Limited.

UN: 1993, The International Bill of Human Rights, New York: United Nations.

UN: 2003, Norms on the Responsibilities of Transnational Corporations and Other Business Enterprises with Regard to Human Rights, s.l.: United Nations.

UN: 2011, Guiding Principles on Business and Human Rights: Implementing the United Nations 'Protect, Respect and Remedy' Framework, s.l.: United Nations.

UNEP: 1992, *Rio Declaration on Environment and Development* [available online at: http://www.unep.org/Documents.Multilingual/Default.asp?DocumentID=78&ArticleID=1163, accessed 22 Sept 2016].

Varadarajan, S.: 1985, *Report on Scientific Studies on the Factors related to Bhopal Toxic Gas Leakage*, New Delhi: Indian Council for Scientific and Industrial Research.

Chapter 6

About the Futile Dream of an Entirely Riskless and Fully Effective Remedy: Thalidomide

Klaus Ruthenberg

Abstract: Hardly any other artificially made medicine has ever had such a dangerous and unexpected potential as thalidomide. The medicine, which started its life as a harmless sleep-inducing substance, caused neurological and teratogenic damage, and, at the moment is a promising candidate for the fight against leprosy, cancer, and AIDS. The present study focuses on the early German history. It takes the bioactivity of that remedy as a chemical property and supports the claim that if a substance is biochemically active, adverse effects can hardly be avoided. The ethical role of an ontological underdetermination of chemicals is discussed, and an analysis of particular responsibilities is presented along the lines of modern principlism in biomedical ethics.

He who is at peace will not waver or become anxious by the constant changes in life, by the 'panta rhei' that is particular to our times. A harmless medicine ensures this steadfastness even in difficult situations.[1]

1. Introduction

The Contergan disaster was possible because of the belief by laypersons *and* experts in the old myth of a remedy without any adverse effects, and

[1] The motto is taken from an advertisement of the German remedies Contergan and Contergan forte by the company Grünenthal, reproduced in Kirk 1999, p. 267, my translation. 'Panta rhei' means 'all is in flux', which is ascribed to the Presocratic philosopher Heraclitus.

the betrayal of that belief. Although the sleep-inducing sedative (or hypnotic) remedy Contergan – containing thalidomide[2] as an active component – was officially launched by the company Grünenthal in Germany only in October 1957, the presumably very first victim of its adverse effects was born at Christmas of the preceding year. Ironically, this victim was the daughter of an employee of Grünenthal, who gave the substance to his pregnant wife, apparently off the record. Their baby girl was born without ears (Stephens & Brynner 2001, p. 19), but the causal relation between the administration of the substance and this situation was far from obvious at that time.[3] Many more humans – estimations go up to about 10,000 – were born suffering from similar malformations and several other conditions.[4] And many more than 10,000 – perhaps five times more – died before any physician could explain why.

All this occurred around 1960, and the teratogen[5] thalidomide was the substance that caused the catastrophe. The world was different after that substance had entered the scene.

'What is a chemical substance?' is a central question in the philosophy of chemistry, and it is just as central in ethics of chemistry, as well. This question is not sufficiently answered by pointing at the compositional or structural formula. All chemicals are long-term epistemic objects, meaning that the amount or realm of properties assigned to these substances can never be closed. Hence, a new substance always carries with it risks for living beings (Schummer 2001). Furthermore, regarding

---

[2] 'Thalidomide' is an artificial name, derived from the older abbreviated name α-Phthalimidoglutarimide; IUPAC calls it 2-(2,6-dioxo-3-piperidyl)isoindoline-1,3-dione. The Chemical Abstract System (CAS) number is 50-35-1.

[3] Though usually very rare (Pschyrembel 1986, p. 389), phocomelic malformations of new-borns have not been excluded from public knowledge, see the 'black painting' named 'Mother showing her deformed child to two women' by Goya (reproduced, for example, in Hoffmann 1995, p. 133, and in Roth 2005, p. 212).

[4] The most typical of these malformations of legs and arms are subsumed under 'phocomelia', which refers to the flippers of seals; the more general medical term for malformations of extremities, however, is 'dysmelia'. There are several other aspects of the impact of thalidomide on new-borns.

[5] Teratology is the science of the phenomena, causation, and mechanisms of morphological and functional developmental disorders in animate beings, see Mutschler & Lemmer 1985. Teratogenicity, carcinogenicity, and mutagenicity are the three main classes of chronic toxicity, contrasting with acute toxicity.

thalidomide, we have to take into account that it is not just one substance, but two (see Fig. 1).

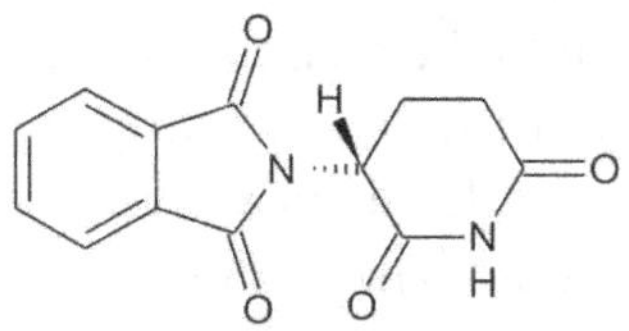

Figure 1. Structural formula of the (R)-(+) enantiomer of thalidomide. The hydrogen atom at the asymmetric carbon atom points up, and the phthalimide group down. In the (S) enantiomer it is inverse.

This chapter is divided into a narrative section and an ethical discourse section. According to its main purpose of introducing a pertinent case to chemists and chemistry students, the narrative shall focus on the main issues necessary for an ethical evaluation. As is appropriate in this specific case, the emphasis of study lies on the early (German) history up to about 1960. More elaborate exposures of the historical material and international aspects with partly different focuses can be found in the references, particularly in Sjöström & Nilsson 1972, Knightley 1979, Kirk 1999, Stephens & Brynner 2001, Zichner *et al.* 2005, and Lenhard-Schramm 2016.[6]

Although there exist some cases of remedies with adverse effects in the history of pharmacy as, for instance, the salts of heavy metals like lead and mercury (and the alloys of the latter) and, more recently, the

---

6 Sjöström & Nilsson 1972 is the most original of the thalidomide monographs, that is, all later publications refer to or are based upon the latter. Both authors were involved in the Astra case in Sweden, thus the book is slanted to a certain degree (and it lacks an index). Knightley 1979 is a brilliantly written must-read with an index, and a very informative appendix on the chemistry of thalidomide (information which is almost entirely missing in the other books mentioned here). Kirk 1999 is a doctoral thesis in the history of pharmacy from the University Greifswald, Germany. It commendably meets the requirements for written scientific work, particularly neutrality, although an index is missing. Stephens & Brynner 2001 has a very useful index, but is not throughout reliable referring to details, and perhaps too speculative in some spots. Zichner *et al.* 2005 does not present material that goes beyond that of the earlier works, particularly that of Kirk. It has its strength in a more sociological and ethical scope, and the integration of orthopaedic aspects (which I do not refer to here). The historical dissertation of Niklas Lenhard-Schramm (2016) is an easy accessible, fresh and substantial documentary, highly recommendable for German speaking readers.

case of Vioxx,[7] the case discussed here is unique in at least the following respects: the pretended and offensively advertised harmlessness, the unexpected emergence of adverse effects (the malformational impact on new-borns and the damage to the neuronal system), the worldwide influence on the vigilance against drugs and drug safety regulations, and its huge catalytic effect on the development of neurology and teratology. Hence, thalidomide certainly ranks among the most influential man-made chemical substances of all time.

Section 2 narrates the emergence of the substance including a critical synopsis of the central (but questionable) scientific paper of Grünenthal scientists, the substance distribution into society, the contact of the drug with human beings, its chemical (stereoisomeric) peculiarity, its temporary extinction and proscription, and, briefly, its rebirth as an active medicine.

2. Main Historical Aspects

2.1 How did the compound come into existence?

From a chemical point of view, it seems likely that thalidomide was synthesized quite early, perhaps well before Grünenthal sold it as a remedy. The preparation path is not very complicated: Starting from glutamine or glutamic acid, the product can be obtained in two or three steps, respectively (Muller *et al.* 1999). However, astonishingly little is known about the early history of thalidomide. The customary or official version, here in the words of the pharmacologist Herbert Keller, who was closely involved, from his witness account at the German Contergan process (the 'Alsdorf Trial', 1968-1970), is the following:

> During works in the field of derivates of glutamic acid, Doctor Kunz and I came across a new compound which was identified as N-Phthalylglutamic acid imide. Besides other pharmacodynamic effects I observed that the substance showed a, however not too impressive, sleep-inducing effect. [Cited after Kirk 1999, p. 52, my translation]

[7] Cf., for example, Schneider 1985, Talbot & Waller 2004. Apart from Contergan, Grünenthal had an intriguing history of pharmacological failures of its own (cf. Sjöström & Nilsson 1972). As to Vioxx, see Rourke 2006 and the references therein.

In their paper from 1956, Kunz, Keller & Mückter claimed that with respect to the increasing consumption of hypnotic medicines there emerged the need for effective drugs without adverse reactions. The most promising candidates seemed to be unsaturated tertiary alcohols on the one hand and piperidine derivatives on the other:[8]

> In our own work devoted to the preparation of peptides, 2.6-dioxo-3-amino-piperidines were synthesized. Their structural peculiarity [*konstitutionelle Eigentümlichkeit*] gave reason to prepare a number of such compounds and test them with respect to their efficacy in animal experiments. [Kunz *et al.* 1956, p. 426, my translation]

Of several prepared substances, 'K 17' (the 17th preparation Kunz made for Grünenthal, as the official version has it) seemed to be pharmacologically the most interesting one. However, none of the other substances are discussed in that paper (and have not been discussed since). K 17 is described as a white solid with a melting point of 271 °C. The solubility is severely restricted: K 17 is insoluble, for example, in water, methanol, ethanol, acetone, diethyl ether, and benzene, and it is only slightly soluble in dioxane and pyridine. Due to this, the administration of the substance was performed in the form of suspensions in most cases. The pharmacological effect was described as the 'reduction of the motility' of the animals shortly after administration, an impact contrasting with the effects of, for example, barbiturates and Doriden, which lead to an initial excitation. The main method the authors applied was the unusual '*Zitterkäfig-Methode*' (jiggle-cage method): Eight male mice, respectively, were held in a cage hanging on a spring. The cage was connected to a platinum wire, which plunged into a bath of sulphuric acid, whenever the mice moved. On contact with the acidic bath the wire completed an electric circuit (24 V) and, through electrolysis, caused the formation of hydrogen which was measured volumetrically. By this procedure, during normal activity, the mice octets produced almost 6 ml/min H_2. When given the sleep-inducing drugs, the hydrogen volume was reduced. Being

---

8 In an extensive survey about the many new pharmaceutical developments in the second half of the 1950s, the author compiles the formulas of disubstituted barbiturates, Persedon, Noludar, Doriden, and Contergan in a row (Kunz 1959). (It is not clear if this W. Kunz from 'Dr. Schwarz Arzneimittelfabrik GmbH, Monheim' is the same who worked with Grünenthal a few years before.)

relevant as calibration standard, the authors defined the status 'sleep' by 50% activity of the base level rather than measuring the hydrogen with actually sleeping mice. Apparently, it is almost impossible to induce sleep in eight mice at the same time in the same cage. Moreover, the dose-dependencies (and, thereby the determination of the smallest sleep-inducing doses) are not presented or discussed at all. Hence, this method is questionable because of major methodical drawbacks regarding calibration:

> Logical enough, perhaps, this [the neglect of calibrationable test methods] led to the evolution of a test for sleep-inducing capacities in which there was no need for actual sleep to be produced. [Knightley 1979, p. 16]

Nevertheless, the paper was accepted and published, and the Grünenthal team came up with comparative results for six substances (Tab. 1), among them Doriden, the market competitor since 1955 (Fig. 2). According to the company's goals, the most important conclusions from these data were the comparable efficacy of sleep-induction and the minimal toxicity.

Table 1. Pharmacological/toxicological comparison of sleep-inducing drugs as inferred from the 'jiggle-cage-method' with mice (after Kunz *et al.* 1956, p. 428).

	smallest hypnotic dose (mg/kg)	start of effect (min)	duration (min)	smallest narcotic dose (mg/kg)	initial excitation phase*	lethal dose** (mg/kg) LD_{50}
Thalidomide	100	5	240-300	>5000	-	>5000
Luminal	40	30	300	150	+++	300
Valamin	50	10	60	375	++	725
Doriden***	75	30	200	400-500	+++	600
Methyl-pentynol	150	15	210	500	++	750
Sodium Bromide	500	80	>360	ca. 7000	-	8000

* 'Initial excitation phase' means the higher activity of the animals directly after administration, an effect well-known, for example, for barbiturates (presumably '+++' means very high and '–' none).

** The actual value of the p.o.-LD_{50} for rats is 113 mg/kg.

*** Note that the competitor Doriden from Ciba is the only other substance of the 'piperidine-family' used here (see Fig. 2).

The *Sunday Times* Insight Team gave the following interpretation of Table 1:

> Without doubt this table is the origin of the illusion that thalidomide was something capable of overturning many decades of pharmaceutical experience: an effective drug and quite without side effects. [Knightley 1979, p. 18]

In the spring of 1955, clinical trials – the official testing of thalidomide on humans – began.[9] No formal application was necessary at that time, and, moreover, no standard procedure for pharmacological trials had been developed yet. Although some side-effects were hinted at, all those reports concluded with similar positive judgments. Peripheral neuritis, in particular, was not described in detail as no thorough searches for side-effects had been conducted at the time, while the clinical studies used were far too short such that any long-term effects had insufficient time to develop. Thus, the severe adverse effects were found by epidemiological observations rather than in the clinical studies.[10]

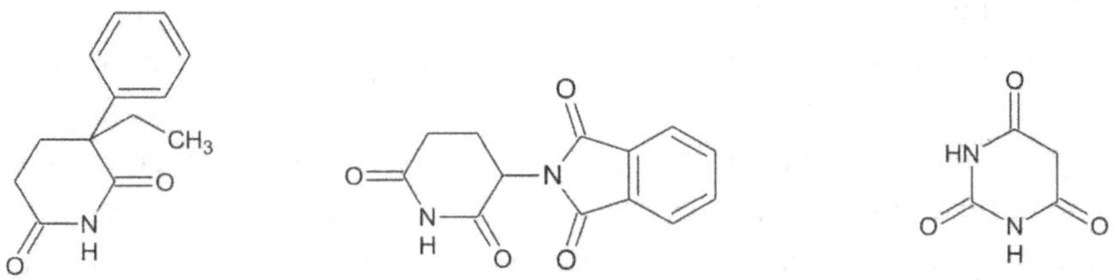

Figure 2. Molecular formulas of Doriden, thalidomide, and barbituric acid (from the left). Note the similar structure of the CO-NH-CO-sections of these molecules (also referred to as imide) which is considered one main locus for interactions with biological targets. Note also that thalidomide has two of these sections.

---

9 Initially, the following four studies sponsored by Grünenthal have been published in scientific journals: Esser & Heinzler 1956, Jung 1956, Stärk 1956, and Walkenhorst 1957. Later, a study on nursing (not pregnant) women followed (Blasiu 1958). None of these went deeper into metabolism and pharmacokinetics.

10 Among the first to become aware of possible neuronal damage caused by thalidomide was the physician Ralf Voss from Düsseldorf. As early as 1959 he contacted the Grünenthal officials, but they dismissed the warning. (See the chapter with the telling title 'How to Sell a Wonder Drug' in Knightley 1979.) Later, he published his observations (Voss 1961).

2.2 How did thalidomide get into the public?

With the backing by a study on nursing (but not pregnant) women (Blasiu 1958), thalidomide was advertised excessively. Two particular aspects were emphasized in this marketing campaign: The lack of toxicity and the naturalness of the support of a 'good sleep'. Furthermore, there seemed to be no risk of addiction. Generally, this led to a very comfortable description, which competitors like Doriden could by no means make a claim on (Kessel 2013). Physicians and apothecaries were the main marketing multiplicators and drug distributors for non-prescription drugs. Referring to the non-barbitural competitor Doriden of Ciba (see fig. 2), Nils Kessel describes the role of these experts as follows:

> The key figure in this arrangement was the physician. He was analyzed as a potential consumer and a 'consumption guide' but officially treated as a partner in a common scientific endeavor. [Kessel 2013, p. 163]

In the case of thalidomide, recommendations ('therapeutic circulars') were distributed to about 50,000 experts, advising them to apply the drug in cases of sleep disorder, fear, contact weakness, nervousness, menopausal problems, female sexual excitability, and more. Obviously, marketing concentrated on female consumers. According to effective law, it was not allowed to address the lay public with advertisements on hypnotic drugs (Kirk 1999, p. 58). Nevertheless, the marketing campaign was very successful, and the revenues were remarkable. In the winter of 1960 Grünenthal made more than 1 million Deutsche Mark monthly with Contergan, an amount that decreased significantly one year later, when the drug was taken off the market. In 1960, however, those numbers represented almost 50% of the total turnover of the company. Clearly, the number of employees rose in parallel with the Contergan sales: from 420 in 1954, to 900 in 1959, and to 1300 in 1961. Understandably – from a plainly economical point of view – the company officials did not want to strengthen the impact of any possible critical statements in their 'consumption guides' or to introduce a prescription for their 'wonder drug'. Yet in 1960, in an internal document from April 14, Mückter wrote:

> Unfortunately, we are now receiving an increasing number of reports on the side-effects of this drug, as well as letters from doctors and

> pharmacists who want to put Contergan on prescription. From our side, everything must be done to avoid prescription enforcement, since already a substantial amount of our turnover comes from over-the-counter sales. [Cited after Sjöström & Nilsson 1972, p. 54]

Eventually, one year later, the management decided to apply for prescription requirement. However, the catastrophe could no longer be stopped.

2.3 How did it get banned?

Only when the previously mentioned presumed first victim of the malformational effects of thalidomide reached her fifth birthday did it become apparent that thalidomide was the cause for her having been born without ears, and for all the other victims. The main figures of this manifestation process were the physicians Widukind Lenz (1919-1995) from Germany and William Griffith McBride (b. 1927) from Australia. Apart from his professional work as a pediatrician and geneticist,[11] Lenz (and his group) performed meticulous epidemiological studies and took the initiative to inform the learned public. He finally called Mückter by phone on 15 November 1961, but the latter was still reluctant to even admit his knowledge about reports on adverse effects. Today we know that Mückter was plainly lying. However, that would not help himself or Grünenthal in any way. On 26 November 1961, the German newspaper *Welt am Sonntag* published a long, scathing article about Lenz's recent complaints entitled 'Malformations from Tablets? Alarming Suspicion of a Physician against a widespread Drug', and the next day Contergan vanished from the German market. In 1962, however, more than 900 new thalidomide victims were registered, and another 14 between 1963 and 1967. The following other countries reported thalidomide cases: Argentina, Australia, Austria, Belgium, Brazil, Canada, Denmark, France, Ghana, Great Britain, Israel, Italy, Japan, Lebanon, Luxemburg, Netherlands, Norway, Pakistan, Sweden, Switzerland, Spain, Syria, and USA. McBride's letter to the editor of the scientific journal *The Lancet* (16 December 1961) is considered the very first publication on the embryopathic effects of thalidomide in humans. In that letter, McBride states

---

11 See, for example, Lenz 1961, Lenz & Knapp 1962, and, also from Lenz's working group, Nowack 1965.

that “congenital abnormalities are present in approximately 1.5% of babies,” and claims: “In recent months I have observed that the incidence of multiple severe abnormalities in babies delivered of women who were given the drug thalidomide (‘Distaval’) during pregnancy, as an antiemetic or as a sedative, to be almost 20%.” (McBride 1961)

The course of the (temporary) decease of thalidomide as a remedy was different in other countries. Even before Grünenthal’s withdrawal of the drug, the Richardson-Merrell Company failed to get approval to register thalidomide (under the brand name ‘Kevadon’) for commercial use in the United States, whereas in Canada the registration was successful. Frances Kathleen Oldham Kelsey (1914-2015), a physician and pharmacologist with the U.S. Food and Drug Administration (FDA), found the submitted material imperfect and unsatisfactory. Her main criticism referred to the obvious differences of the animal test results on the one hand and the impact on humans on the other. Kelsey remained unshaken in spite of heavy contentions with the officials of the company, postponed the approval repeatedly, and the company finally gave up the proposal in March 1962.

Intriguingly, the regulations in the United States and other countries like Germany allowed for ‘blind’ clinical studies on uninformed patients.[12] That weak point in drug legislation lead to the circumvention of the Nuremberg medical code with its central claim of ‘informed consent’, which was compiled after the trials against the ‘Nazi doctors’ (Wiesing 2012, pp. 137-139). The obstetrician Ray Nulsen from Cincinnati, Ohio, for example, one of the large number of physicians Richardson-Merrell mentioned in their FDA application, repeatedly tested drugs for companies in this way. He published his experiences with thalidomide on pregnant women referring to ‘insomnia’ particularly during the third trimester which is just the period in which the fetus is insensitive to malformation (Nulsen 1961).[13] This article was in fact written by Raymond Pogge, the medical director of Richardson-Merrell, rather than by

[12] By ‘blind’ clinical study I here mean the administration of a presumably active substance to humans without their notice or information and before safety has been established.

[13] Widukind Lenz and his group found out that the first trimester of pregnancy is the most critical phase for the impact of thalidomide.

Nulsen, and some of the important aspects were, to say the least, distorted. Even worse is that despite the published material and all his later testimony, Nulsen must have given the drug to women in earlier phases of pregnancy, too: More than a dozen cases of typical malformations and stillbirths he himself delivered are documented, and some of the former went to court. Most of these cases ended in the payment of compensation by Richardson-Merrell. We find a striking comment on the unfortunate role of Nulsen in that history, again by the *Sunday Times* Insight Team:

> The final irony is that if Nulsen had been conducting a *genuine* clinical trial, or if Richardson-Merrell had been really interested in establishing the safe and effective application of thalidomide in a broad spectrum of circumstances, then they both might have won an honored niche in pharmaceutical history for being the first to discover thalidomide's devastating effects on the unborn child. [Knightley 1979, p. 85, emphasis in original]

Hence, although thalidomide was not approved as a remedy in the USA, and therefore a disaster like in other countries did not emerge, even this country has experienced its pertinent cases as well.[14]

Whereas their chapter about thalidomide in the United States carries the heading 'The results of vigilance', the Swedish authors Sjöström and Nilsson (both of whom have been active in the trials mentioned below) call the chapter about their own country 'The results of negligence' (Sjöström & Nilsson 1972). In Sweden, as opposed to (Western) Germany, Neurosedyn and Noxodyn, both containing thalidomide, were prescriptive drugs right from the beginning. Like Grünenthal, the company Astra too was very reluctant to withdraw its products. Because the reports about the withdrawal by Grünenthal referred only to the German trade name Contergan, both physicians and potential consumers in Sweden were perplexed and did not recognize possible risks. The result was that more than 100 Swedish children suffered from phocomelia (in a small country with about 7.5 million inhabitants). In February 1962, a few concerned parents founded an association which finally went to court. Many international experts were invited. The issue ended in a

---

[14] For the thalidomide story in the USA, see Sjöström & Nilsson 1972, pp. 112-130; Knightley 1979, pp. 64-86; Stephens & Brynner 2001, pp. 39-59; and Kirk 1999, pp. 191-205.

settlement between the association for parents of malformed children and Astra. The compensation in Sweden (plus a few Danish cases) amounted to about 15 million USD.

2.4 The two faces of thalidomide

The purity of substances is a central and general topic in the history and philosophy of chemistry, but it becomes particularly intriguing in the instance of thalidomide, as it does for all stereo isomeric substances.[15] Although all its molecules share the same compositional formula (and the same molecular spectra), the bulk material usually consists of two kinds of molecules which can be imagined like our two hands (see Fig. 1). The mixture of these enantiomers is called a racemate, and it is the latter that presumably was used in most experiments, particularly in those of the Grünenthal team, and presumably in the remedy, too.

The biochemical reactivity, however, depends on the particular spatial organization of molecules; in most stereoisomeric pairs we find different effects for each enantiomer (Knabe 1995). Taking into account their general indifference about the real properties of their 'wonder drug', there is no wonder that the Grünenthal researchers did not say anything about this issue. However, after the manifestation of the negative effects of thalidomide several research teams began investigations in that direction, such that thalidomide became the classical drug containing a 'bad' and a 'good' part. I briefly refer to that issue here. In 1965, R- and S- isomers were prepared (Shealy *et al.* 1965), and two years later, biochemists from London published the surprising result that all three entities, R-, S-, and racemic thalidomide, show a teratogenic impact on the New Zealand White Rabbit. This species had recently been specified to be the only species to react equally sensitively to thalidomide as humans do (and it is still used as an 'animal model' in teratogenicity tests during the pre-clinical development of new active substances). Even more surprising than the teratogenicity of any thalidomide species in rabbits was that a very persistent myth emerged and flourished after Köhler *et al.* (1971) had published their study about teratogenic effects

---

15 An informative account of that aspect can be found in Roth 2005.

on mice. According to this myth, the R(+) enantiomer is considered the sleep-inducing but otherwise harmless ('the good') one, and the S(-) enantiomer the embryotoxic ('the bad') one. The results of this study, however, which must rely on at least one misinterpretation, have been falsified by Scott *et al.* (1977), who clearly confirmed the missing teratogenic impact of thalidomide in rodent species other than rabbits. One important pharmacokinetical principle was – and still is – omitted or neglected by the followers of the myth of good/bad thalidomide: the metabolization of active compounds.[16] In aqueous systems like body liquids, thalidomide, although it is almost insoluble, can run through the well-known keto-enol tautomery, which is the dynamic joint between two different isomers. Only the keto-structure carries the asymmetry (Fig. 3).

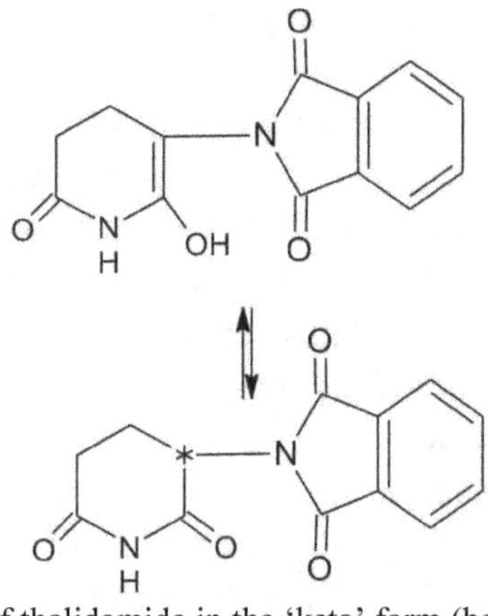

Figure 3. Structure of thalidomide in the 'keto' form (below). The position of the asymmetric carbon atom is labeled. The upper formula refers to the non-chiral 'enol' form. By equilibrizing the racemate is formed. The enantiomers cannot equilibrize directly.

The two enantiomers are transformed into one another, meaning they racemize. The half-life of both R- and S- forms in a neutral phosphate

[16] Whereas pharmacodynamics investigates the impact of incoming substances ('What does the stuff do to the body?'), pharmacokinetics examines what becomes of the incoming substances ('What does the body do to the stuff?'). For a full understanding of the nature of a substance (including adjunct ethical issues) in the biomedical field, it is necessary to take into account both aspects.

buffer is about five hours. The racemization is accelerated if enzymatic (catalytic) effects like serum albumin are present. With albumin, half-times are reduced to 18.5 min (R), and 9.5 min (S), respectively (Knoche & Blaschke 1994). Thus, both enantiomers are transformed/metabolized into the respective other form in the body, resulting after some hours in a racemic mixture. The differentiation of activities, as the myth has it, in fact cannot be tested empirically at all because of the peculiar nature of thalidomide. In addition to racemization, thalidomide is cleaved by hydrolytic reactions, such that it is still unknown which the active molecular agent is.

2.5 The unfinished story

Although it had a complicated and unclear beginning, was introduced to the public as a 'wonder drug', became a 'dark remedy', and has since been withdrawn as an active component, the epistemic and pharmaceutical object thalidomide is still vital and fashionable. At the moment, almost all of the manifest effects including the inhibition of angiogenesis or endocrine disruption are far from being understood,[17] but – or better because of this – the research is booming. The direction of research, of course, is turning away from the sleeping pill to the fight against leprosy, AIDS, and cancer.[18] One of the most astonishing comebacks of a 'dead remedy', the healing of leprosy by thalidomide, has been initialized – by chance: In 1964, Jacob Sheskin (1914-1999), the director of the Jerusalem hospital, gave the medicine, which had already been withdrawn, to a hopelessly ill patient in 1964. After the intended good sleep, the pain and the leprosy symptoms almost immediately vanished. Many other patients reacted similarly. This unexpected positive impact was obtained by a desperate ('blind clinical') move. But still, although there might be some desirable effects with a drug like thalidomide, there will always be a remaining danger, namely the unconscious digestion of the substance. In

---

[17] Stephens *et al.* (2000) give a more recent commentary on the embryopathy explanation hypotheses.

[18] Stephens & Brynner (2001), from whom I borrowed the expression 'dark remedy', tell the more recent part of the history in a fascinating, lively and engaged manner. (Stephens was involved in pertinent scientific work for many years, see, for example, Stephens *et al.* 2000)

countries like Brazil, where leprosy threatens a large number of people, unintended increases in the use of the substance has led to new cases of malformations. Therefore, as long as the substance is available to the public or parts of it, for whatever reason, vigilance must not be neglected (but this is already a point that belongs in the following section).[19]

3. Ethical Discussion

Commenting on the behavior of the scientific decision-makers of companies like Grünenthal (Germany), Richardson-Merrell (USA), Distillers (UK), Astra (Sweden), Dainippon (Japan), *etc.*, Roald Hoffmann writes:[20]

> […] yes, this is abysmal science. And whereas science as a system for gaining reliable knowledge works in spite of instances of poor-quality experimentation – it will easily survive sloppiness, hype, and even fraud – the kind of science that touches on human lives cannot afford to be bad. The thalidomide disaster should not have been allowed to happen. [Hoffmann 1995, p. 136]

In fact, science survived these greed-driven distortions, lies, and slackness, and made significant progress, but many thousands of humans could not do so, and some thousands have suffered unnecessarily. Apparently, the transition from miserable science to 'fraud and deceit in science' is much easier than we would wish, and this has a reasonable influence on the ethical judgment, as we will see below.[21]

3.1 Could the disaster have been avoided?

To start the discussion of pertinent ethical issues, that is to find out the neuralgic points of responsibilities, I shall explore some possible answers to the question used as the heading of this section. A first, admittedly

---

[19] Raza 2002 argues for the strict non-use of thalidomide because of the possible misuse.

[20] At the same spot, however, Hoffmann hints at teratological tests that where performed in the 1950s, but does not refer to the uneasy way to find an adequate 'animal model'.

[21] 'Fraud and Deceit in Science' is the subtitle of *Betrayers of the Truth* (Broad & Wade 1982). In this book, the authors present a list of known or suspected cases of scientific fraud. I have shown here that it would be possible to consider the original thalidomide 'research' as a candidate for that list.

quite simplistic answer would be: yes, if nobody had synthesized the substance, and if nobody had distributed it, and if nobody had needed and consumed these sleeping-pills during the first trimester of pregnancy, the surprising and threatening chemical behavior would not have been realized – at least not in the terrible way it was. A safety-oriented and less trivial positive answer could be motivated by the evidence of neurological impairments, the reports of which commenced in 1959. This less trivial 'yes' could be applied if physicians and scientists, particularly those of Grünenthal, had been much more attentive and careful than they were.[22] The story of the non-licensing of thalidomide to Richardson-Merell in the USA due to the resistance of the FDA physician Frances Kelsey (who in 1962 was honored by President Kennedy) refers to that version to argue for a 'yes'.[23] Of course, Kelsey could not and did not know that thalidomide would turn out to be a teratogen, thus that part of the rejection of the application by the FDA was a lucky outcome. Still, the officials of Grünenthal did a less than mediocre job as scientists and physicians by dropping their scientific curiosity and veracity with respect to their own product. While neglecting the underdetermined chemical character of their product, they decided to follow or tolerate an aggressive marketing campaign based on at least incomplete and distorted results and trivialized or neglected all the anxious reports about side effects.

Another simple but naively whiggish justification of a 'yes' to the question in the heading of this section would be connected to the performance of teratogenicity tests. The problem is: nobody asked for such tests at that time – not the governmental legislators or regulators, and in

---

22 Using mainly the files of the Alsdorf trial, Beate Kirk gives a meticulously penetrating account of that part of the story in 1960-1961, which in fact is an almost classical sociopathological tragedy about practical constraints (*Sachzwänge*) in professional life (Kirk 1999, pp. 60-86). She apparently tends to this interpretation; intriguingly, the results of a study initiated and accepted by Grünenthal in 1957 gave an obvious account of the neurological activity of Contergan – as stimulant (!) for electroencephalograms (Walkenhorst 1957).

23 For detailed accounts of this story see the previously mentioned references. In the German Democratic Republic, thalidomide was not licensed because there were already enough hypnotic remedies, rather than for safety or ethical reasons. Note that the German edition of Sjöström & Nilsson 1972 was published in the GDR. The foreword by the Berlin pharmacologist F. Jung is a practically undisguised piece of propaganda.

particular not the manufacturers – and, furthermore, nobody would have known which experiments to perform or which kind of animal to take. As described above, Blasiu and particularly Nulsen came the closest to the pertinent knowledge, however by dubious methodologies.[24] The missing-question-argument is supported by the switch-on-effect of the number of publications after 1961 (for a few German journals, see Fig. 4): Even the otherwise attentive scientific community did not suspect anything like this, and a worldwide investigation campaign with hundreds of papers was started only after the manifestation of the effect by people like Lenz.

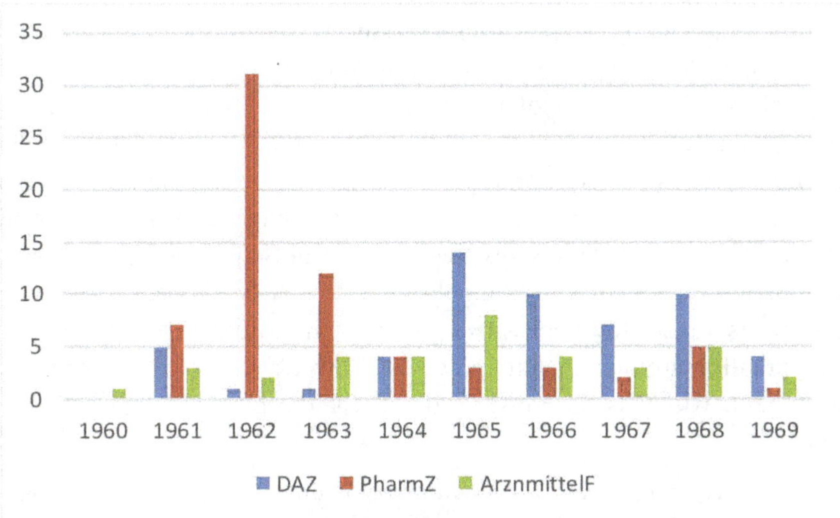

Figure 4. Numbers of articles on thalidomide in three German pharmaceutical journals (*Deutsche Apotheker Zeitung, Pharmazeutische Zeitung, Arzneimittelforschung*) during the main historical phase (after Kirk 1999).

Finally, then, the answer to the question above must be 'no', if someone would refer to the mentioned myth of the 'evil' and the 'good' thalidomide ('distomer' vs. 'eutomer'): there is – unfortunately, perhaps – no practical way to differentiate those two biochemically.

[24] Maio 2005 points out that while it can be immoral not to test medicines on humans sufficiently, it can be immoral to test them on certain groups of humans (here: pregnant women).

3.2 Underdetermination – chemically

The chemical world is in flux. Every new substance can potentially come across many others – those already existing on the one hand, and those which come into the world later on the other. Hence it is impossible to give an exhaustive description of the chemical character and future behavior of chemical entities – in particular not by just pointing at simplifying compositional formulas (like some analytical philosophers still would like to have it). Even a long-standing and intensively researched substance such as water is not fully known up to the present. Chemical substances, even those that are assigned the status of a 'natural kind' by some authors (which is not applicable in the present context, because thalidomide clearly is an artificial product), will never have a conclusive or 'closed' description. Admittedly, underdetermination, in an epistemological perspective, is not a new notion to philosophy of science.[25] I would like to emphasize, however, that, applied to chemistry, underdetermination has a more ontological connotation to it: The material world is changed when a new substance enters the stage, and the prediction of its actual input is severely restricted. Joachim Schummer discusses three arguments for the creator of a substance to justify his or her actions in a moral discourse: the necessity argument, the knowledge argument, and the intention argument (Schummer 2001). With respect to the knowledge argument, which is the most pertinent for the present issue, Schummer claims:

> [...]every substance bears an infinite potential of properties. From that it follows, however, that it is very likely that any new substances can be used to cause harm. Thus, we may expect that our chemist, while being unable to foresee the particular case of harm, knows well about the high probability of possible harm. Therefore, the knowledge argument turns to the contrary and does not help to excuse our chemist. [Schummer 2001, p. 112]

Typically, non-knowledge outmatches knowledge with respect to new chemical entities, which is a good reason to be careful in whatever appli-

---

[25] See, for example, the accounts of that expression in the introductory textbooks Ladyman 2002, French 2007 and Wiltsche 2013. The customary understanding of underdetermination is the lack of data supporting a certain theory.

cation.[26] Hence, the very issue of underdetermination – the knowledge about predominant non-knowledge in chemistry – distinguishes the Contergan case principally from other accidents as, for instance, the Titanic disaster.[27] The only feasible way to eliminate risk would be to stop producing and distributing new substances.

3.3 Who takes responsibility for what?

Because some of the classical schools of philosophical ethics, in their search (or fight) for the last fundament of reasoning, have lost contact with specific practical problems of modern life, particularly in medical and health care, the protagonists of the so-called 'principlism' drew up a list of major moral principles which are acceptable to a large group of professionals in the biomedical realm (*e.g.*, Beauchamp & Childress 2009, Wiesing 2012). These principles can be likened to religious commandments, but do not explicitly refer to any fundamental commitment, and are applied without prioritization or hierarchy. Because that is a modern standard approach, which is applied in many ethic committees, and in order to get an appropriate framework for the pertinent discussion, I assume that the main players of the story investigated in the present study are addressable by these principles (which I simplify for the time being). I try to locate these players (substance creators, substance testers, substance distributors, legislators, regulators, consumers) within that framework with respect to the present case study. Here are the four most customary principles:

(i) *Respect for autonomy*: The medical expert or healthcare professional should respect the decision-making capacities and wishes of the patient. He or she should actively support the enabling of individuals to make informed choices. The informed consent is

---

[26] In a similar sense, the relation between knowledge and non-knowledge is addressed by the Humean induction problem, too: "When we describe a drug as 'safe', therefore, what we should really say is that it is a drug that has not yet been found unsafe."(Knightley 1979, p. 24)

[27] Annas & Elias (1999) compare these two cases. Referring to the re-introduction of the drug in the USA, they say (p. 100): "Like preventing death at sea, preventing thalidomide-affected births will require not only medical technology but also human alertness."

perhaps the most important expression for this respect of autonomy.

(ii) *Non maleficence*: The medical expert or healthcare professional must avoid causing harm. Because all treatments include risks this principle conflicts with (iii).

(iii) *Beneficence*: This considers the balancing of benefits of treatment against the risks and costs; the healthcare professional should act in a way that benefits the patient/consumer. It is necessary to actively support the benefits for the patient/consumer.

(iv) *Principle of justice*: The distribution of benefits, risks, and costs must be fair. This principle is more complicated to interpret. (How do we define justice?)

I have shown above that informed consent has been violated severely in the Contergan case. The groups of creators (chemists, pharmacists), testers (pharmacologists, physicians), and distributors (traders, apothecaries, physicians) have misused their customers (consumers and their babies) as guinea pigs or 'canary birds in the mine of our future'[28] in blind, uninformed clinical trials outside the clinics. Indeed, when it is not known whether a substance being produced is harmful, then the testers and distributors bear the highest amount of responsibility. Later, the producers carry responsibility, too. The legislators must implement and monitor the drug safety framework, which was, as we have seen, quite miserable in most countries during the 1950s, although these vary in detail. I have to stress here that Frances Kelsey was one of the few persons working in administration who took the regulating principles serious, particularly principle (ii), and was very successful with that.

As to the threat for babies, the situation is different. Babies cannot be recipients of any information, therefore principle (i) becomes a respect for future autonomy, and with respect to that, the mother changes into the role of a responsible actor.

Up to this point I have focused on the role of the creators and distributors of thalidomide and their responsibilities. To gain a more adequate and all-encompassing picture for the ethical discourse, however, it seems

---

[28] In a broader sense, a similar canary-bird-metaphor is used by Murphy & Goldkind 2005.

necessary to analyze the consumers, too. Between 1957 and 1961, about 5 million people took 300 million daily doses (100 mg). In most of the scientific and public discussions, new-borns, stillborns, and mothers are solely presented as victims without freedom of action. Of course, for the children this is correct, but is it correct for the mothers, too? Is it necessary to take sleeping pills on a regular basis, perhaps over the whole period of pregnancy? Pregnant consumers take responsibility for the health and welfare of their offspring, independent of the ethical conviction or belief (see principles (ii) and (iii)). There are good arguments not to consider insomnia and nausea as illnesses during the first trimester of pregnancy,[29] and if both are not illnesses, then no medication is needed at all. After all, in present-day pregnancy pharmacotherapy, sleep-inducing drugs are no longer used, and thalidomide (and its relatives) is prescribed only for very particular indications (for example, leprosy, cancer, AIDS) and with detailed safety instructions (Schaefer & Weber-Schoendorfer 2009). Frequently, the general advice is not to prescribe or take new and premature drugs (which is in accordance with the lessons we learned from the Contergan disaster). Hence, we should not forget to take into account the role and responsibility of the drug consumers.

Referring to principles (ii) and (iii): To do no harm to the patient/consumer is an old principle of the medical ethos (*primum nil nocere*),[30] and to do good seems to be almost trivial as a medical principle. Nevertheless, with the present case it is useful to recognize that the former has to do with omitting,[31] the latter with active doing. Ironically, the handling of thalidomide turned those principles upside-down: What was omitted and disregarded by the testers and distributors was particularly principle (ii), and the benefits of some enforced sleep (plus the financial benefits of the companies) could by no means compensate for the malformations of new-borns and stillborns.

---

[29] For a short discussion of that point see Nesse 2012, p. 175.

[30] Note that this principle is explicitly used as an accusing claim against the manufacturers by Voss 1961.

[31] We should reformulate principle (ii) as follows: Leave the patient alone if he or she is well!

The interpretation of the principle of doing justice (iv) goes into a similar direction as the points proposed above. Benefits and risks were not distributed fairly.

3.4 Beyond anthropocentrism

Another important field of ethical dispute addressing pharmacology and toxicology in general and the Contergan case in particular is the field of animal ethics, which I mention here only briefly. After 1961, many countries tightened their regulations regarding the approval of remedies. In the European Community, for example, full toxicology studies, including teratogenicity tests, have had to be conducted on at least two species of mammals since 1975.[32] These species are specified, referring to the teratogenicity tests, as rabbits on the one hand (of a race that is 'sensitive against a proven embryotoxic substance') and rats or mice on the other. Without doubt, this specification is the result of the Contergan case. But who can know in advance whether or not any teratogen reacts like thalidomide? And who can be sure that the chosen animals are of the right species for any risk impact – imaginable or not? The answer to both questions is simple: no one. We still do not know enough about the molecular mechanisms, and it is still possible for a substance to pass the now 'tightened' toxicological controls, and nevertheless cause harm on a different path than expected. What certainly has been tightened is the precarious situation of test animals, which have been sacrificed in significantly increasing numbers since Contergan. The main legitimization for this procedure is the fact that there was no comparable case ever since (but this might lead to a typical induction fallacy). Hence, aside from the general use and consumption of animals as substance testers, the Angst-driven decisions of the legislators and regulators around the world are questionable in both methodological and ethical perspectives.

---

[32] See Hasskarl & Kleinsorge 1979, pp. 206-208.

4. Conclusions

(1) According to Joachim Schummer, those who create and prepare substances carry responsibility for the impact of these substances, and should be aware of providing morally relevant arguments which can support their actions (Schummer 2001, p. 111). Particularly if adverse effects have become known in the case of a remedy, further production becomes questionable from an ethical point of view. In the present context, however, it is not easy to identify and address the very chemist or chemists who prepared the substance for the first time. Along with other disinformational statements, the officials of Grünenthal have not provided trustworthy references about the possible prehistory and the early history of this bioactive substance until the present day. What we have are good arguments that the embryopathic property is not the only pharmaco- and toxicodynamical difference between humans and (most) rodents. Indeed, the question – which is of ethical interest as well – remains: Where did the information about the sleep-inducing impact on humans come from?

(2) As with other pertinent issues of applied ethics – see for example the discourse on active and passive euthanasia – the thalidomide story is foremost about doing and omitting: The main omission of the leaders of Grünenthal – one that is morally inexcusable – has been the continued lack of reports on side effects (as well as other alarming information resulting from the company's studies). Their equally morally wrong action was the misrepresentation of the facts to the scientific community, physician, and the public. Their behavior described in the present chapter was without doubt a scandal.

(3) The embryotoxic effect was a new and unpredictable chemical property of thalidomide; this part of the story was a real accident and no fault (like, for example, the catalytic depletion of stratospheric ozone by fluorochlorohydrocarbons). The characterization of a (new) substance is never finished and sometimes mercurial (*cf.* DDT, cholesterol, laughing gas, fluorochlorohydrocarbons, and many others).

(4) The dream of an entirely harmless but fully effective remedy should never be dreamt again. To paraphrase the old wisdom of Paracelsus: If it is active, then it has side-effects, too.

(5) There has rarely been an artificial substance that came so close to humans with such dangerous, unexpected, and at the same time promising potential as thalidomide. However, it is not the substance that should be blamed, but those humans who do things by halves.

I shall finish by briefly addressing an issue not yet referred to in the present study, and seldom elsewhere in the body of work dealing with thalidomide. That final issue touches on a different aspect of ethical discourse. It is quite common for many institutions and for the public to project disaster or catastrophe on those humans born with the described symptoms, and these humans are considered to be '*Sorgenkinder*' (problem children).[33] But does it go without saying that all thalidomiders lead a miserable life and that they suffer on a daily basis? There are several autobiographical accounts, most of which impressively show that this is not the case (Eistel 2007). Bettina Eistel, to mention only one of these thalidomiders, studied psychology, is a very successful dressage equestrienne with medals from European championships and the Paralympics, and is a well-known television moderator. There is much happiness here and creative power, and there is no need to reduce the lives of these persons to their disabilities. Hence, if there is a problem, it might as well be one of the societal environment rather than that of the thalidomiders themselves.

References

Annas, G.J. & Elias, S.: 1999, 'Thalidomide and the Titanic: Reconstructing the Technology Tragedies of the Twentieth Century', *American Journal of Public Health*, **89**, 98-101.

Beauchamp, T.L. & Childress, J.F.: 2009, *Principles of Biomedical Ethics*, New York/Oxford: Oxford University Press.

Blasiu, A.: 1958, 'Erfahrungen mit Contergan in der Frauenheilkunde', *Medizinische Klinik*, **53** (18), 800.

[33] In Germany the '*Aktion Sorgenkind*' ('Initiative Problem Child') was a fundraising campaign involving a television show and a lottery that was initially motived by the Contergan story. It was first introduced in 1964, and over the years about 1 billion euros have been collected to support the disabled. In part due to this initiative, many negative stereotypes about disabilities and personal differences have strongly and verifiably changed in Germany. Intriguingly, the name of the project was changed to '*Aktion Mensch*' ('Intitiative Human') in the 1990s.

Broad, W. & Wade, N.: 1982, *Betrayers of the Truth: Fraud and Deceit in Science*, Oxford: Oxford University Press.

Eistel, B.: 2007, *Das ganze Leben umarmen*, Bergisch Gladbach: Ehrenwirth.

Esser, H. & Heinzler, F.: 1956, 'Klinische Erfahrungen mit einem neuen Sedativum und Hypnoticum', *Therapie der Gegenwart, Monatsschrift für praktische Medizin*, **95** (10), 374-376.

French, S.: 2007, *Science: Key Concepts in Philosophy*, London/New York: Continuum.

Friedrich, C.: 2005, 'Contergan – zur Geschichte einer Arzneimittelkatastrophe', in: L. Zichner, M.A. Rauschmann, K.-D. Thomann (eds.), *Die Contergankatastrophe*, Darmstadt: Steinkopff, pp. 3-12.

Hasskarl, H. & Kleinsorge, H. (eds.): 1979, *Arzneimittelprüfung und Arzneimittelrecht*, 2nd ed., Stuttgart-New York: G. Fischer.

Hoffmann, R.: 1995, *The Same and not the Same*, New York: Columbia University Press.

Jung, H.: 1956, 'Klinische Erfahrungen mit einem neuen Sedativum', *Arzneimittel-Forschung*, **6**, 430-432.

Kessel, N.: 2013, 'Doriden von Ciba: sleeping pills, pharmaceutical marketing, and Thalidomide, 1955-1963', *History and Technology*, **29**, 153-168.

Kirk, B.: 1999, Der Contergan-Fall: eine unvermeidbare Arzneimittelkatastrophe? Zur Geschichte des Arzneistoffs Thalidomid, Stuttgart: Wissenschaftliche Verlagsgesellschaft.

Knabe, J.: 1995, 'Synthetische Enantiomere als Arzneistoffe', *Pharmazie in unserer Zeit*, **24**, 324-330.

Knightley, P.: 1979, 'Suffer the Children – The Story of Thalidomide', London: A. Deutsch.

Knoche, B. & Blaschke, G.: 1994, 'Investigations on the in vitro racemization of thalidomide by high-performance liquid chromatography', *Journal of Chromatography A*, **666**, 235-240.

Köhler, F., Meise, W. & Ockenfels, H.: 1971, 'Teratologische Prüfung einiger Thalidomid-Metabolite', *Experientia*, **27**, 1149-1150.

Kunz, W.: 1959: 'Eine Übersicht der neueren Arzneimittel aus den letzten fünf Jahren', in: E. Jucker (ed.), *Fortschritte der Arzneimittelforschung*, Basel & Stuttgart: Birkhäuser, vol. 1, pp. 531-607.

Kunz, W., Keller, H. & Mückter, H.: 1956, 'N-Phthalyl-glutaminsäure-imid: Experimentelle Untersuchungen an einem neuen synthetischen Produkt mit sedativen Eigenschaften', *Arzneimittel-Forschung*, **6**, 426-430.

Ladyman, J.: 2002, *Understanding Philosophy of Science*, London & New York: Routledge.

Lenhart-Schramm, N.: 2016, *Die Haltung des Landes Nordrhein-Westfalen zu Contergan und den Folgen*, Ph.D. dissertation, University of Münster.

Lenz, W.: 1961, 'Kindliche Mißbildungen nach Medikamenteinnahme während der Gravidität', *Deutsche Medizinische Wochenschrift*, **86**, 2555-2556.

Lenz, W. & Knapp, K., 1962, 'Foetal Malformations Due to Thalidomide', *German Medical Monthly*, **7**, 253-258.

Maio, G.: 2005, 'Ethische Überlegungen zur Contergankatastrophe und zur Wiedereinführung von Thalidomid', in: L. Zichner, M.A. Rauschmann, K.-D. Thomann (eds.), *Die Contergankatastrophe*, Darmstadt: Steinkopff, pp. 129-134.

McBride, W.G.: 1961: 'Thalidomide and Congenital Abnormalities (letter to the editor)', *The Lancet*, **2**, 1358.

McBride, W. G.: 1977, 'Thalidomide embryopathy', *Teratology*, **16**, 79-82.

Muller, G.W.; Konnecke, W.E.; Smith, A.M. & Kethani, V.D.: 1999, 'A Concise Two-Step Synthesis of Thalidomide', *Organic Process Research& Development*, **3**, 139-140.

Murphy, D. & Goldkind, S.F.: 2005, 'The Regulatory and Ethical Challenges of Pediatric Research', in: M.A. Santoro & T.M. Gorrie (eds.): 2005, *Ethics and the Pharmaceutical Industry*, Cambridge: Cambridge University Press, pp. 48-67.

Mutschler, E. & Lemmer, B. (eds.): 1985, *Pharmakologie: Medizinische Grundbegriffe*, Stuttgart: Wissenschaftliche Verlagsgesellschaft.

Nesse, R.: 2012, 'Warum es so schwer ist, Krankheit zu definieren: Eine darwinistische Perspektive', in: Th. Schramme (ed.) *Krankheitstheorien*, Berlin: Suhrkamp, pp. 159-187.

Nowack, E.: 1965, 'Die sensible Phase bei der Thalidomid-Embryopathie', *Humangenetik*, **1**, 516-536.

Nulsen, R.O.: 1961, 'Trial of thalidomide in insomnia associated with the third trimester', *American journal of obstetrics and gynecology*, **81**, 1245-1248.

Pschyrembel: 1986, *Klinisches Wörterbuch*, Berlin/New York: de Gruyter.

Raza, A.: 2002, 'The third coming – Thalidomide and a final goodbye', *The Biochemist*, February, 21-23.

Roth, K.: 2005, 'Eine unendliche chemische Geschichte', *Chemie in unserer Zeit*, **39**, 212-217.

Rourke, J.S.IV., 2006, 'Merck & Co. Inc.: Communication Lessons from the Withdrawal of Vioxx', *Journal of Business Strategy*, **27**, 11-22.

Santoro, M.A. & Gorrie, T.M. (eds.): 2005, *Ethics and the Pharmaceutical Industry*, Cambridge: Cambridge University Press.

Schaefer, C. & Weber-Schoendorfer, C.: 2009, 'Pharmakotherapie in der Schwangerschaft', *Internist*, **50**, 455-466.

Schneider, W.: 1985, *Geschichte der Pharmazie*, Stuttgart: Wissenschaftliche Verlagsgesellschaft.

Schummer, J. : 2001, 'Ethics of Chemical Synthesis', *Hyle: International Journal for Philosophy of Chemistry*, **7**, 103-124.

Shealy, Y.F.; Opliger, C.E. & Montgomery, J.A.: 1965, 'D-and L-thalidomide', *Chemistry & Industry,* **24**, 1030-1031.

Scott, W.J.; Fradkin, R. & Wilson, J.G.: 1977, 'Non-Confirmation of Thalidomide Induced Teratogenesis in rats and Mice', *Teratology*, **16**, 333-335.

Sjöström, H. & Nilsson, R.: 1972, *Thalidomide and the Power of the Drug Companies*, Middlesex: Penguin (German translation as: *Contergan oder die Macht der Arzneimittelkonzerne*, VEB Verlag Volk und Gesundheit, Berlin, 1975).

Stärk, G., 1956, 'Klinische Erfahrungen mit dem Sedativum K 17 in der Lungenheilstätte und der allgemeinen Praxis', *Praxis – Schweizerische Rundschau für Medizin*, **45** (42), 966-968.

Stephens, T.; Bunde, C.J.W. & Fillmore, B.J.: 2000, 'Mechanism of Action in Thalidomide Teratogenesis', *Biochemical Pharmacology*, **59**, 1489-1499.

Stephens, T. & Brynner, R.: 2001, *Dark Remedy*, Cambridge: Perseus.

Talbot, J. & Waller, P. (eds.): 2004, *Stephens Detection of New Adverse Drug Reactions*, 5th ed., Chichester: Wiley.

Voss, R.: 1961, 'Nil nocere! Contergan-Polyneuritis', *Münchener Medizinische Wochenschrift*, **103** (30), 1431-1432.

Walkenhorst, A.: 1957, 'Das hypnotisch und sedativ wirkende N-Phthalyl-Glutaminsäure-Imid als geeignetes Provokationsmittel bei hirnelektrischen Untersuchungen', *Wiener klinische Wochenschrift*, **69**, 334-339.

Wiesing, U. (ed.): 2012, *Ethik in der Medizin – Ein Studienbuch*, Stuttgart: Reclam.

Wiltsche, H.A.: 2013, *Einführung in die Wissenschaftstheorie*, Göttingen: Vandenhoeck & Ruprecht.

Zichner, L.; Rauschmann, M.A. & Thomann, K.-D. (eds.): 2005, *Die Contergankatastrophe – Eine Bilanz nach 40 Jahren*, Darmstadt: Steinkopff.

Chapter 7

Risk and Responsibility in Chemical Research: The Case of Agent Orange

Claus Jacob & Adam Walters

Abstract: The synthesis of new chemical substances raises a number of ethical concerns frequently overlooked by chemists, such as the risks associated with the production of a new substance, the risks posed by impurities often found in chemical agents, risks related to practical applications, and the question of ultimate responsibility for a new compound, the impurities contained within and wider applications. The case of the synthesis and subsequent use of Agent Orange exemplifies these issues. Risk as well as responsibility for the agent have shifted significantly since its discovery, from the original inventor of a new compound, via the industrial manufacturer of a dioxin-contaminated herbicide, to the user or users of the impure agent as tactical chemical weapon in Vietnam. Analyzing the chain of historical events in the light of moral responsibility allows us to set everyday chemistry into an ethical context and ask a number of important questions, such as who carries responsibility for a new chemical compound, its employment in practice, safety in light of contaminants and impurities, and its proliferation.

1. Introduction

The synthesis of new chemical substances raises several ethical concerns frequently overlooked by chemists.[1] Arguably, the focus resides in the risk associated with the production of a new substance and the ultimate question of responsibility for its manifold consequences. Two special

---

[1] The authors would like to thank Dr. Muhammad Jawad Nasim and Ahmad Yaman Abdin for their assistance during the revision and updating of the original manuscript from 2005.

issues of *Hyle*, published in 2001 and 2002, have initiated a debate of ethics within chemical research (see, for example, Davis 2002, Del Re 2001, Kovac 2001, Laszlo 2001, Schummer 2001a/b). During the years, the substantial interest in this meta-chemical topic has translated into immense efforts to draw attention to – and also adequately address – such controversial ethical issues of risk and responsibility for new compounds and certain impurities associated with them (see Contakes & Jashinsky 2016, Børsen & Nielsen 2017, Kovac 2018, Schummer 2018, Maxim 2017). Given the seriousness of their impact on the future practice of synthetic chemistry, such ethical debates are still in their infancy and often without the desired impact on research chemists. From a chemist's perspective, the importance of an ethical discourse might only become apparent when specific examples from everyday chemical research are used to illustrate where and how ethical issues arise which need to be addressed.

This chapter utilizes the well-documented historical case of Agent Orange and the related twists and turns to discuss, elaborate, and emphasize ethical questions related to risk and responsibility associated with the synthesis of a novel chemical compound. The discourse will focus not only on the two main ingredients of Agent Orange, but also on the unexpected, yet devastating, impurity today referred to as dioxin. As such, it contributes an important historical case study to 'Ethical Cases Studies of Chemistry' and also to the project of 'Green Chemistry'.

The next section briefly sets out the philosophical background of this study. The third section provides a focused - and necessarily incomplete - historical overview of Agent Orange, from the first synthesis of its active ingredients in a research laboratory and the occurrence of the dioxin contamination to the use of the agent as a chemical weapon in Vietnam. Sections four to six discuss the questions of risk and responsibility associated with chemical compounds, from the synthesis of a new chemical agent to its potential uses and abuses, using the case of Agent Orange as one major example[2]. The final section provides a summary and outlook.

---

2 This discussion focuses on risks associated with chemicals. It should be noted that many chemicals have thoroughly positive effects on human health and the environment. Some of the considerations, such as the proliferation of chemical knowledge, apply to both.

2. Risk and Responsibility

Chemistry provides a fertile ground for different ethical analyses. From a chemist's perspective, the ethical issues surrounding the synthesis of a genuinely new compound might be the most interesting one. During this process, the chemist develops a method to produce something intrinsically novel, something which has not existed before. This immediately raises questions related to the risks and responsibilities associated with the compound and its method of synthesis.

As far as risks are concerned, we can distinguish between the risk posed by the first chemical synthesis ever and subsequent repetitions (Table 1). The first ever synthesis is a step into the unknown with practical risks for the chemist and workers in the laboratory. Historical and more recent events teach us that such a step can be very dangerous, especially in cases where the new substances turn out to be toxic or contain toxic contaminants (Abdin *et al.* 2020). Generally, it is good chemical practice to perform such an initial synthesis with small quantities and under secure conditions, but the issues related to the remaining risks are rather of a technical than of ethical nature. Similarly, the first batch of the new compound, in small quantities and safely kept, is unlikely to have a major impact on humanity.

In contrast, the risks arising from the future *availability* of a new compound within – and outside – the scientific community as a result of the first ever synthesis are not confined to the research laboratory, and therefore are considerably more difficult to assess and to control. These risks associated with the coming into existence of a novel chemical are mainly the result of (a) the future use of larger quantities of the compound, (b) the manufacture process of bulk quantities and (c) the dissemination of the synthetic method in the scientific literature. As we will see in later sections, the issues raised by (a) and (b) are usually dealt with at the level of the chemical industry, not research, and are less important here.[3] Nonetheless, there have been instances where up-scaling as part of the manufacturing process has introduced new problems, for instance in form of unexpected impurities (Abdin 2020). The risk posed by (c),

3 They might, for instance, include the use of a chemical as a drug or air pollution as the result of the manufacturing process.

taken by the research chemist at the point of publishing the synthetic method, is frequently underestimated; both the legal and moral assignment of the resulting responsibilities are vague. This risk is directly relevant to our discussion.

Table 1: A selection of different risks commonly associated with a novel compound, in order of occurrence from its first synthesis to use.

Risk	**Actors sharing responsibility for risk**	**Current assignment of responsibility**
Risks associated with first synthesis (*e.g.* explosion)	Synthetic chemist (inventor), employer	Synthetic chemist (inventor), employer
Risks associated with first batch of compound (*e.g.* toxicity)	Synthetic chemist (inventor), employer	Synthetic chemist (inventor), employer
Risks associated with uncontrolled proliferation	Synthetic chemist (inventor), employer, publisher, government	Unclear for the dissemination of synthetic procedure; government (regulator) for shipment of compounds
Risks associated with large scale manufacture (*e.g.* environmental pollution)	Manufacturer, government	Manufacturer, government
Risks associated with bulk quantities of compound (*e.g.* toxicity)	Manufacturer, government, user	Manufacturer, government, user
Risks associated with unknown impurities	Synthetic chemist (inventor), employer, publisher, manufacturer, government, user	Frequently open to ethical and legal discussion

The main actors who might share some of the responsibility for the synthesis, dissemination, manufacture, and application of a new compound are mentioned. This assignment of (shared) responsibility is open to debate, and might differ from the actually accepted responsibility. Please note that impurities might occur at any stage of the synthesis, manufacture, and even during (improper) storage of a compound; moral/legal responsibility for such impurities is particularly difficult to assign (Abdin 2020) .

Since the production of a new compound is always associated with risks, chemists, in order to continue their research activities, require an ethical model which allows them to assess and ultimately to take such risks. Del Re (2001) has addressed issues of risk and responsibility associated with chemical research. As he points out, "in principle, any scientific experiment involves a measure of risk, and therefore of responsibility" – and "responsibility arises from taking the risk" (Del Re 2001). He continues to define the 'choiceworthiness' C of an action as the quotient of the expected *gain* G over the anticipated *risk* R, with G defined as the product of the desirability D of the positive outcome and its probability, and the risk as the product of the gravity W of the negative outcome and its probability.

As we will demonstrate in the case of Agent Orange, this approach encounters some limitations as far as hitherto unknown substances are concerned, especially due to the inability of chemists to estimate the risk associated with *chemical impurities*. As a consequence, conventional risk assessment might not be enough when dealing with new compounds, and we will use it here in combination with a weak version of the *Precautionary Principle*. This principle suggests that it is better not to carry out an action which might possibly be very dangerous, as long as counter-evidence is not available. It implies that the synthetic chemist initially considers any new compound as potentially dangerous and provides initial evidence that it is safe – rather than society having to prove it is unsafe, *before* the compound or its synthetic procedure are released. During the last decade, this changed attitude towards new chemicals has manifested itself in the so-called Registration, Evaluation, Authorisation and Restriction of Chemicals (REACH) exercise of the European Union which aims to identify and contain potentially dangerous chemical substances (Llored 2017).

Since it transcends conventional principles of risk management and incorporates the notion of uncertainty, the Precautionary Principle is therefore useful in dealing with novel chemical compounds and their impurities. It serves as a constant reminder that, though a risk may not be quantified, there may still be reason to believe that damage may occur. Since it is impossible for chemists to conclusively prove the safety of a new compound in practice, we need to apply a weak version of the Precautionary Principle. For instance, an initial estimate of safety and anticipated risk can often be established by looking at chemically similar, already characterized compounds. Similarly, the presence of risky impurities might be predicted by considering the thermodynamics of a chemical process, such as temperature and pressure control. Today, computational chemistry provides tools for such predictions which are often – albeit not always – able to arrive at such predictions.

Indeed, the current state of chemistry does not enable chemists to avoid, or fully rationalize the presence of chemical impurities. Recent scandals such as the one surrounding *N*-nitrosodimethylamine, *N*-nitrosodiethylamine contaminations found in the antihypertensive drug valsartan and the H_2-inhibitor ranitidin confirm this element of uncertainty. This scientific weakness is illustrated further by the frequent failure of so-called structure-activity relationships. Although the latter are good initial approximations, small changes in chemical composition or structure, such as optical isomerism, can significantly affect biological activity. As a consequence, no new substance should be considered as harmless unless it has been carefully tested – and the same applies to impurities, regardless if they are identified chemically in such a substance or not.

The combination of risk assessment and a weak version of the Precautionary Principle is able to address two important aspects of chemical research. First, it allows chemists to reject what they might intuitively feel to be 'immoral' acts, such as chemical weapons research associated with a high value of *W* and often also a high probability of *W*. Second, it allows chemical research, even if associated with risks, to proceed, *i.e.* it does not rule out all chemical synthesis *a priori* because of possible risks.

In the following sections, we will use the notion of risk and responsibility to discuss

- if chemical impurities might complicate risk assessment;
- to which extent the inventor of a new compound carries responsibility for the compound and its synthetic method;
- if the scientific journal publishing the synthesis of a new compound might share responsibility for indirect chemical proliferation;
- if the *Precautionary Principle* should be applied to novel compounds from the point of their first synthesis.

We will start with a brief historical review of the Agent Orange case, with the ethical focus on the issues of risk, responsibility, and contamination, in this case by dioxin.

3. A Brief History of Agent Orange

The history of Agent Orange, and its associated dioxin contamination, is important for the discussion of risks and responsibility. It starts with the discovery of 2,4,5-trichlorophenoxyacetic acid (2,4,5-T), *i.e.* the first ever production of the compound in the research laboratory, and then leads to the large scale manufacture of a 2,4,5-T containing herbicide for agricultural use and the manufacture and use of Agent Orange in Vietnam.

The first synthesis of 2,4,5-T was published by Robert Pokorny (1941). At the time he was employed at C.B. Dolge Company, and working on pesticides for agricultural use. Four years later, the American Chemical Paint Company patented the use of 2,4,5-T as a weed killer along with a plethora of other forms of halogenated phenoxy monocarboxylic aliphatic acids (including 2,4-dichlorophenoxyacetic acid [2,4-D]) and their esters and salts (US Patent 2390941 from 1945). The Chemical Abstract Service decennial indexes show the subsequent rise of scientific interest in the compound, from 13 related papers between 1937 to 1946 to more than 200 papers between 1947 and 1956. The latter period also witnesses a rise in applications on a diverse range of plants, and an increase in the number of patents relating to effective delivery techniques.

Large scale manufacture of the 2,4,5-T herbicide for *agricultural uses* at Dow began in 1950 and ceased in 1979, when Dow was the largest company worldwide manufacturing the compound. In the 1950s and 1960s, large-scale manufacture led to the contamination of the herbicide with the poisonous dioxin 2,3,7,8-tetrachlorodibenzo-*p*-dioxin (TCDD) due to a side reaction (Figure 1) (Heaton 1996). 1,2,4,5-tetrachlorobenzene reacted at high temperatures with hydroxide to form sodium 2,4,5-trichlorophenoxide (2,4,5-TCP, Figure 1, pathway 1). In line with Pokorny's publication from 1941, this then reacted with chloroethanoic acid at 140°C to obtain 2,4,5-T. Temperature control in both processes was essential, because at 160°C the electron deficient 2,4,5-TCP can undergo a condensation reaction (Figure 1, pathway 2) producing the tetrachloro-substituted dioxin. As it was difficult to obtain uniform temperatures in bulk reaction vessels, 2,4,5-T subsequently contained dioxin contaminants in the order of parts per million (ppm). In subsequent years, this contamination level had to be reduced to below 1 ppm in response to legal requirements. This was achieved by better temperature control or the removal of dioxin from 2,4,5-TCP. At the time Dow stopped the 2,4,5-T manufacture in 1979, the legal upper limit of TCDD contamination was 0.1 ppm in the US and 0.01 ppm in the UK (Hay 1982, p. 9).

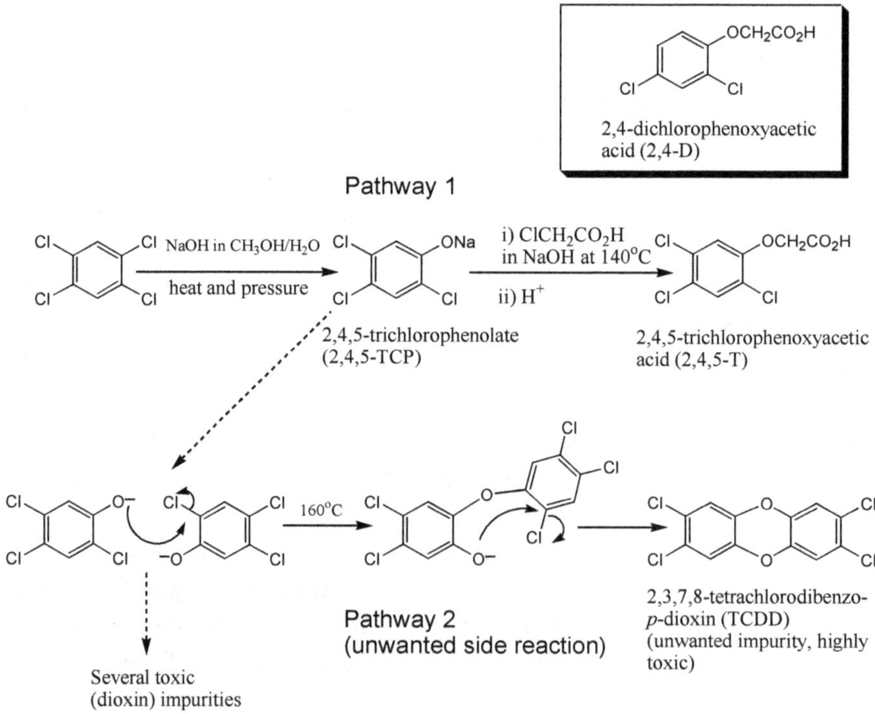

Figure 1. Synthesis of 2,4,5-T. Pathway 1 shows the desired reaction leading to the active ingredient for herbicides. Pathway 2 is a side reaction leading to an unwanted dioxin contamination. In addition to TCDD, a range of other impurities of varying degree of toxicity is formed. Insert upper right: the chemical structure of 2,4-D.

Health problems associated with 2,4,5-TCP exposure, particularly the development of chloracne, were first noticed during industrial accidents involving the compound, such as at Monsanto (1949), Boehringer in Germany (1952) and at Dow (1964). The link between 2,4,5-TCP, chloracne and an – at the time initially unknown – dioxin impurity in 2,4,5-TCP was established by Karl Schultz in 1957 as a direct response to the Boehringer accident (Kimmig 1957). These developments took place several years before the use of 2,4,5-T in Vietnam.

The military use of 2,4,5-T was already considered during World War II and the Korean War, and the British employed small amounts during the 'Malayan emergency'. Nonetheless, the widespread use of Agent Orange for military purposes began only in 1962, together with five other herbicide mixtures.[4] Under the Defense Production Act of 1950, the US government allowed a number of chemical companies to produce 2,4,5-T for Vietnam, including Dow, Monsanto, Hercules Inc. and Diamond Shamrock.[5]

The use of 2,4,5-T in Vietnam had several reasons, most of which were related to the problem of jungle warfare. North Vietnamese soldiers and the Vietcong were adapted to this kind of warfare, using the trees for cover and employing guerrilla tactics. American forces had difficulties to react and their superior firepower rendered immobile in the terrain, so that they sought to remove their enemy's advantage. Operation *Ranch Hand* was launched to defoliate the forests and mangroves, and to destroy crops in order to reduce the enemy food supplies. The operation involved at least 19,900 sorties between 1962 and 1971 (Stellman *et al.* 2003; Hay 1982, p. 147). They employed a range of herbicide mixtures, each known by the color identification band painted on storage barrels. Of these herbicides, Agent Orange was the most widely used blend (45,677,937 liters), consisting of a 50:50 mixture of *n*-butyl esters of 2,4-D and 2,4,5-T.[6] Agent Orange used in Vietnam was contaminated with

4 Refer to Butler 2003a/b; for excellent historical reviews of this and related topics, refer also to Hay 1982 and Gough 1987.

5 Although the following discussion mainly focuses on Dow, companies like Monsanto and Diamond Shamrock share a similar, if not greater, responsibility for the manufacture of Agent Orange.

6 Both 2,4,5-T and 2,4-D were generally applied as either esters or salts of the acids. Long chain (hence low volatility) esters were applied as an oil emulsion and were

varying levels of TCDD, on average around 1.91 ppm or higher (Hay 1982, p. 164; see Figure 1 for chemical structures).

The manufacturer's knowledge about the level of contamination and the toxicity of TCDD at the time of production is hazy. It is complicated by the fact that some herbicide manufacturers purchased already contaminated 2,4,5-TCP. In addition, the determination of dioxins is still complicated and expensive, with only few laboratories able to carry out this kind of trace analysis.

As production levels soared to meet military demands, Dow Chemical and other manufacturers attempted to reduce the TCDD levels. Dow also developed methods to measure TCDD levels more accurately. Nonetheless, dioxin levels in Agent Orange varied dramatically, depending both on manufacturer and production lot. To provide some indication, residues of Agent Orange manufactured at the time were found to contain between 0.05 and 47 ppm dioxin, with considerably higher dioxin concentrations likely in pre-1966 samples, while Dow's product after the construction of its new plant in 1966 was below 1 ppm (Hay 1982, p. 164).

The damaging effects of the TCDD *impurity* to human health turned out to be considerably more serious than the ones of 2,4-D and 2,4,5,-T, both of which have considerably lower toxicity than TCDD. Although the 2,4,5,-T toxicity is still under debate, the main toxic substance in commercial mixtures of 2,4,5-T based herbicide is clearly not the active ingredient itself – it is the dioxin contaminant. For example, neither the West German nor the UK authorities in 1980 were "convinced that 2,4,5-T, if used for the purpose it was intended, presented a health risk" (Hay 1982, p. 177). As a consequence, improved safety of this particular herbicide for commercial use has mostly focused on a reduced dioxin contamination level, and not on 2,4,5-T. This distinguishes 2,4,5-T from

more toxic to plants than the water-soluble salts. The ester's toxicity is due to the lipophilic compound being readily taken up by cuticular lipids. Surfactants were sometimes added to mixtures to increase effectiveness by reducing run-off and by softening lipids (Hassall 1982). In plants esters are hydrolysed back to the biologically active phenoxy acetic acids by carboxylesterase enzymes. The exact modes of action vary, but both herbicides cause lethal, uncontrollable, and grossly distorted plant growth when applied in the correct dosage.

many other commercially used chemicals, where the active ingredient itself, and not an impurity, is toxic.

In 1969, the history of Agent Orange and dioxins took an unexpected turn. While TCDD was known to cause chloracne since Schultz had published his findings in 1957, a study by the Bionetics Research Laboratories provided evidence that (contaminated) 2,4,5-T was teratogenic in laboratory animals (Courtney *et al.* 1970). Agent Orange was therefore not only damaging to the Vietnamese ecosystems, but also potentially dangerous to humans. Its use in Vietnam was phased out in 1970/71, and it is now classified as a Class 1 carcinogen.[7]

The civil use of 2,4,5-T based herbicides did continue, however, and around 58 tons were still used in the UK in 1980 (Hay 1980, p. 177). On the other hand, the U.S. Environmental Protection Agency (EPA) issued an emergency order to restrict the use of the herbicide for agricultural use in 1979, and launched several attempts in the early 1980s to ban it completely. In 1983, 2,4,5-T based herbicides were withdrawn from the U.S. market and replaced by dicamba and triclopyr.[8]

Beginning in the late 1970's, many American veterans sought compensation for a variety of conditions from chemical companies involved in the Agent Orange manufacture. The responsibility for possible damages caused by the agent was a complex and legally tricky issue. These cases were consolidated into a class action lawsuit which was settled out of court. Since then a limited amount of compensation has been awarded by the government and various lawsuits have been filed against the chemical companies. Little has been done to provide compensation to Vietnamese people, although there is some evidence suggesting a high

---

[7] TCDD causes, among others, soft-tissue sarcoma, non-Hodgkin lymphoma, Hodgkin disease, chronic lymphocytic leukemia, diabetes, chloracne, birth defects, fetal death, reduced fertility (in both sexes), modulation of hormone levels, and potentially a wide range of cancers (USEPA 1994, Institute of Medicine 2000, Tuan & Phuong 2003). TCDD is categorized as carcinogenic to humans by the International Agency for Research on Cancer (IARC) and has recently been subject to various legislative controls (Stringer & Johnston 2001). Apart from Agent Orange, dioxins have also played a role in several major accidents, such as in Coalite in the UK (1968) and in Seveso in Italy (1976). Only 39 g of TCDD escaped in Coalite, resulting in a soil concentration of 400 ppb. 1.3 kg TCDD was released in Seveso, leading to soil concentration up to 235 μg dioxin per square meter, and the death of thousands of animals (Stringer & Johnston 2001, pp. 305-34).

[8] See http://www.sciencedaily.com/encyclopedia/herbicide.

rate of birth defects in contaminated areas up to now, which is not surprising because in extreme cases soil samples still exhibit the same concentration of dioxin as Agent Orange (Schecter *et al.* 2003, Tuan & Phuong 2003). The lack of extensive systematic studies on whether birth defects are indeed elevated in the highly contaminated areas poses a major problem.

With the historical development briefly mapped out, the next sections will take a closer look at the risk and the responsibility which result at each step of the development of a compound such as 2,4,5-T, from the first time it is synthesized and the publication of its synthetic pathway, to its large scale manufacture and practical application. It should be emphasized from the start that this analysis aims to stimulate future discussion, and represents just one possible point of view.

4. Risk and Responsibility at the Point of Invention

As mentioned before, there is an extensive, and well-documented debate on the legal and moral responsibilities of the manufacturers and users. The initial parts of the history of 2,4,5-T, *i.e.* its first synthesis, characterization, and publication are often ignored in these discussions. For example, Hay's historical review does not mention Pokorny at all. This is unfortunate, since the inventor of a new compound must obviously share some responsibility for the compound, even if the ultimate use is beyond her/his control. This chapter therefore focuses on the persons and institutions involved with 2,4,5-T at the beginning of the history of Agent Orange. From a chemist's perspective, this raises a number of ethical questions that are all too often ignored when just the final use of a compound is considered.

The history of Agent Orange is a good example of the chain of events leading from the discovery to the application of a new chemical. Along the line, there are several actors who might share responsibility for the compounds involved (Table 1). These include

(a) Pokorny, the research chemist who discovered 2,4,5-T around 1941,

(b) the scientific journal publishing the synthesis of 2,4,5-T in 1941,
(c) the company who patented the use of the chemical as herbicide in 1945,
(d) the companies and their chemical engineers manufacturing Agent Orange,
(e) the American government approving the use of the compound in warfare and
(f) the US Air Force/RVN actually employing Agent Orange as a tactical chemical weapon.

At the point of invention, Pokorny, the synthetic chemist, was the first person involved with 2,4,5-T. It is therefore necessary to ascertain his responsibility for the risks associated with the compound.[9] Schummer has argued that the synthetic chemist developing a new substance will always have some general responsibility for it, as the "first synthesis is the causal step for its existence" (Schummer 2001b). No matter if this is where the chemist's active relationship with the substance ends, and if the damage is caused by the hands of others, without the initial causal step no damage could have occurred. Although this notion of responsibility does not yet include a moral judgment, it implies the obligation of the chemist to respond to moral questions and accept the standards of a moral discourse.

That this may seem bizarre to the average chemist is not surprising. "The fact that the internal norms of [the chemical community] are not in agreement with general moral standards shows that the whole community do not recognize their general moral responsibility and wrongly consider their activity as morally neutral" (Schummer 2001b). Although this is a strong statement, it is indeed curious to realize that research chemists inventing and publishing novel compounds are hardly ever asked by their peers to provide even the most basic information about the potential dangers that might result from these agents.

---

9 Importantly, this is not to pass a moral judgment on the inventor. As we discuss later on, taking responsibility for a new compound does not mean taking the blame for every subsequent uses of it.

As might therefore be expected, and is normal chemical practice, Robert Pokorny's paper from 1941 entitled "Some chlorophenoxyacetic acids" considers his new compounds in the light of successful *versus* unsuccessful synthesis, not under the moral categories of right and wrong. His quarter page report in the 'New Compounds' section of the *Journal of the American Chemical Society* provides the experimental details to successfully make 2,4-D and 2,4,5-T with yields on the 5 g scale and some analytical data. There is neither an introduction to this work, nor a discussion or conclusion. Pokorny does not indicate any future work that could be done with these new compounds, or any potential applications – or safety implications – 2,4,5-T might have.

This does not imply that Pokorny himself was not aware of possible applications. His research at the company was clearly directed toward novel herbicides, and it is likely he considered commercial uses, even if he did not explicitly state them in his publication. The act of the first ever synthesis of these compounds, the production of something intrinsically new, is described from a scientific perspective. At the point of invention of 2,4,5-T, the question of responsibility for this act or the resulting agent is not raised, and it is doubtful that many chemists in Pokorny's situation would consider such a synthetic report from an ethical perspective. It should be mentioned, however, that chemists do speculate about the potential benefits of their novel compounds. They do consider possible consequences, and claim responsibility for future applications, be it mostly for the beneficial ones.

For the sake of argument, let us therefore assume that Pokorny *would* have accepted moral responsibility for the risks associated with his compound. This does not mean, of course, that Pokorny would have to accept responsibility for every use of it or its impurities. Nevertheless, he *could* have performed a risk assessment along the lines suggested by Del Re. Pokorny was obviously aware of the potential agricultural use of 2,4,5-T as a herbicide. Since he could neither have foreseen the TCDD contamination occurring when his reaction is scaled up, nor the dangers TCDD poses, nor the use of his new compound in Vietnam, his risk assessment in 1941 to determine the choiceworthiness of making 2,4,5-T would therefore have been very favorable, *i.e.* the desired benefits of 2,4,5-T as a herbicide would have by far outweighed the known risks associated with the compound.[10] For the same reason, the ethical framework used here would not ascribe any major negative *moral judgment* upon him.

As mentioned already, it is curious that *in practice* a proper risk assessment on new compounds is hardly ever performed by the inventor before the compound is published. In this context, let us look once more at the case of Agent Orange. Any risk assessment made for 2,4,5-T in 1941 would have been severely flawed by the lack of information about the occurrence and implications of the TCDD contamination. In fact, Pokorny does not mention impurities in his publication, and it will be very difficult to find any synthetic chemistry publications explicitly discussing the content and the nature of *impurities* in a new compound.[11] This in itself is an interesting aspect of synthetic chemistry, where numerous 'unidentified' compounds are produced regularly and distributed as impurities.

As far as the ethical discussion is concerned, the presence of such chemical impurities in compounds has serious implications for the risk assessment of novel chemicals. While chemists might be able to draw some conclusions about their desired new compound based on similarities with existing chemicals, they would *initially* be *completely in the dark* regarding the properties of undesired, unknown impurities which are beyond their control – and occur in compounds in an uncontrolled, almost random manner. As the TCDD contamination has shown, minute amounts of impurities (in the ppm range) can completely alter the toxicity of a 99.9999% pure substance, as long as the impurities are a few million times more toxic than the pure compound. In a biological context, the risk posed by impurities is even more serious, since many toxins undergo bioaccumulation in organisms, resulting in biomagnification of their concentrations by thousands or even millions in food chains (Southgate & Aylward 2002).

---

[10] Pokorny's samples of 2,4,5-T might well have contained dioxin impurities. The presence and effects of such dioxins, if present, would have gone largely unnoticed. Laboratory chemicals are, however, frequently of higher purity than chemicals produced on an industrial scale, due to better reaction control and purification methods.

[11] Some chemists do provide percentage purities of new compounds in their publications, but this is not always the case. In addition, impurities frequently vary from synthesis to synthesis, and might consist of yet unknown substances.

While Pokorny might have wanted to consider the toxicity of *his sample* of 2,4,5-T – which might have contained some TCDD – in a few bioassays to rule out major safety risks in handling the compound, he would have been neither able to anticipate the conditions under which impurities might be formed, nor the exact nature and toxicity of these unknowns. Even worse, the impurities in 2,4,5-T might actually have resulted from using contaminated 2,4,5-TCP, in which case the inventor of *that* compound, and not Pokorny, might ultimately share responsibility for the TCDD contaminant in 2,4,5-T.

Although TCDD and other dioxins were later identified as impurities in Agent Orange, other impurities remain unknown, and some might even contain 'yet unknown' compounds, as was indeed initially the case with dioxin in Schultz's studies. Schummer has shown that the increase of knowledge which comes with the synthesis of a new compound is smaller than the non-knowledge generated at the same time (Schummer 1999/2001b). The notion of non-knowledge also applies to such unknown impurities and dramatically complicates risk assessment and the issue of responsibility.

This topic has recently been discussed in considerable detail in a manuscript by Abdin *et al.* in which the authors have introduced a 'Space of Information' to assign coordinates to the information available about the presence, chemical and biological activities of such compounds (Abdin 2020). This allows chemists to visually place compounds and their impurities at the correct places in this multi-dimensional 'Space of Information'.

Such more recent approaches in this field try to address some of the rather difficult questions surrounding risk and responsibility associated with new compounds and their impurities. For instance, how could a chemist like Pokorny be held responsible for the effects of impurities that he was unaware that they even existed, or which were possibly transferred from an impure starting material?

In any case, Pokorny's original publication made 2,4,5-T and 2,4-D available to the chemical community worldwide without a basic understanding if it is safe or toxic, carcinogenic, or explosive. Again, Pokorny acted well within accepted current chemical practice, where the inventor of a new compound is *not* expected to show that the invention is safe.

Admittedly, he could have discussed 2,4,5-T in the light of similar compounds already established to have herbicidal properties, and provided a basic 'risk assessment' based on analogy, but this is difficult and not normally required from chemists at the point of invention.

By publishing the synthetic method, however, Pokorny in principle made 2,4,5-T available to everyone.[12] This kind of chemical proliferation by publication is frequently ignored. Since it is, however, much less controlled than the shipment of substances, the main risk associated with the invention of a new compound might indeed reside with the proliferation of the associated chemical knowledge, and not with the few grams of new compound actually made and safely stored in a secure laboratory. Although this might sound bizarre, Pokorny therefore shares direct responsibility for the worldwide, indirect proliferation of an untested compound. For the inventor, this responsibility might outweigh any responsibility for the subsequent manufacture or use of the agent as discussed above.

The inventor of a new compound is in a moral dilemma caused by potentially conflicting values. On the one hand, the scientific codex requires the publication of results, so that other scientists can repeat, test, and benefit from the new compound. On the other hand, the compound might well be unsafe and cause significant damage – possibly even to the colleagues trying to repeat the synthesis. While here is not the place to discuss this dilemma in detail, chemists should become aware of this issue, and try to address it in the future.[13]

The question of uncontrolled proliferation of chemicals through publication in scientific journals brings us to another, rather unsuspected carrier of responsibility for a new chemical compound – the scientific journals which disseminate synthetic knowledge.

---

[12] Any skilled chemist reading Pokorny's paper would be able to manufacture 2,4,5-T.

[13] Related issues, such as the proliferation of chemicals and commercial interests have been discussed by Laszlo (2001) and Kovac (2001).

5. Publishing as Proliferation

The inventor of a new compound shares responsibility for the dissemination of the synthetic procedure, including the almost uncontrolled indirect proliferation via online journals, Internet postings, and free online services such as *Medline*. It is obvious that without the dissemination of Pokorny's synthetic protocol in the *Journal of the American Chemical Society*, 2,4,5-T would not have become 'freely available' to chemists outside Pokorny's own organization at that particular time. This does not rule out, of course, that someone else might have invented the compound soon thereafter, but this is beside the point.

As a consequence, a journal such as the *Journal of the American Chemical Society* might share moral responsibility for the potentially thousands of kilograms of novel compounds being manufactured on a large scale by whomever is willing to do so. This ethical consideration is in agreement with the fact that journals frequently request the copyrights on manuscripts containing synthetic protocols before publication. Curiously, traditionally there has been little in the *Journal of the American Chemical Society's* 'Ethical Guidelines to Publication of Chemical Research' which would have addressed the ethical issues raised here. The latter deal with good scientific practice and how to provide sound and reliable data, a rather different area of ethics and chemistry. As for safety issues, the synthetic procedure and analytical data of novel compounds can be published without further warning.

In contrast, some journals have recently begun to ask authors to explicitly address such 'safety issues' as part of their publications. For instance, the *Journal of Agricultural and Food Chemistry's* 'Scope, Policy and Instructions for Authors Guidelines' says already in 2001:

> Safety: Authors are required to call special attention, in both their manuscripts and their covering letter, to safety considerations such as explosive tendencies, special precautionary handling procedures, and toxicity.

Such requests might well stimulate a basic discussion of the risks associated with a novel compound. Nevertheless, neither the authors of the manuscript, *i.e.* the inventor(s) of the compound, nor the publisher take *responsibility* for the risks associated with the new agent. From a chemist's perspective, this might look fairly standard, and is in line with Schummer's observation that chemists do not recognize the wider responsibility for their discoveries.

Let us therefore briefly consider how chemically good practice might in the future be aligned with ethically good practice. As the discussion of chemical impurities has shown, a reliable risk assessment for a novel chemical compound is virtually impossible, but some suggestions based on similar chemicals might be possible. As a consequence, a weak version of the *Precautionary Principle* should be applied when dealing with new compounds – and associate unidentified impurities. The anticipated risk can, in certain cases, be described by looking at chemically similar, already established compounds. In other cases, especially when impurities are present and conclusions by analogy are impossible (*e.g.* compare 2,4-D and 2,4,5-T), a basic toxicological screen might be required. The latter would also detect the toxicity caused by unknown impurities.

Although it may be the most suitable approach toward dealing with the risks associated with novel compounds, the *Precautionary Principle* is hardly applied in everyday chemical research. Of course, most synthetic chemists are unable to carry out full toxicological evaluations of each of their new compounds, and to demand such studies would be detrimental to chemical research. Nevertheless, publishers, as a matter of precaution, should follow the lead taken by the *Journal of Agricultural and Food Chemistry* and others and demand at least a brief *discussion* of the potential risks associated with a new compound, such as toxicity or potential environmental impact.[14] The REACH exercise of the European Union has been mentioned already and aims into the same direction.

6. Manufacturers and Users

Responsibility of the manufacturer is a complicated mix of regulatory, legal, and moral issues which cannot be discussed here fully. As in the case of Agent Orange, the manufacturer of a chemical shares responsibility for the resulting product, but no longer for the discovery during the

---

[14] The other American Chemical Society (ACS) journal mentioned here, *J. Am. Chem. Soc.*, does not demand this kind of safety information yet.

first synthesis or dissemination of the synthetic protocol. This is a rather important distinction frequently missed in ethical reflections on chemistry and the chemical industry. Since we are more concerned here with responsibility as part of invention, and since there is ample literature on the legal and ethical issues related to the manufacturers and users of Agent Orange, this section will briefly focus on the issue of unknown risks, and ask if it is possible to deflect the responsibility for impurities.

On its website, Dow Chemical has long denied responsibility for the damages caused by Agent Orange and assigns this responsibility to the users of Agent Orange, *i.e.* the U.S. and Vietnamese governments. This is part of a statement from 2001.

> As a nation at war, the U.S. government compelled a number of companies to produce Agent Orange under the Defense Production Act. Companies supplying Agent Orange to the government included The Dow Chemical Company, Monsanto Company, Hercules Inc., Diamond Shamrock Chemicals Company, Uniroyal Inc., Thompson Chemical and T-H Agriculture and Nutrition Company. [...]
>
> The U.S. military had sole control and responsibility for the transportation of Agent Orange to Vietnam, and for its storage once the defoliant reached Vietnam. The U.S. military controlled how, where, and when Agent Orange would be used. [...]
>
> War damages people, lives, and the environment. Nations, and the militaries of nations, are responsible for war. The U.S. government and the Vietnamese government are responsible for military acts in Vietnam and the use of Agent Orange as a defoliant. The manufacturers feel that in 1984 they took part in a good-faith settlement aimed at healing and bringing closure to this issue. Any future issues involving Agent Orange should be the responsibility of the respective governments as a matter of political and social policy.[15]

This line of arguments shows that at that stage the manufacturer denies responsibility for the use of the chemical in Vietnam, referring to the government's actions. It also uses the controversial idea that the manufacturer is not required to prove the product's safety, and that the injured party must be able to 'prove causation'.

---

[15] http://www.dow.com/commitments/debates/agentorange/background.htm, last visited 10 Oct. 2005.

Our discussion on who shares responsibility for the impact of a new chemical such as 2,4,5-T on society has now reached the point where neither the inventor, nor the publisher, nor the manufacturer intends or is required to accept responsibility. In the case of Agent Orange, this debate is complicated by the fact that the damage was caused by a chemical impurity. Since no one involved with Agent Orange intended the harm caused by TCDD, or initially had any information about the dangers caused by the impurity, this raises a set of very different ethical issues. For instance, while the U.S. and Vietnamese governments might be responsible for using Agent Orange as an herbicide, can the manufacturers also hold them responsible for the effects of the TCDD *impurities* in the herbicide? Even if Dow was not fully informed about the contamination and its associated risks, would it not have been the manufacturer's responsibility to check the quality, *i.e.* chemical composition, of their product? Dow's comment on this issue is rather instructive.

> Much of the source of the resulting public controversy over Agent Orange was an unwanted trace impurity that was present in one of the product's ingredients. The unwanted contaminant was the dioxin compound 2,3,7,8-tetrachlorodibenzo-para-dioxin, commonly known as 2,3,7,8 or dioxin. It should be noted that dioxin was not a commercial product, but rather was an unavoidable manufacturing process contaminant in the 2,4,5-T process. [*Ibid.*]

The description of dioxin in Agent Orange as 'unwanted contaminant' which was 'unavoidable' as part of the manufacturing process is aimed to deflect responsibility and moral judgment from the manufacturer. First, the company did not 'want' the presence of dioxins, and thus acted without bad intentions. Second, the dioxin contamination could not be avoided because of the *chemistry* of the manufacturing process.

From an ethical perspective, Dow's arguments are hardly convincing. The lack of intention to cause bad consequences and the lack of knowledge about the risks do not rule out responsibility. In line with the general concept of responsibility (Schummer 2001b), we therefore propose that the manufacturer of a chemical compound must accept a shared responsibility for both, the alleged (pure) compound and the impurities *usually associated with the compound.* In practice, this would require the manufacturer to estimate the nature, amount, and impact of impurities in a number of samples. While this could still not rule out a significant fluctuation in impurities between samples, it would provide at least some probability for a reliable safety and risk assessment. Such an assessment is possible, for instance, within the 'Space of information' mentioned already (Abdin 2020).

Furthermore, the manufacturer is able to design a suitable manufacturing process for a chemical. In the context of Agent Orange, alternative synthetic routes for 2,4,5-T, avoiding the presence of 2,4,5-TCP, exist and could have been explored by the chemical industry to avoid the presence of TCDD. Responsibility for the decision to produce Agent Orange along the hazardous, yet convenient 2,4,5-TCP route, therefore rests fully with the manufacturers and their chemical engineers.

Within this context, the matter of impurities is complicated by the different quality standards industrial chemicals have to comply with. Pharmaceutical products, such as drugs, for instance, have to comply with considerably higher standards of purity than agricultural products, such as herbicides. As a consequence, there are both ethical and regulatory questions to be addressed.

In any case, the manufacturer clearly shares some responsibility, and there are now stringent rules which govern the commercialization of substances, such as the Toxic Substances Control Act in the U.S.[16] In many cases, however, the burden of responsibility for proving that a substance is toxic still lies with society – simply consider the roles of the Environmental Protection Agency and the British Environment Agency.

Overall, the case of Agent Orange points toward shared responsibility, with the inventor of 2,4,5-T, *i.e.* Pokorny, his employers, the publishers of his 1941 paper, and the manufacturers of the – contaminated – herbicide sharing different degrees of responsibility for the chemical compound and its dissemination. Though ethical responsibility for compounds may be traced back to the persons conducting the initial synthesis and manufacturers, this responsibility in no way detracts from that of the user. The responsibility of the U.S. government is therefore as apparent

[16] As mentioned earlier, the dioxin contamination resulted as part of the synthesis producing the educt 2,4,5-TCP. In many cases, the latter was purchased, not manufactured, by the herbicide company. The responsibility for ensuring the safety of 2,4,5-T might therefore rest with both, the 2,4,5-T and the 2,4,5-TCP manufacturers.

as the responsibility of the US Air Force/RVN for using the product. There is also a long philosophical dialogue about ethics and war, although this is not the place to discuss these issues in detail.[17]

7. Conclusion

The issues surrounding Agent Orange, such as the risk associated with the generation of non-knowledge, *e.g.* caused by impurities, and the indirect proliferation of chemical compounds by publication, have raised important, pressing questions chemists need to address. In the short term, it is unrealistic to expect synthetic chemists to perform in depth risk assessments for each of their new compounds. Nevertheless, the assignment of responsibility for the dissemination of untested substances might sensitize chemists about the implications of their work, and require them to be more cautious when dealing with new substances and the information and uncertainties, knowledge and non-knowledge which come with them.

The attempts of Dow Chemical Company to shift responsibility for its contaminated herbicide first to the user and then to the chemical reaction itself show how controversial this issue is in practice. Traditional concepts of risk assessment might not easily be applied and the question of how a chemist can be held responsible for the action of impurities she/he is not aware of clearly requires further investigation.

From the outset, this case study has primarily aimed at identifying ethical issues in a real life chemistry context, and to stimulate further debate. It is hoped that an ensuing discussion of this case study will make chemists more sensitive and take responsibility for new chemicals as is found, for instance, in the field of 'Green Chemistry'. Only then will it be possible for chemists to follow philosophers and 'recognize their general moral responsibility' and 'align their internal norms with general moral standards'.

---

[17] Some of the ethical issues related to the use of Agent Orange in Vietnam are raised by Hay (1982). Issues of ethics and war are addressed by Coates (1997).

Further Reading

For more further discussions, we particularly recommend Abdin *et al.* 2020, Børsen *et al.* 2020, Kovac 2018, Maxim 2017, and Schummer 2018.

References

Abdin, A. Y., Yeboah, P., Jacob, C.: 2020, 'Chemical Impurities: An epistemological riddle with serious side-effects', *International Journal of Environmental Research and Public Health*, **17**, 1030.

Børsen, T., Nielsen, S. N.: 2017, 'Applying an Ethical Judgment Model to the Case of DDT', *Hyle: International Journal for Philosophy of Chemistry*, **23**, 5-27 [also in this volume, chapter 9].

Børsen, T.; Serreau, Y.; Reifschneider, K.; Baier, A.; Pinkelman, R.; Smetanina, T. & Zandvoort, H.: 2020. 'Initiatives, experiences and best practices for teaching social and ecological responsibility in ethics education for science and engineering students', *European Journal of Engineering Education*, 1-24.

Butler, D.: 2003a, 'Flight records reveal full extent of Agent Orange contamination', *Nature*, **422**, 649.

Butler, D.: 2003b, 'Agent Orange health investigation stuck at square one', *Nature*, **422**, 793.

Coates, A.J.: 1997, *The Ethics of War*, Manchester UP, Manchester.

Contakes, S. M., Jashinsky, T.: 2016, 'Ethical Responsibilities in Military-Related Work: The Case of Napalm', *Hyle: International Journal for Philosophy of Chemistry*, **22**, 31-53 [also in this volume, chapter 4].

Courtney, K.D.; Gaylor, D.W.; Hogan, M.D.; Falk, H.L.; Bates, R.R. & Mitchell, I.: 1970, 'Teratogenic evaluation of 2,4,5-T', *Science*, **168**, 864-6.

Davis, M.: 2001, 'Do the Professional Ethics of Chemists and Engineers Differ?', *Hyle: International Journal for Philosophy of Chemistry*, **8**, 21-34.

Del Re, G.: 2001, 'Ethics and Science', Hyle: International Journal for Philosophy of Chemistry, **7**, 85-102.

Gough, M.: 1987, *Dioxin, Agent Orange*, Plenum Press, New York.

Hassall, K.A.: 1982, The chemistry of pesticides their metabolism, mode of action and uses in crop protection, Macmillan, London.

Hay, A.: 1980, 'Red faces (and hot tempers) on 2,4,5-T', *Nature*, **286**, 97.

Hay, A.: 1982, *The Chemical Scythe*, Plenum Press, New York and London.

Heaton, A.: 1996, 'Pesticides', in: Heaton, A. (ed.), *The Chemical Industry*, Blackie Academic & Professional, London, pp. 238-43.

Institute of Medicine: 2000, Committee to Review the Health Effects in Vietnam Veterans of Exposure to Herbicides, Washington, DC [www.iom.edu]

Kimmig, J. & Schultz, K.H.: 1957, 'Chlorierte aromatische zyklische Aether als Ursache der sogenannten Chlorakne', *Naturwissenschaften*, **44**, 337-8.

Kovac, J.: 2001, 'Gifts and Commodities in Chemistry', *Hyle: International Journal for Philosophy of Chemistry*, **7**, 141-53.

Kovac, J.: 2018, *The Ethical Chemist: Professionalism and Ethics in Science* (Second Edition), Oxford University Press.

Kovac, J.: 2020, 'American Chemical Society Codes of Ethics: Past, Present, and Future', *Hyle: International Journal for Philosophy of Chemistry*, **24**, 79-95 [also in this volume, chapter 18].

Laszlo, P.: 2001, 'Handling Proliferation', *Hyle: International Journal for Philosophy of Chemistry*, **7**, 125-40.

Llord, J-P.: 2017, 'Ethics and Chemical Regulations: the Case of REACH', *Hyle: International Journal for Philosophy of Chemistry*, **23**, 81-104 [also in this volume, chapter 19].

Maxim, L.: 2017, 'Chemists' Responsibility for the Health Impacts of Chemicals: Green Chemistry and Its Relation to Toxicology', *Hyle: International Journal for Philosophy of Chemistry*, **23**, 61-80.

Schecter, A.; Quynh, H.T.; Pavuk, M.; Papke, O.; Malisch, R. & Constable, J.D.: 2003, 'Food as a source of dioxin exposure in the residents of Bien Hoa City, Vietnam', *Journal of Occupational and Environmental Medicine*, **45**, 781-8.

Pokorny, R.: 1941, 'Some Chlorophenoxyacetic Acids', *Journal of the American Chemical Society*, **63**, 1768.

Schummer, J.: 1999, 'Coping with the Growth of Chemical Knowledge: Challenges for Chemistry Documentation, Education, and Working Chemists', *Educación Química*, **10**, 92-101.

Schummer, J.: 2001a, 'Editorial', Hyle: International Journal for Philosophy of Chemistry, **7**, 83-4.

Schummer, J.: 2001b, 'Ethics of Chemical Synthesis', *Hyle: International Journal for Philosophy of Chemistry*, **7**, 103-24.

Schummer, J.: 2018, 'Ethics of Chemical Weapons Research: Poison Gas in World War One', *Hyle: International Journal for Philosophy of Chemistry*, 7, 103-24 [also in this volume, chapter 3].

Southgate, C. & Aylward, A.: 2002, 'Environmental Ethics: Further Case-Studies', in: Bryant, J.; La Velle, L.B. & Searle, J. (eds.) *Bioethics for Scientists*, Wiley, Chichester, pp. 73-83.

Stellman, J.M.; Stellman, S.D.; Christian, R.; Weber, T. & Tomasallo, C.: 2003, 'The extent and patterns of usage of Agent Orange and other herbicides in Vietnam', *Nature*, **422**, 681-7.

Stringer, R. & Johnston, P.: 2001, Chlorine and the Environment: An Overview of the Chlorine Industry, Kluwer, Dordrecht.

Tuan, V.M. & Phuong, N.T.N.: 2003, 'The effects of Dioxin Exposure on the Occurrence of Birth Defects: A Case Control Study at Tudu Hospital, *Organohalogen Compounds*, **60**, 1-3.

U.S. Environmental Protection Agency: 1994, Health Assessment Document for 2,3,7,8-tetrachlorodibenzo-p-dioxin (TCDD) and related compounds III of III, EPA/600/BP, Washington, DC [www.epa.gov].

Chapter 8

When Laypeople are Right and Experts are Wrong: Lessons from Love Canal

Ragnar Fjelland

Abstract: Love Canal, a suburban town in New York State built on a waste disposal site of a former chemical factory, provoked one of the first major environmental controversies. It involved scientists, citizens and politicians, including the US Congress and President. The controversy raises many important problems, and the chapter focuses in particular on the uses of scientific knowledge and the role of scientists. Although the scientists worked for the authorities, they regarded their knowledge as objective and their advice as neutral. However, the residents of Love Canal did not trust them, and engaged their own scientist. At the time of the controversy (1978) the Precautionary Principle had not been formulated, but the controversy involved many issues that have later been related to the principle. One particular issue was the uses of statistics, and the relationship between type 1 and type 2 statistical errors. The chapter relates the controversy to recent debates on the proper use of significance tests and statistics, and argues that context and values have to be taken into consideration. It concludes that in cases like Love Canal it is imperative to inform about uncertainty and to involve all stakeholders.

1. Introduction: Why the Love Canal Controversy Is Important

Love Canal is a suburban town in the US state of New York close to Niagara Falls. In the summer of 1978 it became the scene of one of the first major environmental controversies, a controversy that hit the headlines worldwide. The residents complained that chemicals from a nearby chemical waste site caused health problems. Love Canal was in fact built over the waste disposal site of a former chemical factory. Nearly 22,000 tons of chemical waste (including polychlorinated biphenyls, dioxin, and pesticides) had been deposited. The controversy involved scientists, experts, and politicians, including the governor of the State of New York, the US Congress and the President.

There are many lessons to be learned from the case. Many of them concern the relationship between the authorities, the affected citizens, and the general public: The importance of giving honest information and not downplaying the hazards, and of recognizing the affected citizens as partners and resources, and not just as ignorant and hysterical people. Other lessons concern the uses of scientific knowledge and the role of scientists. I shall in this chapter mainly concentrate on lessons we can learn from Love Canal about the role of science and scientists.

I will focus on the crucial question: Was there a causal relationship between the toxic chemicals in the waste site and health problems among the residents of Love Canal? To put it simply: Was Love Canal a safe place to live in? New York State Department of Health engaged scientists to give an answer to that question. The scientists regarded it as their job to find the facts, and leave it to the authorities to make decisions. Although they reported to the authorities, they regarded themselves as objective and neutral. However, the residents in Love Canal did not trust them, and they themselves took an active part and engaged other scientists.

Were the scientists objective? The question does not only concern the objectivity of scientists, but the deeper question of objectivity itself. The traditional view, held by the scientists working for the Department of Health, was that facts should be separated from values, and that scientists should stick to the facts. I shall in this chapter show that their ideal of objectivity led to bad science, and that it implicitly favored the authorities.

The questions raised in Love Canal are today more urgent than ever, because many of the problems contemporary science faces have similarities to the Love Canal case, characterized by complexity and uncertainty.

2. The Story

The story of Love Canal began in the 1890s, when William T. Love planned to build a hydroelectric power plant and started the construction of a canal to supply water from the Niagara River. However after a section of the canal, which was approximately one kilometer long, twenty meters wide, and three meters deep, had been dug, a financial crisis forced the project to be abandoned. In 1905 Hooker Electrochemical Company was started near Niagara Falls. The factory produced, among other things, chlorine and caustic soda. Between 1942 and 1952 it was allowed to dispose of almost 22,000 tons of chemical waste in the canal, in fiber and metal barrels. In 1953 the canal was full, and was covered with soil and clay, while grass grew on the surface. The school board of Niagara Falls then bought the land, including the canal, for one dollar. One condition of the deal was for a disclaimer to be included in the contract exempting Hooker Company from any future liability (Levine 1982, pp. 11ff.).

The school was completed in 1955, with a capacity of 400 children, who attended every day. Houses were built around the school, most of them modest two- and three-bedroom houses. Although there had been some focus on the chemicals when the school was constructed, most of the residents were unaware of the chemicals that were buried in the ground. And although there were signs of leakage early on, the residents did not pay attention to them. However, things became worse in the 1970s. For example, after heavy rainfall odors from chemical substances became noticeable. The first investigation was carried out in 1976, and the suburb was visited several times by officials of the New York State Department of Health. However, in the summer of 1978, when the newspapers made the general public aware of the situation, the Department of Health was forced to act. In their preliminary investigations they found more than eighty different chemicals at the waste site. Ten of these substances were known to be potentially carcinogenic (*ibid.*, p. 41). The director of the Department of Health declared a state of emergency due to the danger to public health. He justified this on the grounds he was convinced that the toxic chemical substances from the waste site represented a danger to the residents in the area. Soon after that the governor of New York offered to buy the 239 houses situated nearest to the waste site and to help the residents relocate. A fence was put around the evacuated area. The Department of Health began an investigation into the health of the residents, this included blood tests, questionnaires, and an inquiry into the incidences of ill health among the inhabitants. Early in the Fall of 1978 the preliminary results of the study were made public; the health authorities assured the residents that the rest of Love Canal was a safe place to live (Paigen 1982, p. 29).

However, the residents distrusted the information given by the authorities. They felt that the authorities regarded the residents themselves as the problem and suspected that information was being withheld from them. The residents were also never consulted during the process as key contributors of information (Levine 1982, p. 27). One of the residents was Lois Gibbs. She was a young housewife who had neither been interested in politics nor had taken part in any organized political activity. However, she started organizing her neighbors into what became Love Canal Homeowners Association (LCHA), and became its leader. They assisted in the investigations of the Department of Health, among others by calling neighbors and urging them to fill in forms, and by registering medical ailments. One night Gibbs decided to go through all the material that had been collected. She sat down with a map of Love Canal and put a pin on every house that had registered medical problems. It appeared that the pins formed a pattern of narrow paths on the map. Older residents had previously informed her of large stream-beds and swales cutting through the area. They had been filled in when the houses were built. It turned out that photographs from the 1930s displayed the original swales and stream-beds. Gibbs had the idea that there might be a connection between the pattern of the pins on the map and the swales (*ibid*., p. 89).

Because the residents did not trust the authorities, they sought outside assistance. Gibbs' brother-in-law, who was a biologist at State University of New York/Buffalo, assisted her and encouraged her to take a leading role. But even more important, he alerted another scientist, Beverly Paigen, who was doing cancer research at a nearby research institute. She conducted research on the relationship between toxics and cancer and became interested in the problem. At the time she tested out a hypothesis that some families exposed to low concentrations of toxics would be more susceptible to cancer than other families, and she regarded the Love Canal case as an opportunity to test her hypothesis.

Gibbs showed her map to the leader of the Department of Health's research group, but he showed little interest. Then she showed it to Beverly Paigen, who immediately became interested. Members of the home owners' association assisted her in interviewing 1140 of the residents of Love Canal. The result of the investigation showed a clear geographical distribution of ailments, and it appeared to follow a pattern that reminded of the earlier swales. Paigen then divided all houses into two categories: 'Wet homes' were houses built above or close to the swales, and 'dry homes' were all the other houses. The result was striking: Women in 'wet homes' had three times as many miscarriages as women in 'dry homes'. Birth defects, asthma, urinary infections and the number of psychiatric cases were several times higher in 'wet' areas than in 'dry' areas.

Paigen's investigation therefore supported Gibbs' original hypothesis. In addition it had clear consequences for what should be done. One should evacuate residents from the 'wet homes' first. This was contrary to the procedure selected by the health authorities of New York. They assumed that the toxics spread radially from the chemical waste site. Therefore, they had purchased the houses closest to the waste site and evacuated the residents who lived there, and planned to let the rest of the residents remain.

There was a confrontation. When Paigen's results were given to the press, the leader of the research group of the Department of Health said that the investigation was "totally incorrect", and other officials later argued that the evidence was based on "information collected by housewives that is useless" (quoted from Levine 1982, p. 93).

To make a long and dramatic story short: Beverly Paigen and the residents of Love Canal prevailed. In the summer of 1980 the Congress allocated additional funds that authorized the President to use up to 15 million US dollars to relocate the remaining residents. After a number of negotiations between different parties, Love Canal was for the most part vacated in 1981.

3. From Facts to Vlues

Beverly Paigen gave her version of the controversy in an article a few years later. She tells that she had originally believed that the case was a matter of scientific disagreement that could be resolved by having the involved researchers come together and compare the data, experimental design, and statistical analyses. But she was wrong. She was surprised to discover that the facts made little difference, and alleged that "it raised a series of questions that had more to do with values than science" (Paigen 1982, p. 29).

There were two main differences between the research group of the Department of Health and Beverly Paigen:

First, the research group had made the assumption that the toxics spread more or less homogeneously outwards from the waste site. This was in accordance with traditional scientific practice – going back to Galileo and Descartes – to start with the simple and idealized. It followed from this assumption that the area closest to the site contained the highest concentration of toxics, and the concentration would decrease with increasing distance from the site. Paigen, on the other hand, had adopted Gibbs hypothesis that the toxics dispersed along the swales.

Second, they held opposite views regarding the burden of proof and the uses of statistics. The research group had claimed that it had used a 'conservative' scientific approach. Paigen says that it occurred to her that there was a problem when she, in a conversation with a representative from the research group, discovered that they disagreed about how this should be interpreted in every single case they discussed. They both claimed to take a conservative approach, but it turned out that they had opposite opinions about what 'conservative' meant in this situation. The researcher from the Department of Health emphasized that one must be very cautious in concluding that Love Canal was an unsafe place to live. Paigen, on the other hand, maintained that one had to be very cautious about concluding that Love Canal was a safe place to live. She argued that since a mistake could result in severe consequences for the health of the residents, the researchers must be very careful in concluding that Love Canal was a safe place to live. She insisted that underestimating the danger was worse than groundless fear (*ibid.*, p. 32).

Paigen's position is described in the article that I have quoted. Her approach can be expressed in a formulation given by Nicholas Ashford, director of Center for Public Policy Alternatives at MIT. The question should not be "can you publish this in New England Journal of Medicine, but would you let your daughter work with that chemical?" (quoted from Savan 1988, p. 59). We do not have written accounts of the position of the Department of Health scientists, for obvious reasons. When you apply traditional scientific methods and carry out what you and your colleagues regard as good science, you normally do not state this explicitly. However, we do have some quotations from conversations that support this view. Perhaps the best source is Adeline G. Levine's book *Love Canal: Science, Politics, and People*, which has been my main source of information on the Love Canal case. She followed the case closely from the very beginning and had close contact both with residents, politicians and scientists who were involved in the process. Among others she quotes "a high-level official" who emphasized that

> [...] the health department professionals were *scientists*, who did not worry about people's reactions to cautionary statement and recommended actions. They dealt with numbers – with data on physical conditions – and only with these. Political and social matters, the official stressed, were extraneous to the DOH [Department of Health] work. [Levine 1982, p. 40]

Levine also pointed out that the scientists working for the Department of Health were afraid of losing their objectivity, and she quoted one scientist who explained objectivity in the following way: "We deal only with numbers; we are scientists" (quoted in Levine 1982, p. 85).

The positions of the Department of Health scientists and Paigen illustrate two of the categories introduced in Roger S. Pielke' book *The Honest Broker: Making Sense of Science in Policy and Politics* (2007). The book addresses the relationship between scientists and political issues, and it describes different ways scientists may regard their own role as scientists. One category is what Pielke calls 'Pure Scientists': Their role is to sum up the state of knowledge of a limited area, and leave it to the politicians to make decisions. The Department of Health scientists can no doubt be placed in this category. Two other categories are what Pielke calls 'Issue Advocates' and 'Honest Brokers'. They have in common that scientists belonging to these two categories take wider social and political concerns into consideration. The difference is that scientists belonging to the first category take sides in a controversy, and use their expert knowledge to pursue the political agenda that they support, whereas scientists belonging to the second category restrict themselves to pointing to the relationship between various options and political agendas. I think Paigen can be placed in the category of 'Issue Advocates', because she no doubt sided with the residents of Love Canal.

4. The Precautionary Principle: What It Is and What It Is Not

Paigen's fundamental position – that because mistakes could result in severe consequences for the health of the residents, the researchers must be very careful in concluding that Love Canal was a safe place to live – was an application of what has become known as the precautionary principle (PP). I will in this section explain the principle and in the next section look at its preconditions and broader ethical and philosophical context. I will also discuss possible ethical justifications of the principle. Then I will apply it to the Love Canal case.

The account of the precautionary principle in this section follows the report by World Commission on the Ethics of Scientific Knowledge and Technology (COMEST): *The Precautionary Principle* (2005).

The principle, explicitly formulated, is of recent origin. It originated in the 1970s, but today it is best known by its formulation in the Rio Declaration:

> In order to protect the environment, the precautionary approach shall be widely applied by States according to their capabilities. Where there are threats of serious or irreversible damage, lack of full scientific certainty shall not be used as a reason for postponing cost-effective measures to prevent environmental degradation. [Rio Declaration 1992, principle 15]

We see from the text that the term 'approach' is used instead of 'principle'. For our purpose it makes no difference, and I will not go into further details. Let us take a closer look at what the principle implies.

The text uses the expression 'lack of full scientific certainty'. This means that there must be some indications of possible damage.

Ungrounded fear is not sufficient. It is also implied that scientific investigations have been made that make the assumption plausible. It is further required that the results of the scientific investigations are uncertain, and that this uncertainty cannot be reduced or eliminated before a decision is made. In the case of Love Canal both the research group of the Department of Health and Beverly Paigen had been involved, and there was indications of possible damage not only in the evacuated area, but in the surrounding area as well.

Second, the text refers to 'serious or irreversible damage'. This is open to interpretation. If a person dies, it is a tragedy for that person. And in larger accidents, many people may die or be injured. In an airplane crash hundreds of people may die. Therefore, one might argue that according to the precautionary principle we should ban aviation. But we do not ban all aviation because airplanes sometimes crash and people are killed. Uncertainty is a fundamental aspect of the human condition. However, the reduction of uncertainty is an important aspect of modernity, and the theories of probability and statistics were developed as a part of modern science, with the aim of reducing uncertainty. The concept of risk was a part of this endeavor. Risk can be defined as the probability of an adverse quantifiable outcome (therefore, risk is sometimes defined as the product of probability and cost of damage), and risk calculations play an important part in decisions concerning choices of technology. On the one hand it is an acceptance of uncertainty, but on the other hand it is an attempt to control uncertainty (cf. the title of Ian Hacking's book *The Taming of Chance*, 1990).

Third, the text uses the expression 'cost-effective measures'. Again, this is open to interpretation. What is the value of good health, and what is the value of a human life, and what is the value of nature? One way of deciding the value of a harm is the compensation that is paid to the victims, or if the damage is reversible, the cost of cleaning up or repairing the damage. In many cases prevention is more cost-effective than cure, and therefore precaution is cost-effective. If precaution had been taken from the very beginning, the authorities would not have allowed the chemical waste to be buried in Love Canal, and if that had already been done, they would not have allowed the school and the suburb to be built on top of the waste site. However, this could not be undone. As the situation was now, the costs of cleaning up the place, or buying the houses of the residents and aiding them in relocating, paying medical bills, *etc.* would far exceed the costs of safely depositing the chemical waste in the first place. However, the problem is often that benefits and costs are not equally distributed. In Love Canal Hooker Electrochemical Company had got rid of the chemical waste in the cheapest possible way, and sold the land for one dollar on the condition that they should be exempt from any liability.

It was argued that even if Hooker did not have a legal obligation, it at least had a moral obligation. This was denied by the company. In the end the US Congress paid the costs because this had been a political issue. However, to decide in each case what 'cost-effctive measures' implies, is far from easy.

5. The Precautionary Principle: Broader Context and Justification

How can the precautionary principle be justified? I shall not try to give a complete justification, but restrict myself to elaborating some of the philosophical and ethical background that makes the principle plausible.

The precautionary principle may be regarded as a result of a change of scientific paradigm. Newton completed the scientific revolution that started with Galileo, and he made classical mechanics the model of all science. Because classical mechanics is deterministic, they established an ideal of exact predictions that was unprecedented. Using the laws of classical mechanics the motion of celestial bodies could be calculated thousands of years into the past and the future. However, at the end of the nineteenth century the mathematician Henri Poincaré showed that even a deterministic system may become unstable, making calculations difficult. In 1960 Edward Lorenz, who worked on a simplified weather model, showed that even in a simple deterministic system small errors may obstruct exact predictions (sensitive dependence on initial conditions, known as the 'butterfly effect'). His conclusion was that long-time weather forecasts are generally impossible. It was gradually acknowledged that the ideal of classical mechanics only applied to simple, idealized systems. Most natural systems are complex, and therefore uncertainty will in many cases be irreducible. The recognition that nature is complex and that uncertainty normally cannot be eliminated is an important precondition for the precautionary principle.

Another precondition is the recognition that nature is limited and vulnerable. To some extent this recognition represents a return to a view of nature that was dominating prior to the scientific revolution of the 17th century. The dominating view of nature from Antiquity through the Middle Ages was to regard nature as analogous to an organism (an organismic view), where man was regarded as a part of nature. However, this changed in the 17th century. The organismic view of nature was replaced by a mechanistic view, regarding nature as a big machine. The clockwork metaphor became important. Instead of regarding nature as our home, it was rather regarded as a resource that could be exploited for our benefits.

If we jump to our recent history, a highly influential book was Rachel Carson's *Silent Spring*. She brought the public attention to the widespread destruction of wildlife in America by the use of pesticides including insecticides. When the book was published in 1962, Carson was heavily attacked by the agricultural chemical industry and the scientific establishment. Questioning the benefits of the new pesticides was regarded as attacking the faith in scientific progress. However, president John F. Kennedy was so impressed by the book that he ordered the President's Science Advisory Committee to investigate the uses of pesticides in agriculture. When the committee published its report in 1963, it confirmed that Carson was basically right.

Carson's book was not only an attack on the uses of pesticides, but on a scientific and technological development that regarded nature as only a resource to be exploited by us. Although the process has been slow, it is today recognized that nature is both limited and complex.

Another factor is the recognition that we have a responsibility for nature and future generations. I will restrict myself to mentioning one influential book, by the German philosopher Hans Jonas. The book was first published in 1979 in German under the title *Das Prinzip Verantwortung* (*The Principle of Responsibility*) and in English as *The Imperative of Responsibility* in 1984. As the original title indicates, Jonas made responsibility the very foundation of ethics, in opposition to the traditional view that to be a subject of ethics we require autonomy. His point of departure was that all ethical theories up until then had taken as frame of reference that all consequences of human actions are small in 'space and time'. However, the tremendous power of modern science and technology has changed this. The atomic bomb is the most dramatic example. A nuclear war will not only affect all people alive today, but future generations as well. It may even eradicate all life. Jonas therefore set out to develop the foundation of a global ethics, based on responsibility as the fundamental concept.

Responsibility is an asymmetrical concept: A may be responsible for B without B being responsible for A. For example, the parents are responsible for their children, but the children are not responsible for their parents. (When the children grow up and their parents are old, the relationship changes. But that is a different matter.) Therefore, the argument that we should not do anything for future generations because they have not done anything for us, is invalid. We may have responsibility for animals, future generations, and nature even if they do not have responsibility for us. Jonas was particularly concerned that we should not destroy the conditions of life for future generations. He even formulated a categorical imperative to replace Kant's categorical imperative. Like Kant he gave several formulations of his imperative. One of them is this: "Act so that the effects of your action are consistent with a continuing genuine life on earth." (Jonas 1984, p. 11)

At last I want to mention the article 'Asymmetries in Ethics', written by the Norwegian philosopher Knut-Erik Tranøy in 1967. It was written before the precautionary principle became a topic, and has hardly had any influence on the development of the principle. Nevertheless, in the article he introduced some important concepts that can be used to justify the principle. The article was inspired by the philosopher of science Karl Popper and his asymmetry between verification and falsification: Because scientific statements according to Popper are universal statements, they can never be verified. However, they can be definitely falsified. His favorite example was the statement: 'All swans are white.' Observation of thousands of white swans cannot verify the statement, but only one black swan can falsify it. Tranøy applies this asymmetry to some fundamental ethical concepts, like life and death, pleasure and pain, happiness and unhappiness, right and wrong, and good and bad. According to

Tranøy these pairs of concepts are asymmetric, similar to verification and falsification in the sense that the negative terms are more definitive, categorical, and fundamental than the positive.

His first example is the asymmetry between life and death. We know numerous sufficient conditions for death, but whereas we know many necessary conditions for life, we do not know any sufficient conditions. Similarly, there are numerous ways of killing people, but to keep them alive, grow, and flourish is much more indefinite and vague. Therefore, it is easier to find a negative formulation, for example 'you shall not kill', than a positive. Although he does not deal with the pair utility/harm, it is obvious that the same asymmetry applies to this pair as well. Therefore, it follows that it is more important to prevent harm than to promote utility. This is a kind of 'negative utilitarianism' that can be used to justify the precautionary principle.

One might apply Tranøy's asymmetry directly to the Love Canal case. Utility would be the money saved if nothing was done. This money could be used for other purposes, and would increase the total utility in the New York State or in the country as a whole (depending on who actually supplied the money, New York State or US government). The harm would be possible health problems among the citizens of Love Canal. Of course, many good things might be done with the money saved, for example building new schools, or nursing homes for elderly people. But the money might also be wasted, on badly planned projects, corruption, or on an increasing bureaucracy. Although some uncertainty was involved, the harm to the citizens was rather concrete, causing severe health problems. Of course, uncertainty complicates the problem. Let us, therefore, return to the question of uncertainty.

6. The Precautionary Principle and Statistics

Although the precautionary principle was not formulated at the time of the Love Canal controversy, it is interesting to see how it applies to that case.

There was general agreement that the inner, evacuated area was not safe. But what about the remaining part of Love Canal? The researchers working for the Department of Health concluded that it was safe, and their main argument was that the frequency of health problems, for example miscarriages, was not significantly higher than in comparable areas. Therefore, there was no causal relationship between the toxics at the waste site and health problems in the population of Love Canal (except the inner area that had already been evacuated).

One of the standard methods for establishing causal relationships is the use of significance tests. We compare one group which has been exposed to the alleged cause (the experimental group) with a similar group which has not been exposed to the same cause (control group). If we observe a difference in the two groups, we infer that it is due to the alleged cause. However, this inference is only valid if we know that the two groups prior to the exposure to the alleged cause are similar (in all relevant aspects). Therefore laboratory conditions are preferred, because one can control the environment. But even under laboratory conditions it is impossible to control all factors. In particular, in all biological systems there is variation. One technique is to try to distribute the unknown factors equally in the experimental and the control group (randomization). However, all uncertainty cannot be eliminated, and outside the laboratory we have much less control.

Therefore, when we observe a difference between the experimental group and the control group, the difference may be a sign of a causal relationship, but it may also be due to factors beyond our control, or due to chance. But how do we know if it is the one or the other? One standard procedure is using a 'null hypothesis': We assume that the observed difference between the two groups has come about by chance. Only if the probability of this happening is lower than a certain level, called the significance level, will we reject the null hypothesis. A significance level of 0.05 (5%) is normally used. That means that only if the probability of the difference occurring by chance is less than 5%, will we conclude that the difference is not likely due to chance and that therefore we should seek another explanation – for example that the relationship is in fact causal.

7. Type I and Type II Statistical Errors

We may make two kinds of error due to the statistical nature of a problem: A type I error is rejecting a null hypothesis when it is in fact true. This is equivalent to claiming that there is an effect that in reality does not exist ('false positive'). A type II error is failing to reject a null hypothesis that is in fact false. This is equivalent to overlooking an effect that really exists ('false negative').

Although the relationship between a type I and a type II error depends on the specific problem, in general there is a trade-off between them: If we decrease the probability of one, the probability of the other increases. In traditional significance tests the probability of a type I error is controlled (it is set by the significance level), thus leaving the probability of a type II error open. If researchers inform us that they have not found a statistically significant difference, it would be important additional information if the probability of overlooking the difference is large. In statistics text books it is therefore emphasized that the probability for a type II error should always be evaluated. However in practice this is different. We know that this rule is often violated.

One may ask why a type I error is worse than a type II error. One sensible answer is that scientists should have good evidence for asserting something, and when they have not, they should refrain from asserting. It is therefore more detrimental to their reputation to assert something that is not the case than overlooking something that is the case. Sometimes a legal analogy is used to justify the asymmetry. In a criminal case the jury sticks to the 'null hypothesis' of not guilty until the defendant is proved guilty. The burden of proof is on the prosecutor. The justification is that it is worse to find an innocent person guilty than failing to convict a guilty person (Bhattacharyya & Johnson 1977, p. 168).

To put the burden of proof on the one who asserts that there is a difference may make sense in basic science, but it turns out differently when used in applied science. (I shall later show that it does not make sense in basic science, either.) Engineers have treated type I and type II errors differently, because they know that their information is used to make decisions. Type I error is called 'producer risk' because a false positive may harm the producer: He may have to withdraw a product that is not really harmful, or pay compensation for a non-existing harm. Type II error is called 'consumer risk' because failing to detect a harmful effect harms the consumer. The traditional approach of minimizing type I error favors the producer and puts the burden of proof on the consumer or victim of pollution. For example, in environmental questions it is often the case that those affected by the pollution have the burden of proof, while those who pollute have the benefit of doubt. When Beverly Paigen and the residents of Love Canal claimed that the toxic substances from the dump site were the cause of illness, they had the burden of proof.

The scientists working for the Department of Health certainly knew who had hired them. One might suspect that they wanted to come up with results that favored their employer. This may often be the case. As a member of the Norwegian Research Ethics Committee for Science and Technology in the 1990s I attended meetings with researchers who confessed that they had conscientious problems because they knew that if they did not produce the results their contractor expected, future projects would be endangered. Needless to say, these confessions made a deep impression on the members of the committee. However, in Love Canal the scientists might just stick to traditional scientific methods, allegedly being objective and neutral. Therefore, insisting on this kind of method favors one part of the conflict, the 'producer' (or the one who is responsible for the harm).

8. Reversing the Burden of Proof

The Department of Health's research group claimed to do good science, whereas Paigen was regarded as mixing her personal feelings or political agenda into her scientific activity. As we have seen, she used precaution to justify her approach. Today we would just have referred to the precautionary principle to support her claims about the burden of proof, and one might argue that asymmetry between type I and type II errors should be reversed. In cases like Love Canal, where possible harm is imminent, it is certainly better to overestimate the danger than to underestimate it. The philosopher Kristin Shrader-Frechette argued in favor of this position as early as 1991 in her book *Risk and Rationality* (Shrader-Frechette 1991, Lemons *et al.* 1997).

However, Paigen might have taken a more assertive approach. Instead of defending herself, she might have accused the researchers of the Department of Health of doing bad science. To see how, we only have to ask the simple question about significance tests: Why do we use a significance level of $p = 0.05$ (or even $p = 0.01$)? The simple answer is that it represents a difference of two (or three, respectively) standard deviations from the mean if we assume a normal distribution. However, why two standard deviations? It is no more than a rule of thumb. It can be traced back to Ronald Fisher's *The Design of Experiments* (1935), where it looks as if only significant results according to this rule are acceptable. The problem is that Fisher neglected advice from all the other leading statisticians of his time. His simplified version prevailed, though, and became the leading orthodoxy until today. But it has drawn increasing criticism during the last decade. (For the historical background and criticism of traditional significance tests, see Ziliak & McCloskey 2012.)

I said previously that a possible justification for minimizing type I errors in basic science was that it is more detrimental to a scientist's reputation to assert something that is not the case than failing to detect something that is the case. This may be true as a factual description, but it is nevertheless bad science. As Karl Popper pointed out long ago, good science is not characterized by advancing modest hypotheses, but by advancing *bold hypotheses*. This kind of precaution is not a scientific virtue. However, when hypotheses have been advanced, they should be subject to severe tests. Indeed, we shall try to falsify them. If they survive the tests, we should keep them. But only temporarily, because hypotheses can never be verified. However, they can be definitely falsified, and will then be replaced by better hypotheses. We do not have to subscribe to all of Popper's philosophy of science to endorse his view that creativity and boldness are imperative to scientific progress (see for example Popper 1981, p. 96.) We only have to keep in mind the importance of originality in scientific work. This is why plagiarism is regarded as one of the most serious kinds of scientific misconduct.

Many researchers have criticized significance testing. In 2005 the statistician John Ioannidis published the article 'Why Most Published

Research Findings Are False'. The article focuses on the relationship between type I and type II errors. He uses typical values from research fields (for example, in some fields the probability of a type II error may be up to 0.5, which implies that there is a 50% chance of failing to observe a real effect), and some simple calculations to underpin his conclusion: "It can be proven that most claimed research findings are false." (Ioannidis 2005)

This year the American Statistical Association (ASA) issued a statement concerning the uses of significance tests. The conclusion is worth quoting:

> Let's be clear. Nothing in the ASA statement is new. Statisticians and others have been sounding the alarm about these matters for decades, to little avail. We hoped that a statement from the world's largest professional association of statisticians would open a fresh discussion and draw renewed and vigorous attention to changing the practice of science with regards to the use of statistical inference. [Wasserstein & Lazar 2016]

9. Uncertainty and Ignorance

It is important to bear in mind that the significance level is more or less arbitrary, while it would make more sense if it was estimated in each concrete case. Additionally, it is important to keep in mind that significance tests only concerns random errors, due to chance. In most situations systematic errors are much more important: experimental errors, measurements errors, sampling bias, wrong models. *etc.* Love Canal gives a good illustration of this.

All these factors are sources of uncertainty that is much more serious than the risk due to random statistical variation. At least we must make a distinction between three different kinds of uncertainty:

(1) Risk: This is the kind of uncertainty that we deal with in probability theory and statistics. We know possible outcomes, and we know the probabilities of various outcomes.
(2) Uncertainty ('known unknowns'): In this case we know the possible outcome, but we cannot assign probabilities to them.

(3) Ignorance ('unknown unknowns'): In this case we do not even know the possible outcomes and so we cannot estimate their probabilities.

We have seen that the research group of the Department of Health used a model where they assumed that toxics spread homogeneously outwards from the waste site. They expected to find the highest concentrations of both toxins and health problems in the houses closest to the site, and decreasing outwards. However, this was not confirmed. They also took the total population in Love Canal outside the evacuated area and compared it to a similar population. They did not find any significant difference, and this supported their view that the rest of Love Canal was a safe place to live.

This conclusion was, however, based on an erroneous model. The assumption that the toxics spread homogeneously outward from the waste site was wrong, and the hypothesis that it followed the swales, was right. Therefore, they did not see significant differences. However, when Paigen divided the homes into 'wet' and 'dry', she had no problems in establishing significant differences (for example in the number of miscarriages). An important lesson from Love Canal is, therefore, that statistics does not help if one uses a wrong model.

This is an example of ignorance: What actually happened, was not part of the models applied by the scientists.

10. When the Model is Wrong

Experts trained in a field have a tendency to apply the kinds of models that conform with their field. The following example has been taken from Brian Wynne's 'Uncertainty and Environmental Learning' (1992): In May of 1986 a cloud of radioactive material from the Chernobyl accident passed over Cumbria in North Wales. Heavy rains caused a large amount of radioactive cesium to fall over an area used to raise sheep. The authorities in charge assured everyone that there was no cause for concern, but in spite of this, six weeks after the rains a ban against selling meat from sheep that had grazed in the area was imposed because of the high levels of radioactivity found in the meat. Experts claimed however that the radioactivity would rapidly decrease, and that the ban would be lifted in a few weeks. Yet, even after six years the level of radioactivity was so high in some of the affected areas that restrictions had to be upheld.

How could the experts be so wrong? Their predictions were based on extrapolations from the behavior of cesium in alkaline clay soil to the acid peat soil of Cumbria. Measurements showed that the dispersion of cesium in these types of soil was fairly similar, and on that basis they assumed that cesium would sink so far down into the ground that after a short period of time there would be no problem. This was based on the assumption that the radiation would come from the cesium in the soil and would be absorbed by people or animals who happened to be in the area. Under this assumption it was the physical transport of cesium in the soil that was important. However, this assumption was wrong. The sheep received cesium in their bodies through the grass they ate. The important question was therefore not how the cesium was dispersed throughout the soil but if it was absorbed into the vegetation. Here there proved to be a significant difference between alkaline clay soil and acid peat. In alkaline clay soil, cesium adsorbs on aluminum silicate so that it cannot move into the vegetation, whereas in peat it remains chemically mobile and can therefore be taken up by the vegetation. The experts did not consider these possibilities, and that was the cause of their mistaken predictions (Wynne 1992, 121).

Should not a model that takes into consideration for example chemical properties, have been used at the onset? The answer is, of course, yes. But to understand why the experts made such an apparently elementary error we have to take into consideration that they had been trained as physicists. Physicists are used to think in terms of physical transportation and radiation. Chemists are trained to think in terms of chemical reactions and chemical mobility. The problem is that it is not a part of professional training to learn about the limits of the models and methods of a field.

A serious obstacle to coming to terms with this problem is the fact that Thomas Kuhn's description of the scientific community is to a large extent valid. We do not have to accept the more controversial parts of Kuhn's theory in order to agree that scientists are trained within a 'paradigm'. Parts of the paradigm will be the tacit knowledge which is imperative to everyday scientific work. This kind of knowledge cannot be articulated as explicit rules. Kuhn himself uses Michael Polanyi's term "tacit knowledge" (Kuhn 1970, p. 187). When experts deal with situations which fit into their paradigm, this works fine. But when confronted with situations that are not so easy to accommodate to the expert's paradigm, it is a source of error. Because experts in the same field are trained within the same paradigm, they are usually blind to many of their own tacit assumptions. However, experts from other fields may immediately be aware of some of the tacit assumptions of the field. Therefore, in cases involving complexity and uncertainty it is imperative to draw on various kinds of expertise.

11. Conclusion: Lessons

First lesson: Scientists are not outside

In this chapter I have presented two opposite views of scientific objectivity and the scientists' role in giving policy advice in cases where there are different interests involved. On the one hand the scientists working for the Department of Health argued that their role as scientists required that they produce objective knowledge, independent of who might benefit and who might lose from this information. They acted as 'Pure Scientists' according to Pielke's terminology. On the other hand, Beverly Paigen argued that scientists should choose sides, and she applied what is today known as the precautionary principle. She acted as an 'Issue Advocate' in Pielke' terminology.

As pointed out earlier, the researchers of the Department of Health were victims of an erroneous idea of scientific objectivity. This prevented them from cooperating with the residents of Love Canal. If they had entered into a dialogue with the residents, they would have learned about the swales. Lois Gibbs learned about the swales from older residents. When systems are complex, local knowledge may be more useful than mathematical models. The result of this alleged objectivity and neutrality was that the researchers sided with one party, the authorities.

Should the Department of Health researchers have sided with the residents of Love Canal, like Beverly Paigen? That was not required. However, they should have taken the residents seriously, and cooperated with them. The result would have been better science, and better science advice. The first lesson from Love Canal is that scientists must not remain outside a controversy.

Second lesson: Inform about uncertainty

We have seen that statistics was an issue in the Love Canal case. However, I have argued that other kinds of uncertainty are much more important. The following example applies to all kinds of uncertainty.

Let us imagine that a group of researchers is assigned the task of examining whether there is a difference in the incidence of illness between two groups, and they give the answer T0: 'We have not found any (statistically significant) difference.' However for the decision maker essential information is missing, because the answer may be interpreted in two different ways. Either T1: 'We have not found any difference, and we most likely would have found it if it existed' or T2: 'We have not found any difference, but we most likely would not have found it even if it existed.' Needless to say, T0 would carry much more weight if the correct interpretation was T1 than if it was T2. Therefore, it is a serious problem when researchers answer T0, the majority of politicians and others interpret it as T1, and the correct interpretation is T2. This was probably the case in Love Canal.

Even if the scientists of Department of Health regarded themselves as belonging to Pielke's category of 'Pure Scientists', not informing about uncertainty would not only be bad science advice, but bad science. However, it may have serious implications when decisions are based on scientific advice.

Third lesson: Laypeople should be involved

In their book *Uncertainty and Quality in Science for Policy* (1990) Jerome Ravetz and Silvio Funtowicz argue that science has to enter into a 'post-normal' phase to adequately address problems where uncertainty and 'decision stakes' are high. In the book they develop a conceptual scheme to deal with the new challenges. I shall not go into technical details, but will just mention one aspect of 'post-normal' science which is relevant to my discussion: the uses of what they call 'extended peer communities'.

'Extended peer communities' imply an extension of the traditional scientific community to include non-experts as well. However, this does not mean that laypeople should invade the research laboratories and carry out research. It does mean, though, that laypeople should take part in discussions of priorities, evaluation of results, and policy debates.

One reason for including non-experts is that they are sometimes closer to the problem. In Love Canal we saw that the residents had local knowledge that the scientists could not possibly have. Lois Gibbs did not have any relevant formal education, she was not trained in the uses of mathematical models and statistics. However, when she sat down with a map of Love Canal and put pins on every home that had reported health problems, she saw that they formed a pattern, and had the idea that the health problems might be connected with the swales that cut through the area. She thus made a contribution that was much more important than any of the scientists of the Department of Health. When problems are complex, local knowledge is at least as important as mathematical models.

The contribution from laypeople may also be valuable for the opposite reason. In Love Canal the contribution of the residents was valuable because of their closeness to the problems, but laypeople may also be valuable because they have a distance. Experts are often caught in their models, they are victims of 'tunnel vision'. One way of revealing experts' hidden assumptions is to ask apparently stupid questions, which may be regarded as an extension of an important element in the Socratic tradition in philosophy. We know that it was part of Socrates' strategy to pretend that he was more ignorant than he actually was. For example, in the dialogue *Gorgias* he asks the expert in rhetoric, the Sophist Gorgias, what rhetoric is. He then shows that Gorgias' answer is insufficient, and proceeds to 'deeper' or 'wider' questions. Very often the dialogue ends up with the fundamental ethical questions of the right, the good, and justice.

An ethical reason for bringing in common people is that they are affected by the decisions which are made. The questions of global warming, the ozone layer, radioactive waste, and genetically modified food concerns everybody, experts as well as non-experts. These questions are too important to be left only to the experts.

Further reading

The most comprehensive account of the controversy is Levine 1982. A huge collection of original documents concerning the Love Canal case are available at http://library.buffalo.edu/specialcollections/lovecanal/collections/ (accessed 30 March 2016). There is a website for the precautionary principle: http://www.precautionaryprinciple.eu/ (accessed 30 March 2016). From this website you can download among others COMEST 2005. A critical history of significance tests can be found in Ziliak & McCloskey 2012. Many examples of the abuse of mathematical models, in particular in geology, are given in Pilkey & Pilkey-Jarvis 2007. Two recent collections of relevant articles are Pereira & Funtowicz 2015 and Meisch *et al*. 2015.

References

Bhattacharyya, G.K. & Johnson, R.A.: 1977, *Statistical Concepts and Methods*, Hoboken: Wiley.

Carson, R.: 2000 [1962], *Silent Spring*, London: Penguin.

COMEST (World Commission on the Ethics of Scientific Knowledge and Technology): 2005, *The Precautionary Principle*, Paris: UNESCO.

Hacking, I.: 1990, *The Taming of Chance*, Cambridge: Cambridge University Press.

Ioannidis, J.: 2005, 'Why Most Published Research Findings Are False', PloSMedicine, **2** (8), 696-701.

Jonas, H.: 1984, *The Imperative of Responsibility*, Chicago: University of Chicago Press.

Kuhn, T.: 1970, 'Postscript – 1969', in: *The Structure of Scientific Revolutions*, Chicago: University of Chicago Press, pp. 174-210.

Lemons, J.; Shrader-Frechette, K. & Cranor, C.: 1997, 'The precautionary principle: Scientific uncertainty and type I and type II errors', *Foundations of Science*, **2** (2), 207-236.

Levine, A.G.: 1982, *Love Canal: Science, Politics, and People*, Lexington: Lexington Books.

Meisch, S.; Lundershausen, J.; Bossert, L. & Rockoff, M. (eds.): 2015, *Ethics of Science in the Research for Sustainable Development*, Baden-Baden: Nomos.

Paigen, B.: 1982, 'Controversy at Love Canal', *The Hastings Center Report*, **12** (3), 29-37.

Pereira A.G. & Funtowicz, S. (eds.): 2015, *Science, Philosophy and Sustainability*, London: Routledge.

Pielke, R.S.: 2007, The Honest Broker: Making Sense of Science in Policy and Politics, Cambridge: Cambridge University Press.

Pilkey O.H. & Pilkey-Jarvis, L.: 2007, *Useless Arithmetic. Why Environmental Scientists Can't Predict the Future*, New York: Columbia University Press.

Popper, K.R.: 1981, 'The Rationality of Scientific Revolutions', in: I. Hacking (ed.): *Scientific Revolutions*, Oxford: Oxford University Press, pp. 80-106.

Rio Declaration: 1992, *United Nations Conference on Environment and Development* (UNCED), Rio de Janeiro, 3-14 June 1992 (online available at: http://www.unep.org/documents.multilingual/default.asp?documentid=78&articleid=1163, accessed 21 Nov. 2016).

Savan, B.: 1988, Science under Siege: *The myth of objectivity in scientific research*, Montreal: CBC Enterprises.

Shrader-Frechette, K.S.: 1991, *Risk and Rationality*, Berkeley: University of California Press.

Tranoy, K.E.: 1967, 'Asymmetries in ethics', *Inquiry*, **10**, 351-72.

Wasserstein, R. & Lazar, N.A.: 2016, 'The ASA's Statement on p-values: Context, Process, and purpose', *The American Statistician* (online accepted version http://www.tandfonline.com/doi/abs/10.1080/00031305.2016.1154108, accessed 30.3.2016).

Wynne, B.: 1991, 'Uncertainty and environmental learning', *Global Environmental Change*, June, 111-127.

Ziliak, S.T. & McCloskey, D.N.: 2012, The Cult of Statistical Significance. How the Standard Error Costs Us Jobs, Justice, and Lives, Ann Arbor: University of Michigan Press.

Chapter 9

Applying an Ethical Judgment Model to the Case of DDT

Tom Børsen and Søren Nors Nielsen

Abstract: While most chemicals produce, or have produced, direct and intentional effects, many may also cause significant unintended outcomes. In this regard, chemicals are ethically ambivalent. It is not easy to balance the positive and negative consequences, nor is it straightforward to include uncertain future effects within an ethical analysis. Therefore, it is pertinent to develop and implement a method for identifying, assessing, and balancing the ethical issues associated with the production and use of chemical substances. Such a method for accommodating an ethical evaluation is presented in this chapter, being tested and exemplified using the case of DDT. Within the DDT case three major clusters of ethical problems are identified: the first concerning environmental ethics while another relates to the impact on human health, including, in particular, the lethal, carcinogenic, and sub-lethal aspects introduced by suspected endocrine disrupting effects. The third ethical concern focuses on the subject of justice and the distribution of risks, costs, and benefits.

1. Introduction

The Industrial Revolution has been followed throughout the 20th century by an ever-increasing development of new products and technologies. The rapid materialization brings with it an adjacent evolution of new artificial chemicals which often differ considerably to naturally existing compounds, resulting in them being referred to as xenobiotics. Industrial chemical substances have ethical significance because of both their usefulness and the unintentional effects they pose to public health and the environment. This has been recognized and addressed in research journals such as *Hyle* (Davis 2002, Preston 2005) and *Science and Engineer-*

ing Ethics (Parke 1995, Beamon 2005), as well as in reports from international organizations such as the United Nations Environmental Programme (UNEP 2015) and the European Environmental Agency (EEA 2001, 2013).

While it may appear strange that industrial pollutants such as DDT continue to be produced and used today, the initial development of such substances was never driven by the intention to harm human health or the environment. Instead, they were developed to serve beneficial purposes with each pollutant boasting at least one useful quality – the character of which being dependent on the particular chemical in question. Some were developed to increase the production of foods, such as pesticides or growth promoters, while others, such as those administered as medicine, were intended for improving health conditions. Further examples include chemicals used for the extraction and synthesis of other chemicals, or substances used for cleaning processes. In short, the vast majority of these new chemicals have been developed with the intention of contributing to the improvement of life conditions. It is only within the last 30 to 40 years, as a result of increased usage, that the large number of indirect and unforeseen effects have been disclosed to us.

As with any other technology, industrial chemical substances are, under many circumstances, beneficial to their users. They contribute as solutions to societal problems and as such allow political visions to be materialized. However, if used intensively, the chemicals can severely damage the health of both humans and the environment. The problems faced by such bittersweet substances are the consequence of an uncritical faith in human control over nature that cannot be upheld as an underpinning principle for further technological development.

The discovery of secondary effects of the otherwise beneficial use of chemicals was influenced by the publication of Rachel Carson's important book *Silent Spring* from 1962. The book drew public attention to the unforeseen and negative consequences of the use of industrial chemicals, especially DDT. *Silent Spring* criticized both what Georg Henrik von Wright calls the 'myth of progress' and what the Danish biologist, Jesper Hoffmeyer, refers to as the 'technical fix'. The myth of progress names a widespread temptation to sacrifice established ethical orientation systems and functioning practices in exchange for everything that is new, technical, and unconquered, while the technical fix refers to technical problem-solving strategies that apply technical means which temporarily solve a problem without noticing the long-term unforeseen consequences on, for example, the environment or public health. The use of DDT without taking into account the unforeseen and unwanted health effects is an example of a technical fix.

Industrial chemical substances have been named 'chemical boomerangs' (Børsen Hansen 2005, Petrosyan 2011) – compounds developed with the intention of solving individual or social problems yet also containing unforeseen negative (eco)toxicological effects, such as being potential contaminators of air, water, sediments, soil, plants, animals, and humans. In other words, these compounds come back like boomerangs to create new problems that call for further problem-solving. DDT is an example of a chemical boomerang.

The metaphor is striking: as we send something out, we also must expect to get it back. When longer distances are involved, we can even talk about 'chemical sputniks' – chemicals that have been transported through the atmosphere and that condense in colder climatic conditions through 'polar distillation' (Børsen Hansen 2005). This condensation means that indigenous people and animals in the Polar Regions (Siberia and the Arctic) have greater exposures to DDT even though there is no manufacturing in these areas.

Scientists, engineers, and decision-makers can no longer assume that technology overall is purely beneficial to humans. Technology has a dialectical nature – holding the potential both for the improvement and for the degradation of the human condition. This nature can be further explored using the ancient mythology of *Hubris* and *Nemesis*. In Greek mythology, hubris refers to dangerously over-confidence in ones' personal qualities leading to behavior that defies the norms protected by the ancient gods. Nemesis refers to the revenging gods' punishment of the perpetrator of hubris. If one does not respect technology's dialectical nature, and only focus on the technology's positive potential, one commits hubris. The myth can be translated into a relevant lesson for the use of DDT: undesirable consequences for the environment and society – Nemesis – arise as a result of uncritical and intensive use of DDT – hubris (Hard & Jamison 2005).

2. A Method for Proper and Quick Ethical Analysis

This chapter presents and applies a model developed at Aalborg University to conduct a quick and proper ethical analysis of issues and dilemmas related to the use of technology (Børsen & Damborg 2015). In the chapter, we will show that the method can also be applied in the ethical analysis of industrial chemicals. This is achieved by applying it to a case study of DDT, one of the chemicals regulated by the Stockholm Convention (see below).

The model is useful for chemists, chemical engineers, chemistry students, and others who want to make robust ethical judgments of the uses of a chemical substance by splitting the judgment process into four steps:

(1) Identification of intended beneficial consequences, potential misuse, unintended adverse side-effects, and long-term consequences for society.
(2) Linkage of intended consequences, misuse, adverse effects, and cultural implications of the uses of the substance to appropriate ethical values.
(3) Identification of ethical dilemmas related to specific uses of the chemical compound under assessment.
(4) Formulation of appropriate technological and institutional design criteria that can resolve the identified ethical dilemmas.

We will apply the method to the case of DDT to exemplify how industrial pollutants can be analyzed ethically. Ethical analyses of the uses of different chemical compounds will of course not lead to identical ethical estimates. All chemical compounds have their individual properties and uses.

2.1 Ethical values

Ethical values play a central role in the analytical tool presented here. An ethical value is understood as a normative criterion against which one can compare the wider consequences and circumstances of use of a given chemical compound. Do uses and misuses, resulting in both short and long term effects, align with or violate different ethical values? In this chapter, we have formulated a list of nine ethical values against which one can compare the uses of chemical substances. The list was constructed as follows.

The model is inspired by the so-called common-sense morality that can be traced back to Aristotle and Cicero, and through Aquinas to Kant and Ross. The ambition of common-sense morality is to balance ethical concerns which point in different directions and to formulate ethical compromises that all affected parties can accept. Common-sense morality is an ethical theory differing from many other ethical theories by not providing universal answers or decision methods. The ethically correct action is context dependent, where one must independently evaluate their options and choose what seems ethically most correct in a context of conflicting concerns. As the name 'common-sense' implies, this ethical theory cherishes common-sense and believes in the ability of humans to make judgments that are as informed and as reasoned as possible. On one hand, common-sense morality accepts that established ethical approaches reflect legitimate ethical concerns. On the other, it does not insist on only one ethical principle but instead emphasizes ethical reflection and common-sense. Common-sense can be understood as what Aristotle named *phronesis* – practical wisdom and functioning judgment. Phronetic judgment strives after the good life and the individual or collective ability to define actions pointing in that direction, in a context of contradictory but legitimate values.

This ethical approach takes a number of ethical concerns into account. A central aspect is the desire to combine and balance different ethical concerns of involved stakeholders. The founders of common-sense morality Tom L. Beauchamp and James Childress suggest that the fulfillment and balance of four ethical principles – respect for autonomy, utility, minimal harm, and justice – can guide decisions regarding concrete ethical dilemmas (Beauchamp & Childress 2001). Beauchamp and Childress' value of minimal harm is here split up into 'safety' and 'security', which takes into consideration, respectively, unintended and intended harm.

When assessing industrial pollutants, the four original ethical standards of common-sense morality need to be supplemented with ethical values originating from environmental ethics. Hence, we have included three different ethical principles from this branch of ethics: the precautionary principle, which is a well-established concept in regulation of the European Union (EU 2000); stewardship for the Earth which originates in Hans Jonas' philosophy (1984); and respect for nature which is extracted from Arne Næss' writings on deep ecology (1973).

The identification and selection of the ethical values most relevant for this analytic tool has been an iterative process. The combination of the original four principles of common sense morality combined with environmental ethical standards have been discussed at two seminars at Aalborg University and with students at an annual PhD course on ethics and social responsibility for engineers and scientists. These discussions have shown that ethical values inspired from Aristotelian ethics were not present, and therefore, two Aristotelian ethical standards have been added: humility and social stability.

In Table 1, we have included definitions of the ethical values used in the analyses. The list is not complete. New values can be added if users of the model find other values more relevant to their analysis. The table is ordered with the most individualistic and anthropocentric values placed at the top.

Table 1. Ethical values for evaluating of the use of industrial chemicals

Ethical value	**Description**
Autonomy	Everyone has a right to self-determination as long as it does not prevent others from their right to self-determination. Autonomy can be deduced from Kant's categorical imperative: No one must be treated only as a mean and not also as an aim in themselves.
Safety and security	Everybody has the right to be protected from damage, and safeguarded from illness, hunger, accident and other dangers. This value encompasses protection from undesirable events. Sometimes a distinction between safety and security is made where safety refers to the right to be safeguarded from unintentional harm, and security refers to the right for protection against intentional harm (*e.g.* from terrorism)
Justice	Here we include two different definitions: (1) Just actions are to generate the greatest benefit to the least-advantaged members of society. (2) Everybody must be treated according to merit and effort; two people can only be treated differently if their merits or efforts are different. Discrimination and stigmatization are in direct conflict with the ethical value of justice.

Utility	This ethical value has the foreseeable consequences in focus, and states that the ethical correct action is the one that generates the maximal well-being for the highest number of people. Well-being can be defined in different ways: as the feeling of hedonistic pleasure, realization of personal potential, a prosperous life, *etc*.
Humility	This ethical value is the anti-thesis to committing hubris. One commits hubris when one loses contact with reality and over-estimates one's own competencies, does not listen to criticism and thinks one-dimensionally without giving alternatives any consideration. According to Greek myth one will be punished by Nemesis if one commits hubris. One is humble when one is self-restrained.
Social stability	This ethical value focuses on how the various parts of society fit together, and strives for establishing equilibrium by balancing different aspects and interests, and as a last resort forcing out extreme ideas and individuals that disagree with popular opinion.
Precautionary principle	This principle states an action should not be undertaking if there are reasonable grounds for concern, though no scientific evidence, for it having dangerous effects on the environment, humans, animals, or plant health.
Stewardship for the Earth	This ethical value claims that humans are responsible for the world, and therefore are obliged to take care of it by shaping trajectories of social-ecological change at local-to-global scales to enhance and balance ecosystem resilience and human well-being. It has religious origins, as it can be derived from the believe that humans are guardians of God's creation. Nature and natural resources are considered as a gift.
Respect for nature	According to this value all forms of life have intrinsic or inherent value and are to be respected for their own sake. Humans are part of nature and the well-being and flourishing of human beings are not considered more important than the well-being and flourishing of other forms of life. Diversity of different life forms are contributing to the well-being of both individual species and individuals. This value derives from the notion of environmental rights.

David B. Resnik (2012) has constructed a similar list of ethical standards. Both Resnik's list and our list include the ethical values of utility, justice, stewardship, and precaution. In addition, Resnik's includes human rights which overlaps, to a certain extent, with this chapter's listed values of autonomy and safety/security. Resnik's value is broader and includes all human rights, whereas our list focuses on two specific human rights – autonomy and safety/security. Furthermore, Resnik includes animal welfare in his list of ethical values which may be considered as being one element among the many included in this chapter's value of respect for nature – the latter being less restricted to just animals that can feel pain, but instead treasures all life. Finally, Resnik includes sustainability which is a standard including environmental, financial, and social aspects. Our value of social stability focuses on the social sphere whose stability is affected by environmental and financial parameters. Another significant difference is that Table 1 above includes the ethical standard humility which has no counterpart in Resnik's list.

3. Demonstration of the Method: The Case of DDT

DDT was invented in 1873, and has thus been known since the early days of synthetic organic chemistry. Its toxic effect on insects was discovered more than 65 years later in 1939 by the Swiss chemist Paul Hermann Müller who, in 1948, was awarded the Nobel Prize in Physiology and Medicine for the discovery. The chemical was first applied outside the laboratory during World War II, where it was used to fight various insect borne diseases such as malaria and typhus (Bouwman *et al.* 2013). Following the war, DDT continued to be used in the fight against malaria but was also quickly conscripted into use as an agricultural insecticide – offering a potential solution to the issue of feeding the rapidly increasing global population. The application of DDT within agriculture promised increased food production while being seemingly harmless to human health. The future was to tell another story.

In the following, we apply the ethical analytical tool on the uses of DDT, step by step.

Step 1: Identification of intended and beneficial consequences, potential misuse, adverse effects, and long-term consequences for society

Intended benefits

One key application of DDT (as well as other pesticides) has been to maintain crop yields by protecting plants from insects. DDT is acutely toxic to insects and is therefore applied to kill the pests or prevent outbreaks, particularly in intensive farming systems where only a few crops are produced. According to Aktal *et al.* (2009) an almost four-fold increase in Indian food-grain production is estimated to be a result of such pesticide use from the late 1940s to the late 1990s.

Another key application of the chemical is in the prevention of diseases such as malaria, dengue fever, and Zika fever. Insects in these cases act as carriers, or vectors, of life threatening diseases which are typically prevalent in the tropics. The insects most responsible for such diseases are often mosquitoes but can also include flies and ticks. Mosquitoes of the genus Anopheles can transfer Plasmodium sp. parasites to humans by means of their bites, causing several types of malaria which can lead to fever, fatigue, seizures, or even death. The numbers of malaria infections per year is estimated to be around 300 million with a corresponding loss in GDP of 12 billion US$, with most cases affecting developing countries (UNICEF 2004). The dengue virus also causes an unpleasant fever, while the Zika virus is suspected to cause microcephaly in babies whose mothers have been infected during pregnancy. A third group of diseases, represented by the typhoid bacteria, are not very persistent in the environment and hence depend on transfer, by means of insects, for their spread. Typhoid bacteria cause aches and pains, fever, constipation, or diarrhea. DDT use can be very effective in controlling the insects carrying the disease, at least initially.

Malaria, dengue, and Zika virus do not directly discriminate between humans, but there seems to be definite links to the social and geographical conditions in which the individuals live. For example, malaria is typically spread at certain levels of altitude with the culprit mosquitos being attracted to open, still, and fresh water areas. These conditions are commonly associated with areas of low living standards, such as poor communities – typically squatters – situated close to pools of rain and waste water in areas where no sewer systems have been established. Malaria also tends to be most common in developing countries due to a lack of public health infrastructure, and particularly affects the poor who cannot afford protective measures or medicines. In these areas, governments can achieve greater and cheaper levels of malaria protection in principle by spraying DDT in houses and gardens.

Unintended adverse impacts

Extensive evidence shows that exposures to DDT, whether as a pesticide or a disease control agent, can cause ecological and human health effects. DDT belongs to a group of chemicals known as Persistent Organic Pollutants (POPs). POPs are organic chemicals, *i.e.* consisting largely of carbon, that have high molecular weight, are lipophilic, and have a propensity to evaporate and disperse over long distances. These compounds are persistent, *i.e.* hard to break down, and will therefore remain for a long time within the environment or human bodies. They disperse easily and are unfortunately toxic to both humans and wildlife.

DDT is persistent with a half-life time of 2 to 15 years. Its residues may be found for decades in soil, from where they can merge within organisms and later be transferred to consumers by means of food consumption. The chemical's properties make it bind efficiently to organic liquids or solid matter, fats, and fat tissues, and hence offer great potential for bio-concentration in organisms and for bio-magnification by up-concentration through the food chain: Plants treated with the chemical are eaten by plant-eating organism where it accumulates in fat tissues. The organisms are eaten by predators leading to even higher concentrations of DDT in the predator's bodies. Small predators are eaten by other predators resulting in further increase in concentration.

DDT is suspected of having a variety of sub-lethal effects. The chemical is mutagenic and therefore suspected to be carcinogenic, and has had toxic effects on internal organs and on the neuronal system in test organisms. The compound and its many derivatives have an endogenic effect in the human body and interfere with the hormonal system and associated with reproductive and developmental problems, *e.g.* incomplete development of sexual organs. It often hampers reproduction (Borgå *et al.* 2001, Leblanc 1995, Strandberg *et al.* 1998, Wang & Wang 2005) and can lead to abnormal development in offspring. Such problems are also seen in wildlife and likely underpin observations quoted in Rachel Carson's *Silent Spring* (1962), where predatory birds, such as eagles, produce less offspring. This strengthens the case that the ethical aspects of DDT also concern nature. Endocrine disrupting effects are, in general, observed over a wide range of organisms – both vertebrates and invertebrates. The potential danger is that DDT may have an indirect but dramatic effect on the ecosystem structure by breaking essential links and transfers within the ecosystem network.

Misuse

DDT use is historically associated with rapidly growing resistance by pests. In both agriculture and disease control, farmers and public health agencies have incentives to use high doses of DDT to achieve 'better' results more quickly. There are several obstacles to such an approach. The recommended dose alone led to the growth of resistance among populations of 'target organisms' or the insects being controlled. This means that the organisms intended to be killed by the treatment become tolerant to higher and higher concentrations, or can even become insensitive to the compound used. A natural selection of pest organisms takes place: if some organisms can survive being sprayed with DDT, they are more likely to reproduce and pass their traits on to their offspring. Over successive generations, a larger proportion of the pest population can tolerate DDT, which then makes the chemical solution less effective. Furthermore, because DDT is indiscriminate in its effects, it can harm many 'non-target organisms' as well. For instance, soil worms that are in general beneficial to soil conditions or even natural predators of the target pest are also affected by pesticide use. When the undesirable effects of over-intensive use of DDT becomes known, the use becomes referred to as intentional misuse.

A similar type of misuse has been observed in agriculture, and involves an increasing frequency of use of DDT with a precautionary purpose, *i.e.* not suppressing pests as such but preventing potential outbreaks. As a consequence, the crops are under more or less permanent treatment. This misuse is often stimulated by the fact that consultants providing advice to farmers are often financed by companies which have a direct stake in selling synthetic fertilizers and pesticides.

It is difficult to make a strict distinction between uses and misuses of DDT. As a heuristic, we categorize misuse as (1) preventive use for a pest attack, (2) use of overdoses to make sure that a pest is killed, and (3) use of DDT as the primary tool to eradicate malaria. National governments in conjunction with World Health Organization (WHO) – a special division within the United Nations dedicated to public health issues – misused DDT in all three ways in the 1960s and 1970s. All three forms of misuse lead to excessive use, and increase the risk of resistance and uptake in organisms. Misuse of DDT is in this perspective overlapping with overuse.

Societal impact

One characteristic of DDT is that it is a broadly applied insecticide. It attacks many species indiscriminately, whereas some more modern pesticides are targeted at fewer species. As a pesticide, DDT enables the expansion of industrial agriculture systems, with profound effects on land use, biodiversity, and environmental quality. These wider social and environmental changes are often overlooked as people do not appreciate the role DDT played in making industrialized agriculture possible. From the 1930s onward, farmers in the US began intensifying their production by means of machines replacing human and horse labor, high-yielding hybrid seeds, and monoculture crop practices. This combination of farming methods spread around the world after the Second World War, with many developing countries taking up the practices in the 1960s, as part of the Green Revolution.

This type of agriculture is characterized by the removal of natural obstacles to work (*e.g.*, draining water saturated soils, removing hedges) in favor of wide open spaces that could be more easily handled by machinery. Another consequence was that these vast surfaces were often turned into monoculture systems, *i.e.* one crop dominating at first in larger fields but very soon also across the whole landscapes. The increasing demands for food production, not only for human consumption but also for the production of feeds for cattle and pigs, created even higher pressure for increased productivity. Increased mono-cultural farming caused pests to become a more significant problem: with fields planted uniformly containing a single crop and without the natural pest control provided by beneficial insects and birds among diversified crops, the singular crops became more vulnerable to insect attacks. Without pesticide use, industrial agriculture would be less yielding, due to substantial losses from pest infection.

Table 2. Ethical issues and their adjacent short and long term effects

Ethical Issues	**Short term effects**	**Long term effects**
Elimination of pests in order to ensure and increase outcome of production in crops	Successful elimination of pest, Additional elimination and impact on non-target organisms, Acute toxic effects likely to be low in humans but not in nature	Accumulation in soils, Effect at non-targets of the whole ecosystem through bio-concentration and bio-magnification
Prevention of illness: removal of insect-borne vectors in malaria, typhoid control	General improvement of life conditions, Elimination of diseases	Induction of *e.g.* endocrine disruption, Re-appearance of disease vectors
Misuse/overuse	Non-exploited doses build up and left in crops and soils	Resistance of organism leading to no effect of use
Bio-concentration and magnification	Uptakes in tissues affecting farmers and local populations, Uptake in local as well as migrating animals, Binding in food and other organic pools	Residues found everywhere far from use even in pristine sub-arctic and arctic areas, Ever increasing concentration and accumulation until saturation or threshold levels are reached
Mono-cultural food production	Monoculture can produce more food in the shorter term, Gives less diversity in food, leading to nutritional bias and impact on health A more vulnerable economical system Farmers in the developing world are put in debt	Lessened biodiversity, less resilience and buffer capacity, Potential eradication of endangered (red-listed) and protected species, Local agricultural knowledge disappears, Induction of poverty through instable economics

The impacts of use in agriculture are not limited to human and ecological health. Vandana Shiva (1997, 2000) has argued that mono-cultured food production endangers the local agricultural knowledge and indigenous practices that farming communities in developing countries have built up over centuries. Her claim is that modern Western industrial food production does not allow alternative production schemes: they are ridiculed, presented as irrational, or out-competed because so-called external costs are neglected. Shiva further claims that farmers in India can grow enough nutrition to feed the country if they cultivate biodiversity without chemicals.

To sum up, we point to the following intended consequences and short and long term adverse effects for human and ecological health, and societal impact of DDT:

- Increase and ensure production of crop plants,
- Prevention of insect-borne illness,
- Human and ecological health effects, including bio-concentration and bio-magnification,
- Overuse giving raise to resistance among target organisms and uptake in organisms,
- Agriculture based on monocultures that threats self-sustaining local communities.

An attempt to synthesize the considerations that enter our ethical analysis of DDT is shown in Table 2. The next step in our analytical process is to link the short and long-term effects of DDT to ethical values.

Step 2: Linkages between consequences and ethical values

According to the United Nations (UN 2015), the Earth's population will grow to 9.7 billion inhabitants by 2050 leading many to believe that food production will have to be increased correspondingly. In societies dominated by a growth in population – such as India and many nations on the African continent – an argument for using DDT as a pesticide is that it can contribute to increasing food production, and potentially support feeding an increasing number of inhabitants. This argumentation is linked to the ethical value of utility as increased food production can generate more well-being for a higher number of people because fewer will starve.

This reasoning is based on at least two premises: That the use of DDT is, and will be, the best way to increase food production and that the food will in fact reach an otherwise starving population.

A similar line of reasoning can be made for the use of DDT for combating diseases. How can we allow that 300 million people are annually infected with malaria when we have DDT to combat the vector organisms? These two intentional applications of DDT are linked to food and health security which values everybody's right to be safeguarded from hunger or illness. The uses of DDT also relate to the ethical value of justice. In the case of food production this concerns whether the food is distributed fairly, *e.g.* if thc least-advantaged members of society benefit from the increased production, or if they have access only to sprayed food products, whereas the privileged class enjoy organic and high quality meals. Justice is also involved if disease or health risks primarily affect certain vulnerable groups, *e.g.* children or poor people. We observe vector borne illnesses primarily in poorer regions, hence, the use of DDT to combat disease such as malaria can comply with justice.

An argument against the intentional use of DDT as a pesticide, and hence against increased food production, states that it in a long-term perspective violates the values of stewardship of the Earth and social stability. Such use and its long-term consequences supports population growth and a higher world population that is unsustainable both socially and environmentally. From a long-term perspective, societies should not focus on increasing food production, rather they should aim at decreasing population growth and hence minimize the need for growth in the production of food. These arguments address the long-term consequences of intentional use of DDT as a pesticide, whereas the former address short-term issues.

The ethical values of humility and precaution are at stake regarding the overuse of DDT resulting in a build-up of the residues in crops, soils, and animals. Harmful consequences of the overuse of DDT are today well documented, but were not foreseen before intensive over-use began. Both farmers and national governments, encouraged by the WHO, committed hubris when they tried to enhance crop yields or eradicate malaria by very excessive use of the chemical. We see this as a violation of the ethical value of humility as the use of DDT was seen as an omnipotent technology on which the solution to the problems of hunger and malaria was built. This approach also violates the principles of precaution as no preventive measures, such as the establishment of early warning mechanisms, were taken when DDT was used in large scale.

Today we know that overuse of DDT results in bioaccumulation and bio-magnification. The bio-concentration of DDT means that organisms with tissues particularly rich in fats are more likely to suffer adverse health effects whereas bio-magnification signifies that organisms at higher levels of the biological hierarchy, *i.e.* predators, carry a higher risk of being affected. In this instance, the ethical values of stewardship of the Earth and respect for nature are involved. The intensive use of DDT cause ecosystems to be shifted out of balance and neglects their intrinsic value, suggesting that species and their environments must be protected not only directly through wild-life protection but also indirectly by sustained restrictions on DDT.

A less evident strike-back of intensive use of DDT occurs via the endocrine disrupting effect. This effect is not clearly understood and only supported by uncertain scientific evidence. Here, the precautionary principle is at stake and calls for preventive measures even when decisive scientific evidence is not at hand. Endocrine disruption is widespread over the animal kingdom, not only relating to human existence but also posing a threat to many other organisms, striking-back indirectly at society through the loss of biodiversity and diminishing ecosystem services. This aspect potentially risks human health and therefore possibly violates the ethical values of safety and security.

The use of DDT and other pesticides may have societal impacts in communities such as self-sustaining local communities that do not subscribe to the application of modern Western science and technology in food production, and instead practice an alternative, holistic, and low-technological way of producing food. Vandana Shiva (1997, 2000) has argued that the modern Western technologized approach to food production is no better than more diversified approaches. When it comes to forcing industrial and highly technologized food production methods onto poor farmers in self-sustaining local communities, the value of farmers' autonomy is violated. Also disregarded is a respect for nature due to the fact that mono-cultural agriculture practices violate biodiversity and are environmentally unsustainable.

A tentative conclusion based on the above identification of linkages between intentional actual use, potential misuse, adverse effects, and long-term consequences for society and culture and general ethical values may be represented in Table 3. The table is divided in two columns in which the first portrays the ethical issues identified in step 1. The basic ethical values resulting in arguments for or against DDT use are listed in the second column.

Table 3. Main ethical issues of the use of DDT are linked to ethical values.

Issue	**Linkage to ethical values**
To increase food production by protection of crops against pests	In compliance with utility, food safety, justice – if the food is distributed fairly; In a long-term perspective, it violates social stability and stewardship of the Earth
To prevent disease – by killing vectors	In compliance with utility, health safety, justice
Over- and misuse of DDT	Violation of humility, precaution
Bio-concentration and bio-magnification	Violation of stewardship for the Earth, respect for nature, safety and security
Agriculture based on monocultures	Violation of autonomy, social stability, respect for nature

Step 3: Identification of ethical dilemmas in using DDT

To facilitate the identification of ethical dilemmas related to the use of DDT we will introduce a distinction between an unethical situation/action and an ethical dilemma. An unethical situation/action occurs when it violates ethical standards without being justified by reference to other ethical standards. An ethical dilemma is defined as a situation where different ethical standards are in collision or where there is a collision of different interpretations of the same standard.

Use of DDT must be regulated and restricted to prevent overuse

Our quick ethical analysis suggests that the overuse of DDT is unethical and cannot be justified as it violates safety and security, humility, precaution, stewardship of the Earth, and respect for nature, while also casting doubts on the long-term benefits of using DDT. High rates of DDT use must be avoided.

Use of DDT in agriculture is not ethical; modest use of DDT to combat malaria is

A distinction must be made between two purposes of use of DDT. In one case, it was introduced to ensure sufficient food production, the other was due to a wish to increase life quality by eliminating the vectors that transfer harmful disease. This distinction is especially important as the use of DDT must be restricted to avoid overuse, therefore arguing for each case becomes relevant. Alternatives to the use of DDT in malaria prevention is not foreseen to emerge in near future, whereas we will argue that the use of DDT should not be allowed in agriculture as alternative pesticides do exist that can replace the compound in that case.

Anthropocentric versus environmental concerns

We identify an ethical dilemma, which we formulate as a question: Can the utility, safety, and security that a moderate use of DDT provides in our combat against disease be justified at the cost of (i) our stewardship for the Earth and respect for nature, and/or (ii) uncertain potential negative consequences associated with a moderate use of DDT?

The first part of this question reflects two different perceptions of the relationship between humans and nature. Is humanity something different from the surrounding nature, a threat we need to be protected from? Or do we see the natural environment as a part of our humanity?

Killing a target organism is one thing and reflects the first perspective on the human/nature relationship. Target organisms are considered a threat to humanity's well-being. Accidental killing of non-target organisms and all other unintended environmental effects aligns with the second position on the human/nature relationship. Even a moderate use of DDT influences the environment, and if one believes that our natural environment is defining ourselves as human beings, it becomes difficult to justify a moderate use of DDT because we, in that way, jeopardize our humanity.

Most likely this ethical concern is more easily overcome when we consider organisms at the lower levels of the biological hierarchy, such as insects, as they, on a mere physiological basis, have no spine, little brain, and are not able to sense pain. The situation becomes different when the organisms affected are to be found higher up in the hierarchy, such as predatory birds and polar bears, or if the natural environment changes appearance. The suspected effects on various higher level animals are generally sub-lethal and may be considered acceptable when judged from an anthropocentric ethical point of view, such as utilitarianism. On the other hand, when effects are concerned with the ability of organisms to reproduce, such an effect may have severe effects on the natural environment in the first instance and on the human population in a long-term perspective. In such a scenario, the two different perspectives on the human/nature relationship overlap.

Regarding the second part of the question we see that, on one hand, humans are being exposed to an array of *potential* health damages, particularly when DDT is sprayed inside their houses. One type of adverse effect, that of endocrine disruption, can strike anywhere with seemingly no particular social or sexual differentiation. The endocrine disrupting effect has been demonstrated to affect both males and females. However, these effects are not fully understood and might not materialize if the use of DDT is restricted and used with care.

On the other hand, humans – especially those in developing countries with poor public health infrastructure – surely benefit from the control or even elimination of insect vectors and the resulting lessening of their disease burdens. Malaria and other diseases that can be regulated by use of DDT are usually found in regions of the world dominated by 3rd world countries in general, and Africa, South America, and South East Asia in particular. If 300 million incidents of malaria correspond to an additional cost in GDP of 12 billion it means an extra increase in GDP of 4,000 US$ per incident that is avoided. In fact, this number exceeds the per capita GDP of many African countries.

Step 4: Technological and institutional design criteria to transcend dilemmas

Legislation

The presence of DDT in food and the environment was first detected in the 1960s more than two decades after its introduction (Watson 2001). By the late 1960s, countries began to withdraw DDT as a malaria control agent partly due to the environmental effects and partly because of drainage efforts and the removal of wetlands resulting in the reduction of vector survival places. DDT was, in 1972, banned in the US with many other nations implementing similar bans shortly after.

The use DDT is now regulated under the Stockholm Convention (2008) that has been signed – but not ratified – by 180 countries. Today it is illegal to use DDT with the purpose of protecting crops against pests. Through the period of extensive use in agriculture the use of DDT increased to more than 40,000 tons per year (with a peak production of 82,000 in 1963). As a result of the restrictions the production decreased to approximately 3,300 tons in 2009.

Recently, the restricted use of the compound for eradicating disease vectors has been lifted in the Stockholm Convention. Hence, one may identify countries where the usage to fight vector borne diseases is allowed. The WHO's policy is to recommend DDT use as part of a management package in high transmission areas where the degree of control must be particularly high. During the 2000s, some countries, mostly in Africa, resumed using DDT to control malaria as their mosquito populations have become more resistant to other pesticides. Enforcement of the treaty is also a concern.

Our ethical analysis supports the international legislation restricting the use of DDT. We judge that legislation as sound and ethically justified.

Holistic alternatives to pesticides in agriculture

The previous step of our ethical analysis points to expanding available alternatives to the use of DDT in agriculture (and in the combat against vector-borne diseases). A measure to do this is to transform industrialized agriculture methods into more environmentally friendly practices through organic agriculture. Increasing hedges and biodiversity in the landscape would enhance predators of pests which have been found to be equally efficient to the use of pesticides. The development of new pesticides and other technological agriculture practices to prevent pest attack in mono-cultural agriculture is also a possibility if ethical standards (e.g. farmers' autonomy or precaution) are not violated.

There exist many alternatives to intensive forms of agriculture – known under names such as organic agriculture, agroecology, polycultures, permacultures, *etc*., all of which are forms of soil exploitation which involve working closer to natural principles and include higher biodiversity, soil conservation, the input of nutrients by use of leguminous plants, higher diversity landscapes, and even exploitation of the 3rd dimension, *i.e.* production at various levels of height.

There is convincing evidence that advantages exist which are connected to such ways of growing crops. In particular, organic agriculture and agroecology encourage the use of biological and ecological methods for controlling pests. These include maintaining hedgerows that sustain predatory organisms like wasps or birds, planting multiple crops together to reduce the likelihood of pest or disease outbreak, and using methods like the push-pull model developed in Kenya. These methods can greatly reduce the demand for pesticides. Simultaneously, such farming systems have a more continuous demand for labor that may be handled within a family and have a higher crop range that can provide a more nutritious selection of food. Last, but not least, such a system is less vulnerable to economic oscillations in market prices.

On the other hand, alternative pesticides provide us with technological solutions that are easy to apply. As long as the consumption of pesticides is restricted or not over-used but applied with care they can provide alternatives to DDT use. An expansion of available alternatives to include holistic approaches in agriculture will not prevent the use of pesticides in cases where these options are autonomously chosen by farmers and local communities.

Alternatives to DDT in prevention of vector-borne diseases

More challenging is finding ways to reduce the use of DDT for inhibiting vector organisms in order to interfere with the spread of diseases. The argument goes that malaria is, in a number of ways, not only disastrous to human health, but also costly to society in a many developing countries (for a list of viewpoints, see Zelson 2014), and that no other solutions exist. Despite this, the Stockholm treaty calls for DDT to be phased out over time. The history of DDT shows that the insecticide was initially very effective in the 1950s before succumbing to the growth of pest resistance in the 1960s. However, this resistance tends to disappear with time and, as a result, treatments with DDT become effective again. By 2000, when several developing countries, notably South Africa and India, reinitiated the spraying of DDT, the insecticide was proven to be very efficient. However, by 2011, pest resistance had already re-emerged on a widespread scale across Africa.

Other technological solutions might have longer-lasting effects than periodically reviving DDT. These solutions include the deliberate spread of male mosquitoes that have been made infertile through radiation, thus lowering their reproductive success, but not without many obstacles (*e.g.*, there can still be fertile mosquitoes). Research in and development of new 'gene drive' technology is now being tested in Brazil to genetically modify mosquitoes to be infertile (Mendes 2012). This is a radical novel technological intervention that in itself is suitable for being ethically evaluated. Here the questions on long-term effects of eradicating a whole complexion of species will have to be addressed.

Alternatively, insect habitat control measures and behavioral change steps can be taken in combination, without necessarily needing any of these technologies. The question is how far can we go by widespread implementation and dissemination in the use of nets, repair and manipulation of the ambient environment, removal of open waters (sewers, gutters, and rainwater puddles), and regulation of indoor temperature and humidity – all factors that will impact the reproduction and survival in the vector populations.

Early warning mechanisms

The DDT case is an example of a chemical boomerang. On one hand, we are reliant on quite a large number of chemicals offering benefits to us such as the increased availability of food by means of increased crop growth, or improved public health through medicines and disease control. On the other hand, some of these chemicals have at a later stage been found to produce negative impacts. The challenge arises as how to evaluate the positive and negative effects and how to balance those up against each other.

The question is whether we can accept the high number of chemical boomerangs and sputniks and, if so, which ones call for ethical judgment. Should we eliminate the use of pesticides and accept more people dying from hunger or from one of the many diseases having insects as a vector? The number of ethical questions that even one chemical substance raises is endless and although it is possible to group some chemicals, their relating effects tend to be so specific and unique that almost every chemical requires its own assessment. This can be seen in the two volumes of the *Late lessons from early warnings* series issued by the European Environment Agency which consists of an extensive list of case studies relating to industrial pollutants (EEA 2001, 2013). Unfortunately, individualized studies are not likely to be possible if we attempt to make full toxicological screening on the most important high volume substances.

An important point made in *Our Stolen Future* (Colborn *et al* 1993) is that existing risk assessment tools cannot foresee undesired consequences of chemical compounds – such effects may be completely new.

One reaction to this issue is to set up early warning measures with the purpose of spotting new undesirable effects for human health and the environment. These could try to spot weak indicators in the scientific literature. When indicators of new, unwanted, and potentially dangerous effects are recognized, resources to further research in those potential effects are allocated to further research so that it can be decided whether the potential effect is real or not. Early warning mechanisms can be included within existing legal regimes regulating the use of industrial pollutants, *e.g.* into the Stockholm Convention.

4. Conclusion

In this chapter, we have presented a quick and proper method for making ethical assessments of the use of industrial chemicals, and applied it to the use of DDT. The model has helped the authors to form their ethical judgment on the use of DDT. Based on the analysis,

- We recommend that the use of DDT must be heavily regulated to prevent overuse. The Stockholm Convention provides an ethically sound legal framework for the regulation of DDT.
- We do not identify persuasive ethical arguments for the use of DDT in agriculture as alternatives exist both in the form of target-specific pesticides and by means of holistic agricultural approaches. We suggest that it becomes more widely accepted to choose holistic alternatives, and that barriers for making this choice are lowered.
- We find that a modest use of DDT in the fight against malaria is ethically justified until better alternatives are available.
- It is suggested that early warning mechanisms are set up to spot unforeseen effects of the alternatives developed to replace DDT in both agriculture and in the domain of public health.
- An ethical dilemma is identified between the benefits provided by a modest use of DDT in combatting malaria and the effect on nature's integrity and the potential negative consequences related to the moderate use of DDT. The authors encourage students of chemistry and chemical engineering to discuss how to transcend that dilemma.

An ethical estimate is never final and objective. Hence, we encourage our readers to make their own ethical judgment, and challenge our conclusions.

Further Reading

A much more detailed presentation of common-sense morality is found in Beauchamp & Childress 2001. An expanded discussion of most of the ethical values presented here is found in Resnik 2012. For a more detailed and thorough presentation of the history of DDT we refer to Bouwman *et al.* 2013 and chapter two in Widavsky 1995. Carson 1962 and Colburn *et al.* 1993 are influential historical items still worthwhile reading.

References

Aktar, W.; Sengupta, D. & Chowdhury, A.: 2009, 'Impacts of pesticides use in agriculture: their benefits and hazards', *Interdisciplinary Toxicology*, **2**(1), 1-12.

Beamon, B.M.: 2005, 'Environmental and Sustainability Ethics in Supply Chain Management', *Science and Engineering Ethics*, **11**, 221-234.

Beauchamp, T.L. & Childress, J.F.: 2001, *Principles of biomedical ethics*, New York: Oxford University Press.

Bond, J.: 2009, 'Professional ethics and corporate social responsibility', *Process Safety and Environmental Protection*, **87**, 184-190.

Borgå, K.; Gabrielsen, G.W. & Skaare, J.U.: 2001, Biomagnification of organochlorines along a Barents Sea food chain, *Environmental Pollution*, **113**, 187-198.

Bouwman, H.; Bornman, R.; van den Berg, H. & Kylin, H.: 2013, 'DDT: fifty years since Silent Spring', in: EEA (ed.), *Late Lessons from Early Warnings: Science, Precaution, Innovation*, Copenhagen: EEA, pp. 240-259.

Børsen Hansen, T.; 2005: *Teaching Ethics to Science and Engineering Students*, Copenhagen: Center for the Philosophy of Nature and Science Studies [available online at: http://portal.unesco.org/shs/en/files/8735/11289332261 TeachingEthics_CopenhagenReport.pdf/TeachingEthics_CopenhagenReport.pdf, accessed 28 October 2016].

Børsen, T. & Danborg, P.B.: 2015, 'Techno-Anthropological Ethics and Health Information Systems Technologies', in: Botin, L.; Bertelsen, P. & Nøhr, C. (eds.), *Techno-Anthropology in Health Informatics: Methodologies for Improving Human-Technology Relations, Technology and Informatics*, Amsterdam: IOS Press, pp. 83-94.

Carson, R.: 2000 [1962], *Silent Spring*. London: Penguin.

Colburn, T.; Dumanoski, D. & Myers, J.P.: 1996, Our Stolen Future: Are We Threatening Our Fertility, Intelligence, and Survival? A Scientific Detective Story, London: Penguin.

Colburn, T.; vom Saal, F.S. & Soto, A.M.: 1993, 'Developmental Effects of Endocrine-Disrupting Chemicals in Wildlife and Humans', *Environmental Health Perspectives*, **101** (5), 378-384.

Cone, M.: 2009, 'Should DDT be Used to Combat Malaria?', *Scientific American, Environmental Health News*, May 4 [available online at: http://www.scientific-american.com/article/ddt-use-to-combat-malaria, accessed 26 April 2016].

Davis, M.: 2002, 'Do the Professional Ethics of Chemists and Engineers Differ?', *Hyle: International Journal for Philosophy of Chemistry*, **8** (1), 21-34.

EEA: 2001, *Late Lessons from Early Warnings: the Precautionary Principle 1896-2000*, Copenhagen: EEA [available online at: http://www.eea.europa.eu/publications /environmental_issue_report_2001_22 , accessed 28 October 2016].

EEA: 2013, *Late Lessons from Early Warnings: Science, Precaution, Innovation*, Copenhagen: EEA, [available online at: http://www.eea.europa.eu/ publications/late-lessons-2, accessed 28 October 2016].

EEA: 2015, *Priority Substances and Certain Other Pollutants according to Annex II of Directive* 2008/105/EC. Copenhagen: EEA, [update from 8 June 2016 available online at: http://ec.europa.eu/environment/water/water-framework/priority_substances.htm, accessed 28 October 2016].

EU: 2000, *Communication from the commission on the precautionary principle COM 1*, Brussels: Commission of the European Communities [available online at: http://eur-lex.europa.eu/LexUriServ/LexUriServ.do?uri=COM:2000:0001: FIN:EN:PDF, accessed 31 March 2017].

Hård, M. & Jamison, A.: 2005, Hubris and Hybrids. A cultural history of philosophy and science, New York & London: Routledge.

Hellweg, S. & Frischknecht, R.: 2004, 'Evaluation of Long-Term Impacts in LCA', *The International Journal of Life Cycle Assessment*, **9** (5), 339-341.

Johansen, B.E.: 2003, *The Dirty Dozen. Toxic Chemicals and the Earth's Future*, Connecticut & London: Westport.

Jonas, H.: 1984, The Imperative of Responsibility. In Search of an Ethics for the Technological Age, Chicago & London: University of Chicago Press.

Leblanc, G.A.: 1995, 'Trophic-Level Differences in the Bioconcentration of Chemicals: Implications in Assessing Environmental Biomagnification', *Environmental Science & Technology*, **29**, 154-160.

Mendes, H., 2012. *Brazil tests GM mosquitoes to fight Dengue*, *Nature*, News and Comments article [available online at: http://www.nature.com/news/brazil-tests-gm-mosquitoes-to-fight-dengue-1.10426, accessed 28 October 2016].

Næss, A.: 1973, 'The shallow and the Deep, Long-range Ecology Movement', *Inquiry*, **16**, 95-100.

Parke, D.V.: 1995, 'Ethical Aspects of the Safety of Medicines and other Social Chemicals', *Science and Engineering Ethics*, **1**, 283-298.

Petrosyan, P.S.: 2011, 'Chemical Safety Problems', *Conservation Science in Cultural Heritage: Historical-technical Journal*, **11** [available online at: https://conservation-science.unibo.it/article/view/2693/2092, accessed 31 March 2017].

Preston, C.J.: 2005, 'The Promise and Threat of Nanotechnology: Can Environmental Ethics Guide Us?' *Hyle: International Journal for Philosophy of Chemistry*, **11**, 1, 19-44.

Resnik, D.B.: 2012, *Environmental Health Ethics*, Cambridge: Cambridge University Press.

Shiva, V.: 1997, Monocultures of the mind: Perspectives on Biodiversity and Biotechnology, London & New York: Zed Books & Third World Network.

Shiva, V.: 2000, Stolen Harvest: The hijacking of the global food supply, Boston, Massachusetts: South End.

Stockholm Convention: 2008, *Listing of POPs in the Stockholm Convention* [available online at: http://chm.pops.int/TheConvention/ThePOPs/ListingofPOPs/tabid/2509/Default.aspx, accessed 28 October 2016].

Strandberg, B.; Bandh, C.; van Bavel, B.; Bergqvist, P.-A.; Broman, D.; Näf, C.; Pettersen, H. & Rappe, C.: 1998, 'Concentrations, biomagnification and spatial variation of organochlorine compunds in a pelagic food web in the northern part of the Baltic Sea', *The Science of the Total Environment*, **217**, 143-154.

UNEP: 2015, *Stockholm Convention on Persistent Organic Pollutants*. Report from Conference to the Stockholm Convention on Persistent Organic Pollutants. Seventh Meeting, Geneva, 4-15 May, 2015 (UNEP/POPS/COP.7/INF/5).

UNICEF: 2004, *Fact Sheet: Malaria a Global Crisis* [available online at: http://www.unicef.org/media/media_20475.html, accessed 27 October 2016].

United Nations: 2015, *World Population Prospects: The 2015 Revision*, United Nations, Department of Economic and Social Affairs/Population Division.

Wang, X. & Wang, W.-X.: 2005, 'Uptake, absorption efficiency and elimination of DDT in mariner phytoplankton, copepods and fish', *Environmental Pollution*, **136**, 453-464.

Watson, D.H.: 2001, *Food Chemical Safety,* Boca Raton: CRC, Woodhead, vol 1.

Wildavsky, A.: 1995, But is it True? A citizen's guide to environmental health and safety issues, Cambridge, MA & London: Harvard University Press.

Zelson, E.: 2014, 'Rethinking DDT: The Misguided Goals of the Stockholm Convention on Persistent Organic Pollutants and a Plan to Fight Malaria Worldwide', *William & Mary Environmental Law & Policy Review*, 243, [available online at: http://scholarship.law.wm.edu/wmelpr/vol39/iss1/8/, accessed 28 October 2016].

Chapter 10

Applying Utilitarianism and Deontology in Managing Bisphenol-A Risks in the United States

Abigail Martin, Alastair Iles and Christine Rosen

Abstract: We examine Bisphenol-A (BPA) as a case that illustrates key challenges in addressing the public health risks of consumer products in the 21st century. First, we trace growing concerns about the effects of BPA on human health, showing how regulatory approaches can exacerbate the difficulty of dealing with the unforeseen risks of chemicals in consumer products. Second, we highlight the question of who should bear the responsibility – and the cost – of rectifying or preventing unforeseen chemical risks in consumer products. Third, we discuss the challenge of substituting out a potentially hazardous chemical from consumer products in the context of well-established global production chains and consumption patterns. Utilitarian and deontological ethical frameworks have influenced societal debates surrounding each of these three challenges, creating moral dilemmas for actors with different forms of moral agency – both those implicated in the production of harmful chemicals and those pursuing remedies.

1. Introduction

Bisphenol-A is a chemical used in consumer products such as baby bottles, reusable water bottles, and infant formula containers. The substance is found in many other products that require strong, clear glassy materials, such as electronics and food packaging. Some scientific studies have linked Bisphenol-A (BPA) to diabetes, thyroid disease, various cancers, and obesity, but experts disagree over whether BPA is causing harm through its ability to disrupt endocrinal functions. Many chemical and product manufacturers have defended BPA as safe despite concern about the risks of BPA from activists, consumers, and some scientists and researchers. Some companies have voluntarily replaced their products with BPA-free versions. Governments appear to be similarly torn: In Europe and Japan, government regulators consider BPA safe at current exposure levels, while experts advising the Canadian government concluded the opposite.

The case of Bisphenol-A exemplifies societal debate over industrial chemicals in the 21st century. Over the past fifteen years, public concerns about chemicals embodied in consumer products have grown steadily. These chemicals can dissipate from products during their use and disposal, and can be absorbed or ingested into human bodies. In earlier decades, public concern and regulators largely focused on chemical risks created in the manufacturing phase, such as factory pollution and hazardous waste. Such risks are still significant. However, human exposure to consumer products occurs at a much greater order of magnitude: many more people are potentially affected once products leave the manufacturing plant.

Today there is more attention to the health hazards of toxic chemicals in products, but regulators still struggle with the question of how to define toxicity. Traditional toxicology studies used by regulators to determine a chemical's toxicity focus on whether a chemical is carcinogenic (cancer-causing), often overshadowing the question of whether a chemical is estrogenic, or capable of disrupting hormonal processes in the body. Although Bisphenol-A's estrogenic properties have been known since 1938, its risks are still debated – with implications for the numerous chemicals used in consumer products that are suspected of being endocrine disruptors.

The case of BPA exemplifies three key challenges in the chemical industry. First, tracing the growing concerns about human exposure to BPA shows how regulatory approaches can exacerbate the difficulty of dealing with the unforeseen risks of chemicals in consumer products. Second, the case raises the question of who should bear the responsibility – and the cost – of rectifying or preventing emerging chemical risks in consumer products. Third, BPA highlights the challenge of substituting a potentially hazardous chemical for a harmful substance in consumer products in the context of well-established global production chains and consumption patterns.

Utilitarian and deontological ethical frameworks have influenced societal debates surrounding each of these three challenges. This chapter explores how these ethical frameworks raise moral dilemmas for the various actors involved – both those implicated in the production of harmful chemicals and those pursuing remedies: chemists designing molecules; managers devising business models for sustainable products; consumers making purchasing decisions; governments setting health and safety standards; groups pushing for clearer workplaces and products. Each of these actors have moral agency – the power to be morally accountable for one's actions and their consequences. But who has moral agency to advance more sustainable outcomes for the public good? When these actors face a moral dilemma, what ethical perspectives determine what it means to 'do the right thing'?

In the next few sections, we review the history of BPA use in manufacturing products and track the changing science and perceptions of BPA toxicity risks, before turning to discuss the ethical dilemmas of key actors in the BPA production chain. While reading this background, you should reflect on what responses might be appropriate in a situation in which a chemical risk is not yet fully proven but the chemical is commercially lucrative.

2. A Short History of Bisphenol-A

BPA – also known as 2,2-bis-4-hydroxyphenyl – is a synthetic chemical found in numerous products, including automotive parts, water supply pipes, electronics, baby bottles, and other food containers. It is one of the highest production volume chemicals in the world. By the 1980s, the global production of BPA reached almost a million metric tons per year and has grown substantially since then (Fiege *et al.* 2012). According to industry reports, the global demand for BPA was over 6 million metric tons in 2013 – representing a market size of $US 13.87 billion (Grand View Research 2014). Manufacturing capacity was once concentrated in the US, Europe, and Japan, but has expanded to Asia as markets for BPA and consumer end-products made with BPA have become increasingly global.

BPA gained commercial success in the polymer and plastics production with two main end markets: approximately 63% of BPA is used to build polycarbonate plastic resins and 27% goes into formulating epoxy resin monomers. For both polycarbonates and epoxy resins, BPA is an important building block with attributes that industry has found difficult to match with substitute chemicals (Ritter 2011). In the US, just five companies manufacture BPA: Bayer, Dow, Hexion Specialty Chemicals, SABIC Innovative Plastics (formerly GE Plastics), and Sunoco collectively generate approximately $US 6 billion in sales per year (Case 2009).

BPA was first synthesized in 1891, but its commercial production did not begin until the early 1950s, after chemists created the first epoxy resins using BPA in the US and Europe (Vogel 2009). Epoxy resins are produced by transforming liquid polyethers into infusible solids through a special curing process that reacts epichlorohydrin with BPA. These resins are versatile chemicals that can be formulated to have a range of mechanical properties (from extreme flexibility to high strength and harness), chemical resistance, high adhesive properties, and high electrical resistance. Epoxy resins became extensively used throughout the manufacturing sector as protective coatings for metal equipment, piping, steel drums, and the interior of food cans.

In 1957, chemists at Bayer and General Electric began developing another use for BPA as a monomer feedstock in plastics production (Vogel 2013). When polymerized with either carbonyl chloride or diphenyl carbonate, BPA forms a plastic called polycarbonate. Polycarbonate is hard, clear and nearly unbreakable. It is often used to replace glass in a variety of consumer products. The most common trade name for polycarbonate is Lexan.

3. Diverging Opinions on Whether BPA is 'Safe'

BPA is not only ubiquitous in everyday materials but is also prevalent in human bodies. Biomonitoring studies of the American population, for example, have consistently shown the widespread presence of BPA in urine, with slightly elevated levels in children, females, and lower-income populations (Calafat *et al.* 2008, Vandenberg *et al.* 2007). In its

Fourth National Report on Human Exposure to Environmental Chemicals, the US Centers for Disease Control found detectable levels of BPA in 93 percent of urine samples from over 2,500 people, suggesting almost universal and continual exposure (Centers for Disease Control 2009).

Although the amount of BPA found in human bodies is usually relatively low, experts strongly disagree about the levels at which exposure to BPA is harmful. Early toxicology studies in the 1970s found no observable carcinogenic effects in rodents given high doses of BPA for two years, in part because BPA metabolized rapidly in the animals' bodies. Yet, in the last two decades, new scientific research has ignited regulatory controversies over BPA. A growing number of studies suggest that repeated small doses of BPA can disrupt the human endocrine system, especially during prenatal and post-natal development. Fetuses and infants thus have a heightened risk of developmental harm when exposed to BPA transmitted through the placenta or in breast milk, or through containers like baby bottles.

Although environmental and public health advocacy organizations have called on governments worldwide to ban the use of BPA in food packaging, regulators diverge over whether there is enough evidence to justify controls. The challenge is how to decide what this level is, in a situation where scientifically credible studies can substantiate a cause-and-effect relationship between BPA and adverse health outcomes whereas other scientifically credible studies undermine this relationship. Facing apparently conflicting scientific evidence, regulators can reach different conclusions about whether BPA is safe, because of their differing interpretations of the data and because of their assumptions about which risks and harms matter. For example, within Europe, there is a splintering of views underway. The European Food Safety Authority believes that BPA is safe for use in food packages, while France has banned this particular application from January 2015 (Jacobsen 2015).

In the US context, a federal system of government exists, in which federal, state, and local levels of government have their own jurisdictions (or areas over which they can wield legislative and executive powers). Most chemical regulation occurs at the federal level but states and cities can pass their own laws or bans, as long as the federal government has not displaced these with its own. Three federal regulatory agencies are typically involved in overseeing chemical risks: the Food and Drug Administration (FDA) for food; the Environmental Protection Agency (EPA) for toxic substances and pesticides; and the Consumer Safety Protection Commission (CSPC) which has acquired some jurisdiction over phthalates in toys and cosmetics. Each agency must implement federal laws that gives it specific powers, often founded on doing risk assessments to decide whether a chemical must be regulated. In other words, chemical risk regulation is divided between these agencies, leading to many gaps in oversight.

In 1963, FDA approved BPA as "generally regarded as safe" for use in food additives (Turker 2012). Polycarbonate plastics and epoxy resins are regarded as food additives if they come into contact with food. This approval means that regulators have neglected BPA for many years, assuming that it remains safe. It is up to NGOs and citizens to petition FDA to retract its approval. As new evidence has emerged, some state governments have taken precautionary action despite the federal government's hesitancy to regulate BPA more stringently. Starting in 2009, Minnesota, California, Connecticut, Hawaii, Illinois, and other states introduced BPA bans (Barraza 2013). These laws primarily targeted baby bottles and sippy cups and, in some cases, child food containers more generally. Within the federal government, there have been conflicting evaluations. In 2009, Senator Charles Schumer and several other politicians in Congress tried to enact the BPA-Free Kids Act of 2009, which would have eliminated BPA from all child food containers, including cups, bowls, and drinking straws (Barraza 2013). By contrast, the FDA repeatedly re-affirmed between 2008 and 2012 that BPA was safe.

Understanding why the regulatory and policy debate over BPA is so polarized requires looking into how the chemical regulatory process works to evaluate toxicity risks. We need to briefly review how risk assessment has traditionally worked, and how this analysis struggles to accommodate health effects that depart from a standard model of chemical risk. We use the United States as our example, but most industrial countries and some emerging economies have had similar regulatory systems in place.

3.1 Characterizing BPA risks in the United States

In the US, regulators working at FDA, EPA, or CSPC typically assess a chemical's risk in four steps (see Figure 1). Their risk assessments studies are information-intensive and time consuming. They depend greatly on the availability of relevant scientific studies and the protocols that regulatory institutions use to draw conclusions from these studies about what is 'safe' or 'hazardous'.

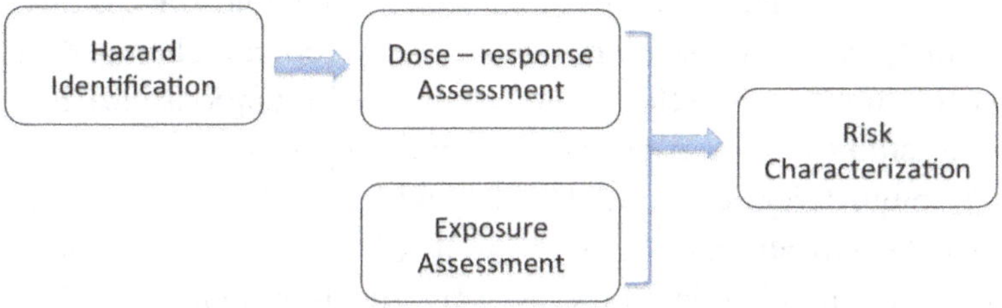

Figure 1. The Four Step Risk Assessment Process.

Hazard identification involves reviewing data from toxicology and epidemiological studies to ascertain what adverse health effects may result from human exposure to the chemical. Toxicology studies use tests of animals in the laboratory to determine whether exposure can cause higher levels of a particular health endpoint – like cancer or infertility – and then extrapolate from these tests what the risk would turn out to be for humans. Such extrapolations, of course, are inherently uncertain because rodents and humans are not perfectly biologically comparable. Epidemiological studies search for disease patterns in human populations, often comparing between two or more groups of people in different geographical locations to uncover environmental contributions to human diseases. Although epidemiological studies can provide suggestive evidence, they cannot provide 'absolute proof' that exposure to a chemical causes a health endpoint.

Especially in the US, researchers have focused on whether BPA causes mortality from cancer. This emphasis is built into the entire risk assessment framework, because legislators, regulators, and scientists have prioritized cancer over all other potential health effects during the past 50 years.[1] Because of this historical legacy, the methodologies and assumptions that have accreted around testing for cancer are not designed to readily detect other types of health effects, or to accept that dose-effect relationships may vary enormously between different types of disease mechanisms. Even cancer – once viewed as a single disease – is now understood to be a fiendishly complex set of diseases with diverse physiological and genetic pathways.

The traditional testing methods used on animals in toxicology studies measure carcinogenic effects over multiple, sequential stages. Beginning with a biological assay, scientists determine whether a chemical in question causes mutagenic effects on bacteria. If mutations are observed, additional studies are carried out on laboratory rats or mice to identify the 'maximum tolerated dose' (MTD), which is the lowest lethal dose that kills the lab animals. A new group of test animals ingest (or are injected with) the chemical at a dose slightly less than the MTD. After two years (or once the animals have died), scientists count the number of tumors that accumulated in the animals' organs and compare the results to a control group.[2]

In recent years, new science has shown that certain chemicals can disrupt the endocrine system – the cells, glands, and tissues that secrete hormones into the bloodstream. Hormones perform numerous essential functions in the human body, and interfering with these can lead to a variety of health effects including breast and prostate cancer, cardiovascular disease, early puberty, obesity, diabetes, erectile dysfunction, and learning and attention-related disorders (Evanthia *et al.* 2009). More generally, humans can suffer from lower fertility rates. Such health effects are not part of the traditional toxicology testing regime, in part

---

1 In 1958, for example, Congress enacted the Federal Food, Drug and Cosmetics Act, which included the now-repealed 'Delaney Clause'. This provision effectively required FDA to prohibit food additives and pesticide residues that were "found to induce cancer when ingested by man or laboratory animals". The Delaney Clause was repealed in 1996, but not before 40 years of toxiciological testing entrenched carcinogens as a fundamental concern.

2 For example, if the control group has an average of one tumor per animal and the test group has an average of four tumors per animal, the chemical is said to increase human cancer incidence by 300 percent.

because it is much harder to screen for them in the laboratory. Within the laboratory, it can be challenging to observe whether a chemical has disrupted an individual's 'normal' hormonal patterns (Olea *et al.* 2002, p. 49). Studies are required to observe how actual human bodies react to chemicals in the environment. Even so, BPA has been a known endocrine disruptor since 1938 (Vandenberg *et al.* 2007).

Dose-response assessment is the second step in traditional toxicity screening. The goal is to connect different exposure levels (dose) with the likelihood that adverse health effects will occur (response). Regulators have generally used experimental animal testing results to generate a 'monotonic' dose-response curve that demonstrates one of two key indicators. Either the 'No Observed Adverse Effect Level' (NOAEL) or the 'Lowest Observed Adverse Effect Level' (LOAEL) is calculated. This calculation allows regulators such as FDA to extrapolate the results to human populations and thus classify a chemical as carcinogenic or not. A dose is considered safe for humans if it falls below the point of NOAEL or LOAEL. Doses that fall higher on the monotonic curve are more toxic. Based on this level, regulators can calculate the reference dose: an estimate of the daily oral exposure level for the human population that is unlikely to result in an adverse health effect over a lifetime.

In the 1980s, EPA and FDA decided that BPA was non-carcinogenic, using this dose-response framework. They based this decision on an early study in 1977 that found no convincing evidence of carcinogenicity. The study was conducted by a private laboratory, contracted by the National Cancer Institute to study BPA for carcinogenic effects (Vogel 2009). After two years of administering high doses of BPA to male adult rodents, researchers found BPA's general toxicity to be low because BPA metabolized rapidly in the animals. Regulators and scientists therefore assumed that BPA must be safe and did not inquire further into its safety for many years.

Yet, the National Cancer Institute and other early toxicology studies of BPA did not test estrogenic compounds. They failed to study female animals, and they did not think about the possibility that humans might be exposed *continually* to these chemicals. Quick metabolism may not mean 'no risk' if it is occurring repeatedly. This lack of scrutiny began to change as endocrinology science matured. Endocrinologists studying *in utero* exposure to synthetic estrogens exposed pregnant mice to low doses of BPA – much lower than in animal studies used in the 1977 NCI regulatory study – and surprisingly observed adverse health effects (Vogel 2009). For instance, a 1990 study showed that male mice embryo exposed to low doses of BPA are more likely to have an enlarged prostates as adults, compared to mice not exposed to BPA (vom Saal *et al.* 1990).

Studying the effects of chemicals crossing the placenta during pregnancy created a new toxicological paradigm that challenged the traditional assumption of 'the higher the dose, the greater the harm'. The dose-response relationship for endocrine disruptors actually follows a U-shape curve, in which low doses and extremely high doses produce the greatest harm (Welshons *et al.* 2003). Scientists call this curve a 'non-monotonic response'. Thus, for endocrine disruptors like BPA, the traditional dose-response assessment fails to accurately assess risk.

Exposure assessment, the third step, involves estimating the actual levels at which humans are exposed to a chemical. For BPA, exposure assessment has tended to focus on food consumption patterns, the occurrence of BPA in foods, and non-dietary sources of BPA exposure. Given the range of consumer products made with polycarbonates and epoxy resins, humans are typically exposed to many sources. Food is considered the greatest source of BPA exposure for most population groups because of food packaging.

Studies indicate, for instance, that BPA can migrate from polycarbonate packaging into food and beverages. This is because polymerization reactions always leave some monomer unreacted. The unreacted portion of BPA stays solid at room temperature, but over time and under higher temperatures, it can leach out (UK Food Standards Agency 2001). Most government exposure assessments have found that BPA migration from polycarbonate into food is very low under typical room temperature conditions, around 5 micrograms per kilogram (μg/kg) or less of body weight per day (Biles *et al.* 1997, Mountfort *et al.* 1997). However, certain populations experience greater exposure, namely infants at 0-6 months age who are fed from polycarbonate bottles that are often heated above room temperature to warm liquid baby formula. For bottle-fed infants, BPA exposures are much higher than for infants fed using non-polycarbonate bottles. BPA can also potentially migrate from epoxy resin can coatings into food and beverages.

Nonetheless, food packaging is not the only major source of BPA: thermal paper, plastics, and electronics could also be other sources, and they are only beginning to be included in exposure studies. People can be exposed to many synergistically interacting sources, thus creating a cumulative low dose that can cross a critical threshold. Regulators, industry, scientists, and environmental NGOs disagree on whether and how to include this dimension.

Finally, **risk characterization** determines what exposure level is 'safe' for the public. In the face of this new research, regulators have struggled to determine what scientific research should be used to define BPA safety. Between 1997 and 2005, at least 115 studies were conducted by public and private research laboratories in the US, Japan, and Europe, which report a range of adverse effects from various BPA exposure levels (Vogel 2009, p. S562). Many government-funded studies conclude BPA could affect human development, even in small amounts. Industry-funded studies cast doubt on these findings, questioning whether studies that test for estrogenic activity are methodologically sound (raising complaints of lack of reproducibility, poor design, potential confounders unaccounted for, inappropriate manipulation of data, flawed statistical analysis, and so on). These claims hinge on the assumption that traditional toxicology science is the only legitimate way for assessing risk.

To provide apparent clarity on the state of BPA science, the Harvard Center for Risk Assessment published, in 2006, a review study of all published studies on BPA, and concluded that only two studies – both of which were funded by industry – provided 'reliable' data, in large part because many studies diverged from the traditional monotonic dose-response assessment paradigm. However, the Harvard Center has a history of favoring industry perspectives on risk. By contrast, in 2007, the National Institute of Environmental Health Sciences released findings from two government-sponsored review studies of the same scientific literature. A special expert panel told the institute that BPA concentrations in the human body are associated with "changes in the prostate, breasts, testis, mammary glands, body size, brain structure and chemistry, and behavior of laboratory animals" (vom Saal *et al.* 2007, p. 134).

Thus, the conflicting interpretations of the science continued to play out in regulatory circles.

4. Who is responsible for Acting on BPA Risk?

In the US context, there have been ongoing debates about who should take responsibility for reducing any BPA risks. Should it be government with its regulatory authority and public protection role? Or should it be industry with its practical capacity to change products and its interest in profit? Should intervention be left to consumers to decide through the market whether they want to bear what might be a speculative risk? In the 1970s, the answer would have been clear: the federal government had a political stake in defending its population from environmental degradation, and was expanding its regulatory apparatus to enable strong precautionary action. But by the 1980s, when the Reagan Administration's deregulatory agenda was in full flow, the answer would have been: let industry and the market choose. In the 2000s, regulation was more welcome politically, but some government agencies still favored industry interests after decades of lobbying and Congress pressures.

FDA's story illustrates the entrenched institutional cultures and constraints that shape US chemical regulation. Historically, FDA was meant to protect Americans from adulterated and contaminated foods (as well as assuring pharmaceutical safety). But FDA became caught between its dependence on traditional risk assessment methods and its closeness to industry. FDA retained considerable discretion to interpret scientific research to support its regulatory cases. Nonetheless, in the 2000s and 2010s, fissures have opened up inside FDA, resulting in confused positions that reveal much about its organizational thought processes.

Given the inertia of a regulatory system founded on traditional toxicology principles, it was little surprise that in 2008, FDA agreed with the Harvard Center's conclusion that BPA was safe at current exposure levels. While some retailers voluntarily removed products containing BPA from their shelves, FDA cited the lack of validity of low-dose studies that rely on questionable scientific methodologies that may not be reliable enough for regulatory toxicity testing. This finding undercuts attempts by federal legislators to intervene more vigorously. The Consumer Product Safety Modernization Act had just passed in Congress. Inspired by state-level and European Union bans on phthalates, this law empowered the Consumer Product Safety Commission to ban, for use in children's soft toys, a handful of phthalates that were suspected of being significant endocrine disrupters. This law would have included a similar ban on BPA in children's food containers.

From 2009 onwards, FDA began to revise its position on BPA. It is not a coincidence that the Obama Administration entered power in 2009. Like other government agencies, political appointees oversee FDA, altering with each Administration. Thus, politically contentious regulatory decisions may change according to the values of each Administration. The agency requested that an independent science advisory board review BPA findings (yet again) and make recommendations. By 2010, FDA officially began expressing concerns about BPA safety. In 2012, FDA rebuffed a petition from the Natural Resources Defense Council to ban BPA in food packaging and containers as no longer 'generally accepted as safe'. Oddly enough, the agency announced that baby bottles and children's drinking cups should no longer contain BPA. FDA explained that this voluntary call was not based on scientific evidence but on a request by the American Chemistry Council (the chemical industry's main trade association) to implement the ban in order to boost consumer confidence in the midst of regulatory confusion (Tavernise 2012).

The controversies over BPA research continue on. A 2012 review of more than 800 studies on BPA found that even extremely small doses of BPA can be toxic and that low doses of BPA are linked to higher rates of obesity, diabetes, thyroid disease, breast cancer, prostate cancer, and other illnesses (Vandenberg *et al.* 2012). The study authors conclude that "fundamental changes in chemical testing and safety determination are needed to protect human health" (*ibid.*). Yet, in February 2014, FDA scientists concluded that low-level exposure to BPA is safe, and that low-dose studies did not show adverse effects (Delclos *et al.* 2014). However, some scientists not affiliated with the FDA study criticize the FDA's study design for not investigating all relevant health endpoints like changes in brain behavior (Bienkowski 2014). Researchers from another government study organized by the National Institutes of Health charge that FDA did not use the most up-to-date quality control methods in use among university researchers, nor did it incorporate new scientific findings about how chemicals affect human bodies through endocrine disruption (Blake 2014).

Academic researchers and advocacy groups criticize the FDA of being too accommodating to the scientific standards of industry scientists, whose real aim is to institute a standard of proof for risk that is unattainable – a tactic that the tobacco industry used to fight regulation. Proponents of BPA's safety respond that academic scientists relying on large government grants have a vested interest in keeping BPA on the publicly-funded research agenda, and that their research is overly influenced by personal goals of doing 'advocacy research' that supports activists' claims (Miller 2014). In these highly adversarial circumstances, scientific evidence will remain contested, and it seems unlikely that a decisive ban on BPA is forthcoming unless government strongly favors a precautionary stance.

In the absence of federal regulation, and with a patchwork of a few states restricting BPA through bans or labeling laws, what can be done? Some companies have attempted to address consumers' concerns about BPA through product innovation.

5. Substituting BPA for Alternatives in Well-Established Production Chains

In response to consumer concerns, NGO campaigns, and imminent regulation in the late 2000s, manufacturers quickly introduced new product lines, marketing them as containing BPA-free materials. As scientific evidence pointed to BPA's endocrine-disrupting effects, consumer safety, environmental health, and disease-advocacy NGOs built campaigns targeted against retailers and brand name products. These campaigns provided information to consumers about BPA risks and urged them to boycott products known to feature BPA. For example, the Breast Cancer Fund advised against buying plastic-coated toys and cooking utensils altogether, with 'BPA-free' plastics permissible if necessary. Some consumers began using stainless steel bottles and glass bottles as alternatives to plastic bottles. The SIGG scandal in 2009 further intensified the debate over whether BPA should even be allowed to be present in bottles

(Baker 2009). SIGG, a stainless steel bottle manufacturer, had claimed that its products were BPA-free, only for independent laboratory testing to discover that the bottle lining contained trace amounts of BPA. A widespread consumer backlash against SIGG was the result.

In 2008, the Canadian government imposed a ban on certain uses of BPA in consumer products. Even before the law came into effect, many consumer product manufacturers and retailers removed BPA products from their offerings in Canada. Because of a common industry chain supplying both countries, Wal-Mart Canada, CVS, Toys'R'Us, Playtex, Nalgene, Whole Foods, and other companies in the US and Canada voluntarily stopped selling baby bottles and water bottles made with BPA. Nalgene Outdoors Products, based in Rochester, exemplifies this response to market pressures and regulatory risks. Steven Silverman, the firm's manager, said: "Based on all available scientific evidence, we continue to believe that Nalgene products containing BPA are safe for their intended use. However, our customers indicated they preferred BPA-free alternatives, and we acted in response to those concerns" (Austen 2008, p. C1). To stay in the bottle business, Nalgene introduced a line of bottles made from Eastman Chemical Company's Tritan copolyester.

As this action suggests, product manufacturers began investigating BPA substitutes in response to retailers pulling BPA products from their shelves. A number of companies have turned to existing functional equivalents of BPA to develop their own BPA substitutes. Bisphenol S (BPS) is a popular BPA substitute, which boasts similar product attributes as BPA but with less risk of leaching from with heat or sunlight.[3] BPS was first made in 1869 as a dye, but did not find commercial application until 2006 as a substitute for BPA in paper products such as cash-register receipts, airplane luggage tags, and boarding passes. BPS can now be found in products made from recycled paper, like pizza boxes and food buckets.

BPA-substitutes also come from outside of the bisphenol family. In 2002, the Eastman Chemical Company began working on a new heat-resistant plastic called Tritan, which it released in 2007. Many product

[3] A BPA molecule consists of two phenol groups connected by a branched three-carbon group, whereas a BPS molecule has two phenol groups connected by a sulphone group ($-SO_2$).

manufacturers quickly began using Tritan instead of BPA-containing polycarbonate to make infant products, water bottles, and food containers. To buttress its claims to safety, Eastman released third-party test results showing that Tritan monomers do not bind to oestrogen or androgen receptors, and is therefore free of estrogenic activity (Eastman Chemical 2010).

Yet some recent scientific research indicates that Tritan, BPS, and other BPA substitutes *do* induce estrogenic activity and therefore pose endocrine disruption risks similar to BPA. In 2011, researchers at CertiChem, a chemical screening firm, reported that 92% of 102 commercially available plastic products (*e.g.* 'BPA-free' plastic cups marketed for children and purchased from Target, Walmart, and Babies R Us) leached chemicals with estrogenic activity (Yang *et al.* 2011). Scientists at the University of Texas Medical Branch in Galveston found that BPS, like BPA, disrupts the endocrine system at extremely low doses, noting the similar size and structure of both chemicals, which both have the potential to bind to natural oestrogen receptors inside cells (Viñas *et al.* 2013). These studies are troublesome in light of biomonitoring studies that show that even though BPS has only been in use for less than a decade, it is already pervasive in human populations: 81% of 315 urine samples from men and women in the United States and seven Asian countries contained BPS (Liao *et al.* 2012).

In 2014, EPA's 'Design for the Environment' program, which seeks to identify safer alternative chemicals, released a report assessing 19 chemical alternatives to BPA used in receipt paper, including BPS (US EPA 2014). The report found that BPS poses similar risks to public health as BPA, concluding that all of the BPA alternatives are associated with some trade-offs. Because the major US chemicals policy, the Toxic Substances Control Act (TSCA), has allowed new chemicals to enter the market without being tested for safety, BPA substitutes have been offered to consumers with no or little public research available on the real or potential risks of such substances.

6. Analyzing the Ethical Situations of Moral Actors Regarding BPA

As BPA has emerged as a chemical of concern over the past 15 years, many actors in the chemical industry, government, and civil society in the US have wrestled with the difficult dilemmas that using possibly harmful chemicals in products can pose. In deciding what to do, these actors have most often taken three major ethical stances.

First, actors can employ *deontological reasoning*. Deontological theories determine moral action according to intention, or whether an action is done for the right reasons. The word 'deontology' derives from the Greek words for duty (*deon*) and science (*logos*). Deontologists assert that people have a duty to do the right thing no matter the consequences. Yet, deontologists differ in how they define what constitutes 'good' or 'right' action. For example, the German philosopher Immanuel Kant (1724-1804) asserted that because what is virtuous is not always identifiable, we must instead do 'right' (Hinton 2002). For Kant, to determine what 'right action' is, one must consider whether the action would be desirable as a 'universal law'. For example, the precautionary principle is sometimes invoked as a deontological base for environmental laws (Kysar 2010). According to deontologists, societies should prioritize human and environmental welfare ahead of economic profit as a universal norm of justice. Protecting human well-being is a fundamental societal value. If there is a possibility of serious, irreparable injury to human health occurring, even if scientific proof remains uncertain, then societies should intervene to prevent that harm.

Second, actors can rely on *consequentialist ethics* to guide their choices. This ethics holds that the consequences of an action will determine whether it is morally permissible or not. Moral action, then, is that which produces 'right consequences'. Utilitarianism, the most influential form of consequentialism, applies the principle of utility to determine 'right consequences'. The morally right action is the one that produces the most utility ('the greatest good for the greatest number'). A strict utilitarian gives equal weight to each person's well-being in assessing utility, ignoring whether an action has a negative impact on specific subgroups of society. However, John Stuart Mill (1806-1873) proposed balancing utilitarianism with a set of fundamental rights that defend individual liberties, namely the right to protection from harm, free speech, free association, and self-determination (Mill 1859). Mill's merger of utilitarian moral theory with liberal political philosophy suggests that a moral society is comprised of individuals who are free to pursue their personal goals and interests, unless such pursuits harm another person.

Over the last century, utilitarian calculus has come to dominate decision-making in government and business organizations. The practice of cost-benefit analysis (CBA) operationalizes Mill's liberal utilitarianism by measuring all of the costs and benefits of a policy in economic terms, applying weights to different costs and benefits, and comparing the totals, with the overall objective being to maximize the ratio of total costs to total benefits. This utilitarian calculus requires that the calculator adopt an impartial 'view from nowhere' in assessing all the costs and benefits for a range of possible actions. Thus a chemical industry consultant suggests: "The really prudent step is to make the best scientific evaluations of the risk from the product as compared to the risks and loss of benefits associated with removing it from the market before any actions are taken" (Entine 2011, pp. 19-20).

Finally, actors can choose to ostensibly *sidestep ethical questions* and instead engage in technical arguments over how to interpret the scientific evidence. Companies, regulators, and legislators frequently portray science as an arbiter of objective, factual truth. By doing so, they remove social and political values from debates over chemical risks, and emphasize issues of methodological rigor, data quality, and expert credibility as the fundamental ones (Kinchy 2012). If other actors attempt to raise concerns about the underlying priorities of chemical manufacturing, or about the observation that chemical exposures are widespread, they are attacked as lacking in scientific rigor and biased. Actors who take a 'scientized' perspective fail to recognize that they, too, are often cloaking social and political values under a veneer of objectivity (*ibid.*).

We will now consider several key actor groups and summarize their potential reasoning. Our approach is to consider the practices and statements of these actors as they respond to the BPA situation.

6.1 Business managers and corporations

Business executives must ponder whether their company should eliminate BPA or continue using it. At this time, they have access to incomplete, uncertain information about BPA risks, so they must necessarily be speculative. Because of their organizational culture, managers are likely to use utilitarian reasoning to evaluate the costs and benefits to their company in taking a particular course of action. This analysis can vary between managers as well as different types of companies.

A firm manufacturing BPA stands to lose billions of dollars worth of sales over a few decades, since the substance has one of the highest production volumes worldwide. The firm could switch to producing substitutes for BPA to replace this lost market. But finding suitable alternatives and re-engineering product design will require millions of dollars in R&D. Moreover, once alternative ingredients are found, new testing must be done to ensure those substances are safe, and new supply chain partners must be cultivated and vetted. Whereas BPA is already grandfathered under the Toxic Substances Control Act, new chemicals may face more searching scrutiny from EPA. The company could undertake all of these activities only to lose its market-share to competitors who are not taking similar action. To oppose any change, managers could point to the existence of weak scientific evidence that shows low-level exposure to their product causes future adverse health effects. From their perspective, hypothetical harm is not enough to warrant real financial risk. Exiting the BPA market could violate a company's fiduciary duty to its shareholders, to whom they are obligated to make sound financial decisions. From this view, it would be morally permissible for a company to wait for clearer scientific evidence, new regulations, or sizeable consumer demand for safer alternatives.

In contrast, a retailer or a consumer product manufacturer might face a more complicated calculus. On one hand, 'downstream' firms can decide to trust that the chemical industry has public health interests at heart. There is no need to worry about possible BPA risks, since the chemical has already been on the market for decades, and any latent health consequences would have appeared by now if they were going to. Moreover, BPA has been approved by US government regulators for use in products. Many firms appear to require stronger government positions on BPA before they are willing to intervene. Such a stance effectively sidesteps the company's ethical responsibility, if any, by assigning it to scientists and governments. For example, the Coca Cola Company's position on BPA maintains the *status quo* in the absence of regulatory action:

> While we are very aware of the highly publicized concerns and viewpoints that have been expressed about BPA, our point of view is that the scientific consensus on this issue is most accurately reflected in the opinions expressed by those regulatory agencies whose missions and responsibilities are to protect the public's health. The consensus repeatedly stated among regulatory agencies in Australia, Canada, Europe, Germany, Japan, New Zealand and the United States is that current levels of exposure to BPA through food and beverage packaging do not pose a health risk to the general population. [Coca-Cola 2012, p. 38]

The downstream company could also decide to apply utilitarian reasoning in the same way that a BPA manufacturer might. There would be costs from changing a product to be BPA-free, such as sourcing new raw materials, reformulating designs, and adjusting manufacturing equipment to use BPA substitutes. Manufacturers might encounter retailers which are reluctant to change their product lines. In evaluating the merits of eliminating BPA, firms could skeptically ask: how many human lives will be affected by health problems? Are adults dying from cancer, or are children suffering from development problems such as undescended testes? Or are children afflicted by 'only' small cognitive impairments like memory fragility? Companies may argue that too few people will be seriously injured to warrant the considerable costs of transition to safer alternatives. They may treat their ethical conundrum as a business decision.

Yet downstream firms might benefit from maintaining or even expanding their markets as consumers demand safer products. They could acquire a reputation for protecting consumer well-being that translates into higher share prices, better employee morale, and government goodwill in setting standards. The firms may avoid legal and regulatory liability for continuing to use BPA. Even if courts are not yet ruling in lawsuits that BPA products harm consumers, there is arguably enough scientific evidence of BPA risks now to find that companies should have known about these risks and were negligent in failing to act to prevent them.

Nonetheless, business managers and companies do not have to adopt utilitarian reasoning. They may use deontological reasoning to declare that the risks of BPA call for precautionary action as soon as a threshold of sufficient scientific evidence of potential harm is passed. That is, they do not have to await a decisive ruling or regulatory order before acting voluntarily to invest in innovation or to phase out BPA. Still, there can be wide differences of opinion as to what the appropriate threshold of scientific evidence should be. It could vary from 'probable risk' to 'reasonably likely risk' and to 'plausible but not likely risk'.

6.2 Chemists and designers

Most chemists and designers who work with BPA and other endocrine disrupting chemicals in consumer products do so within an industry or corporate context. Therefore they are subject to similar pressures and considerations that managers have to face. They may adopt utilitarian reasoning along the same lines as discussed just above. They may decide that they ought to help maintain their employer's business, ahead of any ethical issue. Indeed, they may think that they could lose their jobs if their employer does not thrive, or frowns on their 'whistleblowing' as regards chemical risks. They can put their personal welfare ahead of public welfare.

Nonetheless, chemists might believe that they have a duty to protect vulnerable populations such as children and pregnant women when designing molecules. Like business managers, they may have their own families and children to worry about, and they do have the moral agency to make fundamental design choices. Here, chemists can decide whether they have a deontological obligation to implement green chemistry principles, which include designing out toxicity from chemicals where practicable, designing safer products, maximizing atom efficiency, using renewable feedstock, designing chemicals to degrade readily (Anastas & Warner 1998). However, the vast majority of chemists are not trained in environmental health, green chemistry, or even toxicology, so they can struggle to carry out green chemistry principles. They may not even know that the green chemistry field exists. Thus their moral agency is inhibited by their professional and organizational conditions.

In practice, a number of product designers likely have raised significant concerns about BPA from within chemical and product manufacturers. They can point to emerging scientific evidence of estrogenic activity as a reason to intervene. However their ability to do so may rests on the availability of chemical substitutes that are reliably safer. In this regard, product managers can support the development of improved information tools like the EPA's Safer Chemical's list and independent or non-profit organizations that offer testing services to identify which materials leach chemicals with estrogenic activity.

6.3 Regulators and legislators

In the US, chemicals regulation is based on a utilitarian framework that requires government officials, companies, and other actors to focus on the question of whether material convenience outweighs physical harm. The Toxic Substances Control Act has hamstrung officials at EPA when it comes to trying to regulate harmful chemicals. On one hand, regulators were obliged to prove that a chemical posed substantial risk to human health, the benefits of a regulatory action outweighed the burden to industry and society, and the action was the most reasonable possible. In practice, regulators had to compile massive amounts of evidence, using cost benefit analysis and risk assessments, to justify any ban. This situation resulted, in part, from an US Court of Appeal ruling in 1989 that EPA did not provide enough evidence to warrant an asbestos ban. On the other hand, regulators lacked power to require companies to carry out even basic toxicity screening of chemicals, so they were usually unable to prove that a risk existed. Thus otherwise sympathetic officials were unable to escape their regulatory confines. By contrast, since the 1990s, EPA officials have become more willing to enforce the new substances review program, in which new chemicals are evaluated more closely for their health effects. But the number of existing chemicals far exceeds new chemicals: some 62,000 were already permitted under TSCA in 1976.

In recent years, EPA's leadership has ordered the agency to tackle chemicals more aggressively, a call grounded in the precautionary principle. For example, in a 2009 speech at the Commonwealth Club in San Francisco, the head of the EPA, Lisa Jackson, declared: "We need to review all chemicals against safety standards that are based solely on considerations of risk – not economics or other factors – and we must set these standards at levels that are protective of human health and the environment" (Jackson 2009). Even so, existing chemicals – of which BPA is one – readily evade review. In response, regulators could decide to create voluntary programs to coax industry into doing more toxicity testing of chemicals and possibly phasing out especially harmful ones. Whether or not regulators actually engage in this work is an ethical choice. EPA officials may rely on utilitarian reasoning: is trying to build a BPA program likely to be politically or economically costly, compared to the health benefits? Will there be strong push-back from powerful companies? Or, the officials could invoke deontological reasoning and prioritize the voices of those people who are most affected by exposure to BPA, like children and mothers.

In turn, legislators are elected in a political environment in which donations and lobbying from industry interests have been rampant since the 1980s. This is particularly true at the federal government level where the chemical industry has pervasive influence. Members of Congress thus face ethical quandaries: should they reject donations from the chemical industry and jeopardize their chances of re-election? Or should they heed calls from parents who are becoming more worried about chemical risks? Should more ecologically conscious Senators compromise their values and reach an agreement with industry-friendly Senators in order to reform the Toxic Substances Control Act? Or should they hold out for more stringent rules that are less likely to become law?

In many cases, legislators (and regulators) simply decide to follow the *status quo*. They are unwilling to abandon their dependence on campaign finance. They can follow the lead of many companies in ignoring ethical questions altogether and emphasizing the need for more science. This behavior is seen in the 2008 Congressional hearing on BPA, in which most industry and government witnesses avoided debating the ethical issues at stake. Instead, they sparred over whether or not scientific evidence existed to justify regulatory action. As Marian Stanley of the American Chemistry Council testified during the 2008 Congressional hearing: "In the past 2 years comprehensive scientific assessments from the European Union, the U.S. National Toxicology Program, Health Canada, NSF International, and the European Food Safety Authority have all been undertaken, and these assessments support the continued safe use of consumer products containing BPA" (US Congress 2010, p. 81).

In other words, few participants at the hearing (apart from Dr. Ted Schettler, a NGO scientist) brought up ethical questions such as: why are humans being exposed to harmful chemicals at all? Who is responsible for putting us into this predicament? Why not make safe products to begin with? What are the views of those people who are harmed the most? They remained trapped within a risk assessment paradigm that favors seemingly objective and neutral analysis.

Yet, legislators can decide to take moral leadership by introducing new laws that force industry to remove BPA and other substances from products. In the US, this has tended to happen more often at the local and state government levels, which can be less subject to industry influences. As of 2015, at least 12 states from California to Massachusetts had passed some sort of restriction on BPA use (National Conference of State Legislatures 2015). Most of these laws banned the manufacture and sale of drinking bottles that contain BPA, if they were intended for use by young children. One law, in Minnesota, banned all uses of BPA in food containers for use by children. Moral leadership can also happen at the federal government level. In 2015, lawmakers introduced the BPA in Food Packaging Right to Know Act, which would make it illegal to sell food in containers with BPA unless the container is labeled with the statement: "This food packaging contains BPA, an endocrine-disrupting chemical, according to the National Institutes of Health."[4] This law, however, simply creates information aimed at consumers, who are obliged to decide whether they are willing to accept the risk of buying a

4 The bill has been introduced to committee but no hearings have been conducted. See: https://www.congress.gov/bill/114th-congress/senate-bill/821.

potentially harmful product. It arguably shifts moral agency from firms to consumers.

7. Conclusions

By studying the BPA case, chemistry students can learn several critical lessons for use in their future careers. The full health and ecological impacts of chemicals may only appear long after widespread use in products begins. BPA use in products has become ubiquitous and many millions of people are exposed to BPA. Although BPA's estrogenic properties have been known since the early 1900s, whether its estrogenicity is harmful remains hotly debated. Faced with scientific controversy, government and company decision-makers may struggle to recognize and act on the emerging risks. Failure to address the potential risks of BPA has implications for the numerous chemicals used in consumer products that are suspected of being endocrine disruptors.

Companies are not inherently unethical in having helped disperse harmful chemicals into society if they could not foresee the health consequences. However faced with new knowledge about product risk, companies may act unethically if they persist in making or using these chemicals after evidence emerges signaling cause for concern (European Environmental Agency 2001). With BPA, sufficient warning signals arguably exist now to justify the trade-off between ethical action and financial costs.

As we have seen, difficult ethical dilemmas exist when deciding whether sufficient scientific evidence exists to warrant actions such as reformulating products, removing chemicals, or regulating substances. Should precautionary action be taken, even though this might jeopardize a firm's markets and profits? Or should the *status quo* be retained even though it might cause harm to some people? Deontological and consequential ethics theories are often applied to resolve these debates. The diverse values that people hold can color their thinking; calculations can vary widely. For example, Nalgene used utilitarian analysis to decide to remove BPA from its bottles, reasoning that the benefits of doing so would outweigh the costs. Yet, other firms did not withdraw BPA, arguing that the economic costs of changing to an alternative substance would be too large. And the American Chemistry Council, joined by many companies, argued that more science was required to prove that a BPA risk existed, or that regulators had certified BPA as safe, implying that ethical questions were irrelevant.

Numerous actors involved in chemicals production and use have moral agency because of their ability to influence chemical product design. They can be held ethically accountable for their actions and the consequences of those actions. They can be inside firms, in government departments, and scattered across civil society. Chemists are not the only actors who can make ethical choices that matter, and they can find allies in many places inside and outside industry. These allies can be sympathetic managers, other firms, regulators, and environmental NGOs who want to make chemicals safer. Civil society involvement can also increase scrutiny of the choices of corporations and chemists regarding toxic chemicals. The work of activists such as the Breast Cancer Fund and Clean Production Action can help improve decision-making through critically appraising why companies are using BPA in products.

The BPA case is only one example of a growing number of chemical risks calling for an industry response. Other substances of concern include phthalates, flame retardants like polybrominated diphenyl ethers (PBDEs), and perfluorinated compounds such as perfluorooctanoic acid (PFOA) and perfluorooctanesulfonic acid (PFOS). These chemicals raise similar concerns through their ability to damage human development and reproduction; they can also cause certain kinds of cancers. Phthalates are commonly found in soft toys and cosmetics; flame retardants are everywhere from electronics to furniture; PFOA is used in teflon-coated pots and Goretex clothes. As we saw, in trying to substitute apparently safer chemicals for known or suspected toxins, firms can choose substances that are little better due to lax regulatory screening.

As this case makes clear, the way forward is difficult. Removing toxic chemicals from consumer goods is a vast political, economic, and design challenge that poses many complex ethical dilemmas for the diverse stakeholders involved. Better understanding of the ethical dimensions of creating safer consumer goods can help lay the foundation for opening up better communication between regulators and legislators, company managers, chemists and product designers, consumers, and the public. By helping all actors develop deeper analysis of the ethical values at stake and the underlying moral commitments that they share with one another, they can create a safer, more sustainable chemical industry, even with imperfect science. You can play an important part in this process through whatever you end up doing.

References

Anastas, P.T. & Warner, J.C.: 1998, *Green Chemistry: Theory and Practice*, New York: Oxford University Press.

Austen, I.: 2008, 'Bottlemaker to Stop Using Plastics Linked to Health Concerns,' *New York Times*, April 18, C1.

Baker, N.: 2009, 'Why I'll Swig From My Sigg Bottle No More', *Huffington Post*, Sept 26, available at: http://www.huffingtonpost.com/nena-baker/why-ill-swig-from-my-sigg_b_269603.html, accessed 21 Oct. 2016.

Barraza, L.: 2013, 'A new approach for regulating bisphenol A for the protection of the public's health', *The Journal of Law, Medicine & Ethics*, **41** (1), 9-12. [not cited in text]

Bienkowski, B.: 2014, 'New BPA Experiment Finds No Low-Dose Effects', *Environmental Health News*, February 13. Available at: http://www.environmentalhealthnews.org/ehs/news/2014/feb/bpa-low-doses, accessed 21 Oct. 2016.

Biles, J.A.; T.P. McNeal, T.H. Begley & Hollifield, H.C.: 1997, 'Determination of Bisphenol A in reusable polycarbonate food-contact plastics and migration to food simulating liquids', *Journal of Agricultural and Food Chemistry*, **45**, 3541-3544.

Blake, M.: 2014, 'Scientists Condemn New FDA Study Saying BPA Is Safe: "It Borders on Scientific Misconduct"', *Mother Jones*, March 24, available at: http://www.motherjones.com/environment/2014/03/scientists-slam-fda-study-bpa, accessed 20 April 2016.

Calafat A.M.; Ye X., Wong L.Y.; Reidy J.A. & Needham L.L.: 2008, 'Exposure of the US population to bisphenol A and 4-tertiary-octylphenol', *Environmental Health Perspectives*, **116** (1), 39-44.

Case, D.: 2009, 'The Real Story Behind Bisphenol A', available at: http://www.fastcompany.com/1139298/realstory-behind-bisphenol, accessed 21 Oct. 2016.

Centers for Disease Control: 2009, *Fourth National Report on Human Exposure to Environmental Chemicals*, available at: http://www.cdc.gov/exposurereport/, accessed 21 Oct. 2016. .

Coca-Cola Company: 2012, 'Sustainability Report 2011/2012', available at: http://www.coca-colacompany.com/sustainabilityreport/me/product-safety-and-quality.html#section-cocacola-and-alcohol-ethanol, accessed 21 Oct. 2016.

Delclos, B.K.; Camacho, L.; Lewis, S.M. & Vanlandingham, M.M.: 2014, 'Toxicity evaluation of bisphenol A administered by gavage to Sprague-Dawley rats from gestation day 6 through postnatal day 90', *Toxicological Science*, **139** (1), 174-197.

Eastman Chemical: 2010, 'New Third-Party Test Results Confirm Eastman Tritan Copolyester is Free of Bisphenol A and Estrogenic Activity', *ChemInfo.com,* May 13, available at: http://www.chem.info/news/2010/05/new-third-party-test-results-confirm-eastman-tritan-copolyester-free-bisphenol-and, accessed 21 Oct. 2016.

Entine, J.: 2011, 'Scared to Death: How Chemophobia Threatens Public Health', available at http://www.pavementcouncil.org/pavementcouncil/americancouncil.pdf, accessed 21 Oct. 2016.

Evanthia, D-K:; Bourguignon, J-P.; Giudice, L.C.; Hauser, R.; Prins, G.S.; Soto, A.M.; Zoeller, R.T. & Gore, A.C.: 2009, 'Endocrine-Disrupting Chemicals: An Endocrine Society Scientific Statement', *Endocrine Re*views, **30** (4), 293-342.

European Environmental Agency: 2001, *Late lessons from early warnings: the precautionary principle 1896-2000*, available at: http://www.eea.europa.eu/publications/environmental_issue_report_2001_22, accessed 20 April 2016.

Fiege, H.; Voges, H-W.; Hamamoto T. *et al.*: 2012, 'Phenol derivatives', in: K. Othmer (ed.), *Ullmann's Encyclopedia of Industrial Chemistry*, Weinheim: Wiley-VCH, pp 521-586.

Grand View Research: 2014, 'Global Bisphenol A (BPA) Market Demand Is Expected To Grow At CAGR of 4.7% From 2014 To 2020', available at: https://globenewswire.com/news-release/2014/11/12/682387/10107790/en/Global-Bisphenol-A-BPA-Market-Demand-Is-Expected-To-Grow-At-CAGR-of-4-7-From-2014-To-2020-New-Report-By-Grand-View-Research-Inc.html, accessed 21 Oct. 2016.

Hinton, T.: 2002, 'Kant and Aquinas on the Priority of the Good', *The Review of Metaphysics,* **55**, 825-846.

Kinchy, A.: 2012, Seeds, Science, and Struggle: The global politics of transgenic crops, Cambridge, MA: MIT Press.

Kysar, D.: 2010, Regulating from Nowhere: Environmental law and the search for objectivity, New Haven: Yale University Press.

Jackson, L.: 2009, 'Remarks to the Commonwealth Club of San Francisco', September 29, available at: https://yosemite.epa.gov/opa/admpress.nsf/8d49f7ad4bbcf4ef85 2573590040b7f6/fc4e2a8c05343b3285257640007081c5!OpenDocument, accessed 21 Oct. 2016.

Jacobsen, H.: 2015. 'EU's food safety agency gives green light to Bisphenol A', available at: http://www.euractiv.com/section/agriculture-food/news/eu-s-food-safety-agency-gives-green-light-to-bisphenol-a/, accessed 21 Oct. 2016.

Liao, C.; Liu, F.; Alomirah, H.; Loi, V.D.; Mohd, M.A.; Moon, H.B.; Nakata, H. & Kannan, K.: 2012, 'Bisphenol S in urine from the United States and seven Asian countries: occurrence and human exposures', *Environmental Science and Technology*, **46** (12), 6860-6866.

Mill, J.S.: 1859, *On Liberty*, London: Parker.

Miller, H.: 2014, 'BPA Is A-OK, Says FDA', *Forbes,* March 12, available at: http://www.forbes.com/sites/henrymiller/2014/03/12/fda-research-confirms-bpa-is-a-ok/#6f08bee2714d, accessed 21 Oct. 2016.

Mountfort, K.A.; Kelly, J.; Jickells, S.M. & Castle, L.: 1997, 'Investigations into the potential degradation of polycarbonate baby bottles during sterilization with consequent release of bisphenol A', *Food Additives and Contaminants*, **14**, 737-40.

National Conference of State Legislatures: 2015, 'NCSL Policy update: State restrictions on Bisphenol A (BPA) in consumer products', available at: http://www.ncsl.org/research/environment-and-natural-resources/policy-update-on-state-restrictions-on-bisphenol-a.aspx, accessed 21 Oct. 2016.

Olea, N.; Fernandez, M.F. & Olea, M.F.: 2002, 'Human Exposure to Endocrine Disrupters, an Overview', in: L. Chyczewski, J. Niklinski & E Pluygers (eds.), *Endocrine Disrupters and Carcinogenic Risk Assessment*, Amsterdam: IOS Press.

Ritter, S.K.: 2011, 'BPA Is Indispensable For Making Plastics', *Chemical and Engineering News*, **89** (23), available at: https://pubs.acs.org/cen/coverstory/89/8923cover4.html, accessed 21 Oct. 2016.

Tavernise, S.: 2012, 'F.D.A. Makes It Official: BPA Can't Be Used in Baby Bottles and Cups', *New York Times*, July 17, A15.

Turker, M.S.: 2012, 'Banning Bisphenol A in the United States and Canada: Epigenetic Science, the Precautionary Principle, and a Missed Opportunity to Protect the Fetus', *Journal of Health & Biomedical Law*, **8**, 173.

UK Food Standards Agency: 2001, 'Survey of Bisphenols in Canned Foods', March 2001. Available at: http://www.food.gov.uk/science/surveillance/fsis-2001/bisphenols, accessed 21 Oct. 2016.

US EPA: 2014, 'Bisphenol A Alternatives in Thermal Paper', available at: https://www.epa.gov/sites/production/files/2014-05/documents/bpa_final.pdf, accessed 21 Oct. 2016.

US Congress, 2008: 'Safety Of Phthalates And Bisphenol-A In Everyday Consumer Products', *Hearing Before The Subcommittee On Commerce, Trade, And Consumer Protection*, June 10, available at: http://www.gpo.gov/fdsys/pkg/CHRG-110hhrg56091/html/CHRG-110hhrg56091.htm, accessed 21 Oct. 2016.

Vandenberg L.N.; Hauser, R.; Marcus, M.; Olea, N. & Welshons, W.V.: 2007, 'Human exposure to bisphenol A (BPA)', *Reproductive Toxicology*, **24** (2), 139-77.

Vandenberg, L.N.; Colborn, T.; Hayes, T.B.; Heindel, J.J.; Jacobs Jr., D.R.; Lee, D.-H.; Shioda, T.; Soto, A.M.; vom Saal, F.S.; Welshons, W.V.; Zoeller, R.T. & Myers, J.P.: 2012, 'Hormones and Endocrine-Disrupting Chemicals: Low-Dose Effects and Nonmonotonic Dose Responses', *Endocrine Reviews*, **33** (3), 378-455.

Viñas, R. & Watson, C.S.: 2013, 'Bisphenol S disrupts estradiol-induced nongenomic signaling in a rat pituitary cell line: effects on cell function', *Environmental Health Perspectives,* **121** (3), 352-358.

Vogel, S.A.: 2009, 'The Politics of Plastics: The Making and Unmaking of Bisphenol A "Safety"', *American Journal of Public Health*, **99** (S3), S559-S566.

Vogel, S.A.: 2013, Is it Safe?: BPA and the Struggle to Define the Safety of Chemicals, Berkeley: University of California Press.

vom Saal, F.; Quadagno, D.; Even, M.; Keisler, L.; Keisler, D. & Khan, S.: 1990, 'Paradoxical Effects of Maternal Stress on Fetal Steroids and Postnatal Reproductive Traits in Female Mice from Different Intrauterine Positions', *Biology of Reproduction,* **43**, 751-761.

vom Saal, F.S.; Akingbemi, B.T.; Belcher, S.M. & Birnbaum, L.S.: 2007: 'Chapel Hill bisphenol A expert panel consensus statement: integration of mechanisms, effects in animals and potential to impact human health at current levels of exposure', *Reproductive Toxicology,* **24** (2), 131-138.

Welshons, W.V.; Thayer, K.A.; Judy, B.M.; Taylor, J.A.; Curran, E.M. & vom Saal, F.S.: 2003, 'Large effects from small exposures. I. Mechanisms for endocrine disrupting chemicals with estrogenic activity', *Environmental Health Perspectives*, **111**, 994-1006.

Yang, C.Z.; Yaniger, S.I.; Jordan, V.C.; Klein, D.J. & Bittner, G.D.: 2011, 'Most Plastic Products Release Estrogenic Chemicals: A Potential Health Problem That Can Be Solved', *Environmental Health Perspectives*, **119** (7), 989-996.

Chapter 11

Undoing Chemical Industry Lock-ins: Polyvinyl Chloride and Green Chemistry

Alastair Iles, Abigail Martin, and Christine Meisner Rosen

Abstract: We examine polyvinyl chloride (PVC) as one example of the ethical challenges that the chemical industry faces when putting green chemistry into practice. Green chemistry has emerged as a powerful new philosophy for designing molecules, reactions, and products to be intrinsically non-toxic and sustainable. We consider three issues: Should the chemical industry overcome the inertia of path dependent technologies and introduce safer, more sustainable technologies? What will motivate companies and their employees to practice green chemistry under conditions where changing technologies and businesses can create substantial economic, market, and technical risks? How should the precautionary principle be applied in terms of the real-world complexities of manufacturing chemicals? To do so, we look at examples of environmental and health harms in the feedstock and PVC manufacturing lifecycle stages, along with green chemistry solutions that could be employed. PVC suggests how difficult it could be to adopt green chemistry solutions; nonetheless, these solutions may make significant contributions across the chemical industry generally.

1. Introduction

Polyvinyl chloride (PVC) is one of the oldest and most ubiquitous plastics in the world, dating back to the 1920s. In terms of global production volumes, PVC is second only to polyethylene. In 2015, manufacturers produced 43.6 million tons, worth $US 57 billion (Zion Research 2016). Over 50 percent goes to make infrastructure materials like water pipes, wire coverings and window frames used in buildings and automobiles, with the rest used to create durable consumer products such as toys, credit cards, and vinyl curtains. PVC has replaced many traditional materials like textiles and wood because it offers longevity and strength. While PVC presents an 'old' chemistry issue, in that its environmental harms have been fought over for well over 40 years, it still presents widespread risks to health and ecosystems through its myriad contemporary uses.

In the past 20 years, green chemistry has emerged as a powerful new philosophy for designing molecules, reactions, and products to be intrinsically non-toxic and sustainable. Many green chemists such as Terry Collins and John Warner envisage a world where chemists and engineers – along with company managers and government regulators – take the ethical lead in diminishing the exposure of human societies to harmful chemicals. These scientists argue that green chemistry is founded on the precautionary principle. Put briefly, this ethical principle holds that companies, engineers, and chemists should act to prevent chemical risks even if these are scientifically uncertain, and even if regulation does not require action. As a result, companies can benefit in many ways: they no longer need end-of-pipe technologies to control pollution from chemical processes, and can avoid exposing people to toxic risks through consumer products. Nonetheless, companies and scientists can reject the precautionary principle, or construe it in divergent ways.

The chemicals used to make PVC – from chlorine to mercury and from phthalates to vinyl chloride monomers – are known to cause cancers, neurological disorders, reproductive and development problems, and other deleterious health effects. Some effects are well-established while others are debatable. Some chemists say that if PVC had been developed more recently than the 1930s, it would never have been commercialized. Still, we live in a world with PVC, whose impacts must be dealt with somehow. Green chemists are creating solutions to tackle PVC impacts (or even to replace PVC altogether) – yet they have run into a fundamental barrier: the pervasive technological and economic lock-ins of PVC. As PVC has matured, various production processes and consumption patterns have become intertwined with each other, to the point where they are hard to unwind.

The PVC case thus raises three major ethical issues. Should the chemical industry overcome the inertia of path dependent technologies and introduce safer, more sustainable technologies? What will motivate companies and their employees to practice green chemistry under conditions where changing technologies and businesses can create substantial economic, market, and technical risks? How should the precautionary principle be applied in terms of the real-world complexities of manufacturing chemicals?

To address these issues, we examine how green chemistry can help generate new path-shaping opportunities throughout the PVC lifecycle. We briefly review the historical evolution of the PVC production chain to show where and how the many lock-ins that characterize this chemical have materialized over time. We then review two key stages in the PVC lifecycle – feedstock production and PVC manufacturing for end-uses – to illustrate some of PVC's environmental and health impacts, along with green chemistry solutions that could be adopted. We conclude with analysis of the considerations that key groups of companies may face when choosing whether to do green chemistry research and development. As you read through the PVC background, you can reflect on the ethical considerations that should come into play when deciding whether and how to change a deeply entrenched chemical chain.

2. The Precautionary Trajectory of Green Chemistry

The precautionary principle can be traced back to the *Vorsorgeprinzip* idea that developed in Germany during the early 1970s (European Environmental Agency 2001). When protecting water from pollution, German officials thought, there ought to be 'forward-looking' planning to prevent environmental damage. In retrospect, much of the US environmental law framework that emerged around 1970 was founded on the precautionary principle (Raffenberger & Tickner 1999). For example, the Clean Air Act requires the Environmental Protection Agency to protect public health and welfare when making ambient air quality standards for pollutants such as lead and nitrogen dioxide. In contrast to subsequent laws, regulators cannot consider economic cost in setting standards. European countries were slower to put the principle into practice but it is now firmly established in EU environmental laws. By the 1990s, the precautionary principle was being widely cited in international environmental treaties and declarations.

Within the US, the Reagan Administration greatly weakened the emphasis on prevention during the 1980s, by requiring a favorable cost-benefit analysis (a quantitative method to determine whether benefits exceed costs) before approving new rules (Ashford 2005). Its officials prioritized voluntary industry action over regulation. By the 1990s, government agencies and courts were repeatedly accepting the arguments of industry lobby groups (*e.g.*, the Chemical Manufacturers Association) that policy interventions without a strong economic case needlessly damaged businesses. In this context, a group of 32 lawyers, activists, and scientists met in January 1998 at the Wingspread Conference Center in Racine, Wisconsin to discuss how to define the precautionary principle for environmental health decision-making. These participants issued a consensus declaration known as the *Wingspread Statement on the Precautionary Principle* (Tickner *et al.* 2003). The statement said:

> When an activity raises threats of harm to human health or the environment, precautionary measures should be taken even if some cause-and-effect relationships are not fully established scientifically.

The conference affirmed that whoever wants to create the potential harm must have the burden of proving that it is not injurious (because the actor often has much more capacity and knowledge to do so compared to, say, the public; and may benefit lucratively from the activity). If more scientific evidence is needed to determine whether a proposed activity is safe for the public, the actor must generate this knowledge. A reasonable range of alternatives to the activity (including no action) should be evaluated, while decision-making should be "open, informed and democratic, and must include potentially affected parties" (*ibid.*).

Advocates commonly invoke reasons such as the following to justify the precautionary principle (European Environmental Agency 2001). Given data limitations, scientists may not be able to conclusively prove that something causes an effect. Achieving corroboration may require decades of scientific research and in the meantime, substantial, irreparable harm could be caused to societies and ecosystems. Yet, taking early action may save human lives, not to mention large sums of public funds through avoiding health care and environmental clean-up costs. Existing regulatory practices like risk assessment have failed to adequately protect humans and animals.[1] Traditionally, risk assessment has assumed that chemicals can have 'safe' levels of exposure, only to be confounded by emerging scientific knowledge. Moreover, uncertainties may be due to the existence of 'undone science', or gaps in technical knowledge because of a long-running failure of industry and scientists to inquire into, for example, the health effects of pesticides on farm workers (Frickel *et al.* 2009).

In industrial countries like the US, dominant policy frameworks tend to be utilitarian. Consequentialist ethics emphasizes looking at the consequences of an action to decide if it is morally permissible (Martin *et al.* 2016). When invoking utilitarianism, people say that the morally right action is the one that produces the most utility ('the greatest good for the greatest number'). Cost-benefit analysis (CBA) is premised on the idea that government rules or industry decisions should proceed only if significant benefits (calculated in monetary terms) will result. For example, a company should only choose to remove a chemical if a large number of people will have their health protected without excessive cost or loss of profit. Policies that force industry to spend many millions of dollars per human life preserved or enhanced are economically 'inefficient'. Critics have pointed out that CBA conceals numerous problems, from imposing artificially monetary prices on environmental health to relying on fallacious, often ideologically driven assumptions about the value of human life and health (Ackerman 2008). Nonetheless, this way of thinking continues to be tremendously influential in the green chemistry arena, since it meshes with traditional business worldviews. Companies can, and do, interpret the precautionary principle in a more utilitarian sense. They say that there are limits on what should be done – that precaution only 'works' beyond a certain threshold of danger that would warrant costly interventions. For them, many sorts of uncertainty do not pass this line – because of their underlying organizational decision-making criteria.

---

1 For explanation of risk assessment and its limitations in the context of US chemicals regulation, see Martin, Iles & Rosen 2016 (reproduced as chapter 10 in this volume).

By contrast, strong interpretations of the precautionary principle invoke deontological ethics, particularly in the European Union. That is, moral action depends on the intention: whether an action is done for the right reasons. Deontologists assert that people have a duty to do the right thing no matter the consequences. The Wingspread Statement bases precaution on a paramount duty to prevent harm to humans and ecosystems, and does not qualify precautionary measures as worthwhile only if they are cost-effective. Crucially, *who* decides on what constitutes an actionable peril is not industry but the societies and peoples who are being potentially harmed. This shift of judgment power recognizes that corporations tend to prioritize their own existential interests over those of societies. Companies should, then, eliminate risks that *societies* see as particularly damaging (*e.g.*, harming the cognitive development of children) even if conventionally calculated monetary benefits may not be large,[2] and even if substantial uncertainties exist.

Do chemists, engineers, managers, designers, and many others involved in chemical production and use have a special responsibility to repair the toxicity and sustainability of molecules? Must they be constrained by perceptions of their own capacity for action and by the findings of cost-benefit analyses when deciding how far they can go? Must they integrate societal views into their decision-making? In short, how can the precautionary principle be applied to chemistry?

In 1998, chemists Paul Anastas and John Warner proposed one way to operationalize the precautionary principle: green chemistry. They defined green chemistry as "The utilization of a set of principles that reduces or eliminates the use or generation of hazardous substances in the design, manufacture and application of chemical products" (Anastas & Warner 1998, p. 2). While green chemistry originally began as a specific pollution prevention practice (*i.e.*, molecular design to reduce waste), it quickly evolved conceptually into a much broader set of approaches targeting toxicity and sustainability (Geiser 2015). Green chemistry seeks to inject ecological and health values into the otherwise technical process of designing and making chemicals. While many green

---

[2] In many cases, the 'real' benefits of action may be large but be excluded through CBA's selective omission of facets like performance in school and longer term employment prospects on the basis these cannot be measured, or given monetary value.

chemists underline the fact that no chemical can ever be fully benign, they prioritize these values much more than had been the case before.

The 12 Principles of Green Chemistry

Provides a framework for learning about green chemistry and designing or improving materials, products, processes and systems.

1. Prevent waste
2. Atom Economy
3. Less Hazardous Synthesis
4. Design Benign Chemicals
5. Benign Solvents & Auxiliaries
6. Design for Energy Efficiency
7. Use of Renewable Feedstocks
8. Reduce Derivatives
9. Catalysis (vs. Stoichiometric)
10. Design for Degradation
11. Real-Time Analysis for Pollution Prevention
12. Inherently Benign Chemistry for Accident Prevention

Table 1. The Principles of Green Chemistry (from Anastas & Warner 1998).

Anastas and Warner (1998) drafted what became known as the 12 Principles of Green Chemistry (12 GC, see Table 1). Arguably, this framework is grounded in the deontological version of the precautionary principle. The traditional model of risk suggests that Risk = Hazard x Exposure. Instead of simply containing exposures to risk, Anastas and Warner argued, chemists should focus on preventing hazards to begin with. They suggested that prevention 'at the source' is superior to simply controlling pollution and waste, since if there is no hazard, then there is no risk. The 12 GC principles are meant to inspire molecular designers in developing methods and technologies to create inherently benign materials and energy. These principles can also guide the decisions of a broad community of process and product designers, business managers, regulators, and advocacy groups. The 12 GC principles call on chemists to engage in various practices, such as choosing less hazardous reagents and solvents, or designing reactions to have 'atom economy' (or efficient conversion). Chemists should design molecules to degrade readily in the environment, draw on renewable feedstock such as agricultural crops, and integrate catalysts to improve reaction productivity. Importantly, the principles have been updated to say chemists should design safer chemicals. This implies replacing harmful chemicals with better ones and conducting alternatives analysis to know which ones these are.

Early on, leading green chemists acknowledged that they have special moral agency to reshape the molecules that they help make. Anastas and Williamson (1996, p. 1) wrote:

> For those of us who have been given the capacity to understand chemistry and practice it as our livelihood, it is and should be expected that we will use this capacity wisely. With knowledge comes the burden of responsibility. Chemists do not have the luxury of ignorance and cannot turn a blind eye to the effects of the science in which we are engaged.

Academic chemist Terry Collins considers it an ethical imperative for his peers to create what he calls a 'sustainable civilization' (Collins 2001).

Many green chemists point to the terrible, known harms of lead, polychlorinated chemicals, asbestos, and other substances as validating their work (Tickner & Geiser 2005). They believe there is an overwhelming case for the removal of such substances. By contrast, green chemists (who are still only a small minority in the chemistry profession) can differ as to whether more scientifically uncertain effects should be enough to justify using specific green chemistry choices. Even so, by 2001, some chemists were already saying: "Our present knowledge strongly suggests that anthropogenic [endocrine disrupters] should be identified and eliminated altogether" (Collins 2001, p. 49). They feel responsible for having collectively helped create a planet where households are filled with chemicals, where wildlife bear heavy burdens of persistent organic pollutants, where girls are maturing much younger due to endocrine disruption. More generally, scientists often must work with some degree of uncertainty as they make particular green chemistry decisions. For example, using computer modeling tools to predict the toxicity of a new substance means accepting a level of uncertainty in these tools because of inadequate health data and modeling assumptions (Faulkner *et al.* 2017).

Such ethical questioning is also entering the downstream consumer product industry. Product developers at the Seventh Generation firm argue: "From the perspective of a green household cleaning formulation the risk of harming human health and/or the environment far outweighs the benefit of providing consumers with the latest innovation in household cleaning" (Bondi 2011, p. 430). They suggest, however, that green chemistry does not mean that companies cannot innovate: green chemistry is *intrinsically* innovative since it creates safer cleaning products that exclude endocrine disrupters. Products must be judged according to a new, socially made criterion of safety, alongside performance and cost. This criterion should include precaution where necessary. If firms are wavering about what to do, the precautionary principle offers a way to rank priorities for change (Bondi 2011), depending on the gravity and magnitude of the potential danger, populations who may be affected, and the reversibility of the danger. The emphasis is on hazard, not on cost. To accomplish this ranking, interest in the analysis of alternatives is growing: firms can compare between different solutions, including green chemistry, to choose the safest alternative (Tickner & Geiser 2004).

Importantly, American and European actors can diverge in their understandings of whether and how the precautionary principle underlies green chemistry, because of their political, cultural, and historical conditions (Wilson & Schwarzman 2009). In Europe, precaution is partially built into the REACH regulatory framework that has governed chemicals in this region since 2006. Companies must prove that their products, whether new or existing, are safe; substances that are likely to pose significant perils can be de-registered after regulatory review. To some degree, then, firms have strengthened their green chemistry efforts as a regulatory compliance approach. In the US, green chemistry has largely occurred through voluntary industry actions, because a dysfunctional toxics regulatory state prevailed until July 2016, permitting lax oversight of chemicals already on the market. Here, the precautionary principle has likely played a greater role in motivating ethical industry action in the absence of stringent regulation. In this chapter, we look more at the latter situation because here ethics becomes central.

3. Historical Development of PVC Industry Lock-ins

To understand how PVC became widely used and locked-in, we look at the history of industry choices in manufacturing PVC. More generally, the ability of companies to use green chemistry principles may depend on the ways in which technologies and reactions have evolved.

3.1 Early PVC production

The first generation of PVC polymers emerged in Germany but was unsuccessful in the market due to performance problems. In 1872, a German chemist named Baumann first discovered how to polymerize vinyl chloride monomers (VCM) into PVC. Commercial production, however, did not accelerate until 1913, when another German chemist called Klatte patented a new method for producing VCM. This method obtained acetylene feedstock by reacting calcium carbide with water in special generators, and adding hydrogen chloride gas to acetylene gas using a mercury chloride catalyst (Wilkes *et al.* 2005, see Table 2).

Klatte's approach to VCM production was particularly economical because it made use of acetylene and chlorine – raw materials that had encountered problems of overproduction. Beginning in the late 19th century, companies had built chlor-alkali factories to meet rapidly growing demand for alkali (caustic soda, soda ash, sodium hydroxide, and baking soda) (Thornton 2000). Through brine electrolysis, chlor-alkali facilities produced alkali with chlorine and hydrogen as by-products. While chlorine was used as a bleaching material in the textile and paper industries, alkali was the more valuable industrial product due to interest from manufacturers of glass, soap, paper, textiles, and other products. However, for every ton of caustic soda made, 1800 pounds of chlorine were also generated (Thornton 2000). If producers could not find uses for chlorine, they would be forced to slow caustic soda production or store dangerous chlorine gas. Thus, the industry sought markets for chlorine-based products, including VCM made from acetylene and chlorine as raw materials.

PVC production in Germany grew during World War I (Mulder & Knot 2001). PVC was an attractive construction material because it offered a longer life for products traditionally made from corrosion-prone metals. Yet performance problems were soon apparent: PVC degraded when exposed to heat and light, turning brittle. With competition from low-cost, durable natural materials like rubber, interest in PVC collapsed after WWI.

3.2 A second generation of PVC production

In the 1920s, a second generation of higher quality PVC emerged through innovations in polymer science and process engineering. Many companies founded polymer divisions, igniting the nascent field of polymer science (Chandler 2009). One innovation in particular transformed the performance value of PVC. In 1926, Waldo Semon, a polymer scientist at the US tire manufacturer B.F. Goodrich, discovered that formulating PVC with additional substances turned it into a flexible, water-proof, and fire-resistant material that could bind to metal and be readily molded into stand-alone products. By 1930, B.F. Goodrich was producing PVC commercially, with other companies following in the US, Germany, and Japan. In Germany, I.G. Farben researchers developed a co-polymerization method that promised to soften PVC. Polymer scientists continued to experiment with PVC formulations, adding plasticizers (to make it softer, less brittle) and stabilizers and hardeners (to make it more durable, more rigid). In addition, new developments in PVC process engineering improved the injection molding capabilities of PVC manufacturers.

As a result, PVC production expanded once again during and after World War II. For materials that were hard to attain in a wartime economy like electrical wire coatings, PVC became a desirable substitute. After the war, to solve the problem of excess PVC production, PVC manufacturers aggressively marketed their product as a cost-effective, higher performing alternative to other plastics and materials like woods, metals, glass, rubber, and ceramics (Mulder & Knot 2001). Initially, consumers perceived PVC as a sub-standard plastic, since it deteriorated with use. Using additives therefore became the dominant method for controlling PVC characteristics. Additives emerged as a sub-sector in its own right, as industry invested in developing hundreds of additives.

Rigid PVC formulations suited mass-produced construction and consumer products, including windows, doors, pipes, combs, toothbrushes, and eyeglass frames. Flexible PVC formulations included flame-resistant cable insulation, water-proof raincoats, shower curtains, artificial leather, and phonograph records. By the 1970s, the average consumption of PVC per person in industrialized countries exceeded 20 pounds yearly.

3.3 The shift from acetylene to ethylene

In the 1950s, many companies chose between acetylene and ethylene as feedstock for PVC production (Mulder & Knot 2001). The primary method for making VCM was still the hydrochlorination of acetylene gas. In the early 1950s, a new route for producing VCM from ethylene appeared, at a time when ethylene was abundant due to cheap oil supplies. Although ethylene is the raw material for most basic petrochemical products, the industry searched for additional commercial processes to consume ethylene, and turned plastics including PVC into a new huge end-use market (Spitz 1988).

The ethylene-EDC-VCM route is depicted in Table 2. Most ethylene comes from petroleum production, although it can also be obtained from natural gas and biomass. This VCM route involves two major processes. The first is the direct chlorination process which reacts chlorine (obtained from salt electrolysis) and ethylene to form the intermediary ethylene dichloride (EDC). The second is the oxychlorination process: hydrogen chloride (HCl) is obtained as a by-product from the direct chlorination process and used to create more EDC by reacting HCl with ethylene in the presence of catalyst and air (or oxygen). The EDC from this process is then dehydrated and thermally cracked using pyrolysis (a thermochemical process of decomposing organic material in high temperatures in the absence of oxygen or halogens) to yield VCM. EDC and VCM can be sold as commodities to PVC producers, but large chemical companies like Dow can house the entire EDC-VCM-PVC chain.

Value Chain Stages	Acetylene Route			Ethylene Route (90% of global capacity)			
Raw material extraction	Limestone ($CaCO_3$) is ↓ converted into lime (CaO or calcium oxide).	Coal is ↓ refined into coke (3C), a hard, high carbon fuel.	Natural gas ↓ Combusting methane + oxygen yields ↓ Hydrogen (H) + **Acetylene gas (C_2H_2)** ↓	Salt (NaCl) ↓ Electrolysis of brine creates alkali, chlorine gas (Cl_2), hydrogen (H_2). ↓ Hydrogen (H_2) + Chlorine(Cl_2) ↓ **Hydrogen chloride (HCl)** ↓	Oil ↓	Natural gas (including shale gas) ↓	Biomass ↓
Raw material conversion to obtain Acetylene, Ethylene and Hydrogen Chloride	Lime (CaO) + coke (3C) reacted at high heat yields ↓ Calcium carbide (CaC_2) + Carbon monoxide (CO). ------- Calcium carbide (CaC_2) + water (H_20) combined in special generators yield ↓ Hydrated lime + **Acetylene gas (C_2H_2)** ↓				Ethylene obtained from thermal cracking		Ethanol dehydration on acid catalysts leading to ethylene (in development)
Manufacturing VCM	Acetylene (C_2H_2) + Hydrogen Chloride (HCl) via mercuric chloride catalyst ↓ VCM			ethylene dichloride (EDC) ↓ VCM			
Manufacturing PVC including polymerization, processing, and specialty products manufacturing	Depending on the intended market, PVC plants formulate PVC with plasticizers (e.g. phthalates) and stabilizers (e.g. lead, organotins) to give PVC the flexibility and durability desired by their customers.						
	Construction (over 50% of total market volume in 2013).		*Electronic*	*Automotive*	*Packaging*	*Other, including*	
End uses	*End user segments include automotive, electronic, construction, packaging, medical and consumer goods like toys, bottles, and eyeglasses. There is little value added by consumers in this stage beyond how consumer use affects the ability of PVC materials to be absorbed by one of the below "end of life" processes.*						
End of life	**Disposal:** *Most PVC is disposed of in landfills.*	**Reuse:** *PVC building material (pipes, sheets) is being reused in urban agriculture and gardens, among other uses.*		**Recycling**: *PVC is difficult, namely due to the many additives that make PVC useful.*			

Table 2. The Global Value Chain for PVC.

Today the ethylene route accounts for more than 75 percent of global PVC production capacity (Wilkes *et al.* 2005). It is used mostly in the US, Europe, and other countries with well-developed petrochemical industries and strict environmental regulations. Yet, the acetylene route remains attractive in places where oil supplies are costly or scarce and a readily available supply of lime and coke exists to convert calcium carbide into acetylene. Calcium carbide production requires substantial energy resources as well. Most VCM plants in China, Russia, and other parts of Eastern Europe therefore use this process. China's VCM production is now at least 85% from acetylene (ICIS Chemical Business 2012). Asian PVC factories are particularly polluting and energy-intensive, compared to PVC factories in other regions.

Today, the two production pathways for VCM continue to straddle the global value chain for PVC, as shown in Table 2. Each route requires combining a hydrocarbon feedstock (such as ethylene or acetylene) with hydrogen chloride (made from hydrogen and chlorine gas). Over time, both routes have been refined for increased scales of operation and efficiency. Yet, fundamentally new and commercially viable pathways for making PVC have not been developed since the 1950s.

Ironically, PVC may have become prevalent because it was one of the first successful commodity plastics back in the 1930s. A polymer scientist wrote in 1966: "Had this polymer been discovered at the present stage of development of the plastics industry, it would almost certainly have been eliminated as useless because of its general instability to all common degradative agents" (Grassie 1996, p. 647). But PVC now drives chlorine production, rather than serving as an outlet. PVC is now the largest downstream product of chlorine gas, accounting for 41% of chlorine demand in the US and 38% in Europe (Thornton 2000).

PVC is widely used because of its cheapness and technical advantages. The chemical industry has built a vast production system around the two pathways and many manufacturing industries depend on PVC as a material for their own products (Mulder & Knot 2001, Knot *et al.* 2001). Many downstream production processes rely on specialized formulations of PVC with modern additives. Numerous users of PVC – for example, builders, water engineers, and home renovators – prefer it for its ease of use and durability (Mulder & Knot 2001). To change or abandon PVC use, companies would have to invest in new technologies, molding machines, and feedstock. They would have to reformulate their products to remove PVC or to use different PVC formulations. Companies are also unwilling to relinquish a highly lucrative market, while their customers are reluctant to switch away from a substance. In these ways, PVC shows how the chemical industry features many lock-ins that reinforce each other and make it difficult to escape from a technological path. However applying green chemistry ideas can open new scope for making PVC production more malleable again.

4. Opportunities for Green Chemistry in the PVC Chain

Despite PVC's many useful applications, labor and environmental advocates, academic researchers, and others have argued that the PVC value chain causes many hazards (Thornton 2000). PVC is one of the most heavily scrutinized chemicals of all. Numerous critics have assailed this substance since the 1970s and industry claims to have studied it thoroughly for its toxicity. Nonetheless, as we will see, extensive evidence of toxicity associated with PVC does exist, though significant uncertainties exist around the additives used in PVC. What, then, is the scope for using green chemistry? What moral agency do various actors within the PVC chain have to use the precautionary principle and therefore justify using green chemistry?

We employ global production chain analysis to help identify some opportunities for putting green chemistry principles into practice. The PVC chain begins with raw material extraction. In between, there are a number of manufacturing stages. The end-of-life stages may extend the chain further, depending on how PVC products are disposed of, recycled, or reused. Generally, the further downstream we go, the greater the potential flexibility that actors have to change the technology (Mulder & Knot 2001). At all stages, a number of context-specific factors mediate environmental and health risks and how actual impacts occur. These factors include local infrastructure and energy resources, access to specific technologies, the enforcement of government regulations, and actions by company executives, workers, government officials, advocacy groups, and consumers. Moreover, what happens at one stage can affect other stages profoundly. For example, activists may persuade a toy manufacturer to mandate the removal of a toxic additive, which feeds back upstream into the decisions of PVC manufacturers.

We focus on the toy industry to examine the agents of change who have the capability to reduce harms created in the global PVC chain. PVC is widely used in children's toys because of its durability, low cost, and ability to be molded (Tickner 1999). In the 1950s, the toy industry introduced a number of PVC-based products that became consumer icons: Mr. Potato Head (1952), Lego (1955), and Barbie (1959) (Meikle 1995). Injection molding and PVC offered a cheaper manufacturing route than more traditional materials. Prior to PVC, the toy company Mattel fabricated its commercial doll-heads with porcelain and doll-bodies with leather. To replace this design, Seymour Adler, the head plastics engineer of the Mattel Barbie product team, chose a soft PVC formulation to give the doll more physical detail, such as crevices between Barbie's fingers and toes, and to strengthen its ability to withstand child play (Lord 2004). Today, Barbie is made with diverse plastic components: its arms are ethylene-vinyl acetate (EVA), its torso is acrylonitrile-butadiene-styrene (ABS), and its bend-leg armature has polypropylene. Its outer legs are still made of PVC, albeit with a different formula requiring less plasticizer.

Since the late 1990s, advocacy groups and concerned scientists have criticized the toy industry for its use of PVC. Toy manufacturers from Hasbro to Lego, along with large toy retailers including Toys'R'Us, Walmart, and Target, have faced intensifying calls to make toys safer for young children, who are particularly susceptible to carcinogenic and endocrine-disrupting substances during their physical and cognitive growth (Iles 2007). In response, some toy companies are trying to remove PVC from their products in favor of safer plastics, or to reformulate PVC with non-toxic additives. In turn, PVC manufacturers continue to face long-running pressures from workers and fence-line communities to reduce their exposure to harmful vinyl chloride monomers and to hazardous chemicals like mercury used in producing chlorine (Thornton 2000). Chemists and executives in these firms can try to switch to alternative feedstock to make PVC-like plastics, or to redesign production processes early in PVC's life cycle. Such decisions can ripple downstream but may be difficult to make, given existing industry structures.

We will analyze the ethical arguments that chemists, engineers and business managers may consider when deciding whether and how to use green chemistry solutions. To do this, we concentrate on the PVC production and downstream product design/production lifecycle stages. (The end-of-life stage also poses many concerns that green chemistry could address, notably poor PVC recycling rates and high resistance to degradation in landfills.)

4.1 Raw materials and PVC manufacturing

Impacts

PVC exemplifies the complex web of material inputs and manufacturing steps that characterize the chemical industry. Commonly, a plastic is made by polymerizing a monomer that is in turn produced from a number of intermediate chemicals, all of which have their own feedstock sources. The initial choice of feedstock and basic chemicals can cause a cascade of multiple environmental and health damages along the lifecycle chain. For PVC, two major feedstock concerns exist: chlorine and petrochemicals. Moreover, the choice of intermediate chemicals and processing pathways can create further deleterious impacts. In this case, turning vinyl chloride monomer into PVC is associated with high cancer risks for factory workers.

In terms of toxic effects alone, PVC use raises grave concerns. In both the acetylene and ethylene routes, VCM producers must use hydrogen chloride, which is formed by reacting chlorine gas and hydrogen gas at temperatures above 250° C (Wilkes *et al.* 2005). Chemically, elemental chlorine and hydrogen chloride are toxic at high concentrations; chlorine is also vigorously reactive. They are logical candidates for replacement by safer alternatives. Moreover, manufacturing both chlorine and acetylene uses mercury, an extremely dangerous substance. Traditionally, chlorine is produced by electrolysing brine salt in chlor-alkali factories. Not only does this electrolysis consume copious energy, it also depends on the use of mercury as a cathode (Wilkes *et al.* 2005). In the last century, chlor-alkali production has been second only to the fossil fuel industry in mercury releases to the environment. Humans can ingest mercury by eating fish and wildlife contaminated by industrial emissions. Even in minute amounts, mercury can cause numerous health effects on human bodies, including impaired neurological development in children (Thornton 2000). Many mercury-cell factories have been decommissioned (in favor of membrane and diaphragm technologies) due to concerns about mercury. Between 2005 and 2010 alone, global mercury-cell capacity decreased by about 30 percent (Global Mercury Partnership 2012). Still, over 100 mercury-cell factories continue operating worldwide.

Further downstream, the acetylene route to make VCM uses a mercury catalyst-mediated addition of HCl to acetylene. In 1952, the Chisso Chemical Company began manufacturing acetylene at its factory on Minamata Bay in Japan. It discharged mercury sulphate effluents into the bay, eventually resulting in over 20,000 local inhabitants suffering or even dying from neurological diseases and birth defects (Hylander & Goodsite 2006). Other acetylene factories have been associated with similar health damages for neighboring communities.

The ethylene route for manufacturing VCM does not involve mercury and is less energy intensive. However, ethylene production emits extremely hazardous organochlorine by-products generated during the oxychlorination process from the synthesis of EDC to make VCM (Thornton 2000). These substances include chlorinated dioxins, chlorinated furans, and polychlorinated biphenyls (PCBs). They can travel into the environment through various pathways, such as incinerating PVC waste, recycling vinyl-containing metal products, and accidental fires in buildings, warehouses, or landfills. These organochlorine substances often display persistent bioaccumulative toxicity: they strongly resist degradation, accumulate in fatty tissues in humans and wildlife, and can be widely dispersed by air and water. Dioxins in particular are carcinogenic, cause reproductive and developmental problems, and damage the immune system. Waste from ethylene-based VCM facilities in the US and Europe have some of the highest known dioxin concentrations (Thornton 2000).

Further downstream, a larger number of companies are involved in polymerizing VCM into PVC and formulating hard and soft varieties of PVC for specific applications. They include firms like Shin Etsu Group, Formosa Plastics Corporation, Solvay, Ineos and Oxy Vinyls. Some of this PVC may require further processing before sale to distributors, retailers, or end-use consumers. The health risks of PVC have been known – and deliberately obscured – for at least 70 years (Soffritti *et al.* 2013). From 1938, European manufacturers observed that VCM was toxic in animals at low levels of exposure but kept these studies secret (Markowitz & Rosner 2002). By the 1950s, firms such as Dow Chemical and B.F. Goodrich were monitoring their workers through collecting urine and blood samples. They discovered that VCM exposure led to an increased probability of developing liver angiosarcoma and to statistically significant excesses of brain cancer and neurological effects not only in workers but in their families. Seeking to weaken calls for regulatory action, these firms publicly claimed that PVC products were benign. Starting in the 1970s, though, US regulators discovered the hidden data and imposed strict workplace exposure standards that forced US factories to steam-strip PVC in closed polmerization vessels (Soffritti *et al.* 2013). This step reduced VCM levels in PVC resins by 99%. Unfortunately, many older factories continue to operate globally.

Solutions

How might green chemistry solutions be used to address the raw material and initial manufacturing impacts of PVC? Chemists and business executives could contemplate actions like substituting new membrane technologies instead of brine electrolysis; replacing hazardous catalysts like mercury-based ones with safer catalysts; abandoning acetylene in favor of ethylene; or using different feedstock such as biomass to make ethylene while reducing the use of fossil fuels. Alternatively, companies can stop using PVC altogether as a plastic; they could use safer and sustainable plastics instead, thereby eliminating chlorine altogether. Here we consider two examples of green chemistry solutions: catalysts and biomass feedstock.

A core green chemistry principle is to use catalysts rather than traditional stoichiometric reactions because these can accelerate reactions, increase efficiency of conversion, and reduce energy inputs with lower temperatures and pressures needed. In many cases, existing reactions may need to be redesigned to accommodate catalysts. In making acetylene, a catalyst is already used, but it may not be as efficient as modern catalysts. Many catalysts, like mercury-based ones, are toxic (mercury catalysts are short-lived because of their rapid loss of mercury, Johnston *et al.* 2015). The principle also suggests the substitution of toxic and polluting catalysts with benign ones. This green chemistry technique has been particularly popular in industry because it is a drop-in solution that does not call for extensive change yet can lead to substantial efficiency and economic cost gains.

Interestingly, a new international mercury treaty requires VCM factories to switch to a mercury-free catalyst by 2017, assuming there is an economically available alternative. This has incentivized academic and industry researchers to search for new catalysts in anticipation of the treaty's implementation (Zhang *et al.* 2011). One viable alternative is a gold nanoparticle catalyst technology that researchers at the Johnson Matthey Company (a leading catalyst maker) are commercializing now (Liu *et al.* 2014). This idea has already been around for 30-plus years. In 1985, Cardiff University chemist Graham Hutchings discovered that gold could catalyze the hydrochlorination of acetylene to VCM; later, he found that gold performed far better than mercury (Perks 2010).

But only in the past few years has a chemical firm chosen to develop this idea into an actual technology. One reason is that China is one of the few countries where the acetylene route is still used intensively, and where mercury pollution is a pressing concern. Johnson Matthey identified a potential new market in China for a green acetylene catalyst and thus invested in its R&D for eight years to reduce the use of *aqua regia* in making the catalyst and to increase longevity and turnover. Its scientists screened many cationic gold-carbon complexes to identify the ones that performed the best in producing acetylene. A small pilot plant was installed at a Chinese PVC manufacturer, resulting in 99% selectivity, lower energy use, and less toxicity (Johnston *et al.* 2015). The manufacturer has now built a full plant, implying it thinks the technology is economical, while Johnson Matthey has located a new gold catalyst factory in China. Replacing mercury catalysts, however, cannot address the problems that choice of feedstock and intermediate chemicals can create. This is a limited process change. Moreover, gold nanoparticles may also be toxic to workers: the scientific evidence to date is contentious but suggests some potential harmfulness to cells (Frattoni *et al.* 2015).

Another green chemistry principle that VCM and PVC manufacturers may consider to reduce the health and environmental costs of using fossil fuels is to use renewable feedstocks or inputs. They can choose to use biomass feedstocks – such as agricultural crops, grasses, and crop harvesting waste – in making PVC. In general, biomass feedstocks can yield dramatic decreases in greenhouse gas emissions, compared to petrochemical feedstock like naptha. These feedstock can also help avoid the larger social, geopolitical, and environmental impacts of extracting fossil fuels.

Bioplastics are a small but fast-growing segment of the polymer industry, with significant progress being made in the past decade to develop bio-based pathways to supply major existing chemicals or novel plastics (Iles & Martin 2013). Examples include manufacturing polyethylene from sugarcane, polylactic acid (PLA) from corn, and polyethylene terephthalate (PET) from corn. This work draws on the long history of lipid and carbohydrate chemistry to enable the use of materials such as fatty/oily or woody plant materials. Fermentation is a traditional bioconversion technology that is now widely used to make bioplastic precursors. In some cases, microbes (including yeasts and bacteria) are genetically modified to preferentially metabolically convert raw materials in ways that naturally occurring species could not perform. In other cases, naturally available enzymes and microbes have been identified to perform such transformations.

Regarding PVC, significant industry research has been underway for some time to develop a biobased version of PVC, using biomass to make ethanol (a biofuel) that can then be converted to ethylene. Bioethanol is produced by liquefying and fermenting sugary materials (primarily corn and sugarcane but potentially other crops like sugar beet), and distilling ethanol from the resulting mixture. It is also possible to obtain ethanol from hydrolyzing starch into glucose, or from using cellulosic technologies with corn stover (DuPont has opened a commercial scale plant in Iowa) or sugarcane bagasse (currently only at the pilot stage). From the 1950s, the Brazilian chemical producer Salgema used an old technology to make 100,000 tons of PVC from sugarcane ethanol annually, before going out of business due to competition from petrochemicals (Grushkin 2011).

More recently, in the 2000s, the Brazilian subsidiary of Solvay, a Belgian-based chemical producer, began experimenting with new, more efficient techniques for making PVC from sugarcane ethanol. Plans were underway in the early 2010s for building a small factory but have not yet been realized, because of technical challenges and the difficulties of competing economically with petrochemical PVC. By contrast, Braskem, the leading Brazilian chemical firm, has successfully made ethylene from sugarcane ethanol for processing into polyethylene for some years already. The firm could extend this process to manufacturing PVC but has prioritized its polyethylene market. At present, bio-based chemicals tend to be more expensive than their oil-based counterparts, so they must either have superior technical attributes or be capable of attracting sustainability price premiums, to justify their manufacture (Iles & Martin 2013). This does not yet seem to be the case for bio-PVC, which is likely to be more costly than oil-based PVC.

4.2 Consumer Product Use and Phthalates Exposure

Impacts

Further downstream, PVC suppliers and product manufacturers work together to develop formulations of the plastic for use in goods. Their choices regarding what to include in PVC can lead to consumers and factory workers being exposed to toxic substances. Suppliers mix additives with PVC resins to achieve performance characteristics according to customer specifications (Wilkes *et al.* 2005). Hundreds of chemical additives are now used to give PVC products various combinations of characteristics; they include stabilizers, lubricants, plasticizers, impact-modifiers, fillers, flame retardants, and reinforcing agents. Without these additives, PVC would be less durable and flexible, limiting its market.

As PVC products are used, they can leach, flake or outgas plasticizers and stabilizers. In 2006, 5.8 million metric tons of plasticizers were used, far dwarfing stabilizers at 670,000 tons, let alone other additives (Makarian 2006). We will therefore focus on plasticizers as an example.[3] In this life cycle stage, manufacturers theoretically have much greater capability to change between different formulations, such as switching from phthalate plasticizers and lead stabilizers to safer alternatives. Yet removing the chemicals is complicated because of their role in making ubiquitous PVC use feasible.

Plasticizers are a particularly important additive because they transform brittle PVC into a soft, malleable yet strong and stable plastic. Plasticizers and PVC resins are heated and mixed until they dissolve into one another. The plasticized material is molded into the product and cooled. The PVC-plasticizer relationship is unique because few polymers can retain high concentrations of plasticizers. PVC, however, responds favorably to plasticizers, which explains why PVC accounts for between 80-90% of all plasticizer consumption (Wilkes *et al.* 2005). Around 70 plasticizers are on the market. Product designers and manufacturers choose from diverse plasticizer options depending how soft, flexible, or heat-resistant the final product should be. The type and amount of plasticizer varies depending on the application. For example, PVC pipes require little to no plasticizers. By contrast, plasticizers can constitute 30% of the weight of PVC medical products (*e.g.* blood bags) and up to 80% of PVC recreational equipment (fishing tackle, children's toys and bouncy balls).

Phthalates are a popular plasticizer family that is compatible with PVC-based product and personal care products. Phthalates are typically formed by reacting a phthalic acid with an alcohol. Compared to other plasticizers, they are a general-purpose plasticizer that offers performance advantages not found in more specialized plasticizer families. These include strong solvency, low volatility, low diffusion, and flame resistance (Wilkes *et al.* 2005). Three phthalates account for about 75%

---

[3] Historically, heat stabilizers frequently made use of lead and other heavy metals. Around 2006, the European PVC industry committed to phasing out lead stabilizers by 2015 in favor of mixed metal stabilizers (Makarian 2006).

of all plasticizers used in PVC: diisononyl phthalate (DINP), diisodecyl phthalate (DIDP) and di-2-ethylhexyl phthalate (DEHP).

For decades, phthalates were thought by industry and regulators to be safe substances because they did not reveal any cancer-causing properties. It was not until the 1990s that scientific research began documenting that phthalates are potential endocrine-disrupting substances. Phthalates can interfere with the hormonal system, causing developmental and reproductive problems in both humans and animals. Numerous toxicological studies have now shown that phthalates can affect the growth of laboratory animals. A study of premature breast development in girls in Puerto Rico showed significantly higher levels of DEHP among girls with premature breast development compared with girls developing normally (Raloff 2000). Many toxicologists now agree that at least several commonly used phthalates pose substantial health risks to humans. Most phthalates, however, still have not been scientifically proven to be harmful even though they are suspected of being endocrine disrupters.

PVC releases phthalates throughout its lifecycle, especially when it becomes overheated, agitated, or is stored too long and begins to breakdown. The larger population can be widely exposed to phthalates through ingestion, inhalation, and skin absorption. Phthalates can also cross the human placenta during pregnancy, thus affecting foetuses. Human biomonitoring surveys, such as those of the Centers of Disease Control in the US, suggest that virtually all Americans and Europeans contain phthalates in their blood and urine. Environmental health NGOs have used the CDC surveys to show that young children have particularly high levels of phthalates (Iles 2007). They attribute this disproportionate exposure to the fact that children play with soft PVC toys and put them into their mouths. Phthalate exposure from PVC toys is especially concerning because children are more vulnerable due to their smaller bodies and critical periods of physical and cognitive development.

In response to this emerging new science, during the mid-2000s, a number of governments began eliminating a few commonly used phthalates from use in cosmetics and soft toys targeted at children (Iles 2007). Starting in 2004, the European Union took precautionary action and imposed short-term bans on three phthalates that were later permanently extended through the 2007 REACH chemicals regulation. Since

2008, the US federal government has also restricted these phthalates nationwide as part of a law reforming the Consumer Product Safety Commission. Chemical firms and many toy manufacturers have denied either that phthalates pose risks or have claimed that children's products contain safe levels of phthalate plasticizers (Iles 2007). Nonetheless, a number of retailers and toy companies have announced they will phase out PVC altogether, or will eliminate phthalates in their products. Such corporate decisions are driven not just by regulatory bans but by consumer demands, NGO campaigns, and retailers' own fears about legal liability, loss of market share, and damage to public reputation.

Solutions

How might green chemistry solutions be used to address the impacts of PVC product design choices? A fundamental green chemistry principle is to design safer chemicals. Toy companies and their designers can specify the chemicals they want to use in products, and can work with their suppliers to come up with new formulations or feedstock. Accordingly, one key green chemistry solution designers can use is to reformulate PVC so as to remove its toxic additives in favor of safe additives. They can also choose safer and more sustainable plastics *other* than PVC.

As part of the toy industry's growing response to criticism of PVC and phthalates, chemists and PVC makers have begun researching plasticizer alternatives. A minimal approach by numerous manufacturers is to 'drop in' other phthalates with safer health but comparable performance/cost profiles. One example is DINP (diisononyl phthalate). Because scientists disagree strongly over whether DINP is safe for children, the Consumer Product Safety Commission has declined to regulate DINP to date, arguing that few children are at risk. Some tests have shown that ingesting DINP causes kidney and liver damage in animals (Ackley 2000), which has led to the European Union restricting its use. In 2015, California listed DINP as a carcinogen on its Proposition 65 list. This means that retailers and manufacturers must label DINP-containing products with toxicity warnings, and hence now face greater regulatory pressure to abandon DINP.

A more demanding approach is to introduce different substances as plasticizers and remove phthalates altogether. These substances include citrates, sebacates, adipates, and phosphates (Tickner 2011). For example, acetyl tributyl citrate, bis(2-ethylhjexyl)-1,4-benzenedicarboxylate (Eastman 168), and dioctyl terephthalate (DOTP) are being used in some PVC toys. Some companies are exploring biobased alternatives to plasticizers. Metabolix, a renewable chemicals firm, has been developing polyhydroxyalkanoate (PHA) co-polymers that can substitute for some (not all) of the (fossil fuel-derived) phthalates used in PVC. Such co-polymers can impart greater durability and flexibility to PVC. PHA is made through bio-fermentation of sugars, where water-insoluble inert co-polymer builds up inside microbes that are pulped into a broth for extraction. Toxicological testing currently suggests that PHAs are non-toxic to humans. Similarly, Dow Chemical has introduced 'BioVinyl', which is a phthalate free plasticizer made from ill-defined 'plant by-products' and which promises a 41% decrease in greenhouse gas emissions compared to fossil fuel plasticizers.

Unfortunately, for most of these phthalate alternatives, toxicological and environmental safety data does not yet exist (Tickner 2011). Companies were not obliged under the old Toxic Substances Control Act (the major US chemical regulation) to carry out studies prior to marketing them. Such alternatives can still leach out of products. For DOTP, toxicity testing suggests that it is slightly irritating to eyes and may cause dermatitis; it is likely to biodegrade readily and should not build up in ecological food webs. By contrast, Eastman 168 lacks any health and environmental data. If companies choose phthalate-free alternatives, they risk creating new health hazards that may take time to manifest and be recognized.

The most radical approach is to eliminate PVC and use another plastic (or even different materials like metal or wood). Researchers looked at a set of 55 plastics on the market and concluded that PVC is easily the worst plastic because of its unique combination of chlorine, VCM, and additive toxicity (Lithner *et al.* 2011). Many other plastics would be significantly safer, if not necessarily sustainable. Starting in the late 1990s, NGO and academic analysts have repeatedly pointed out that toy manufacturers do have a wide variety of plastics options to choose from if they desire to jettison PVC. In 2011, scientist Joel Ticker surveyed these options and reported that, for example, high-density polyethylene (HDPE) could be substituted in, with relatively few effects, primarily the use of flammable chemicals in making it. While HDPE is recyclable, it cannot be biodegraded or composted. Nonetheless, most of the petrochemical-based plastics still posed serious toxicity and ecological concerns (Tickner 2011). At present, most potential biobased plastics are still not ready for use, and the main safer possibilities already on the market – PHAs and thermoplastic starch – demonstrate several problems. While PHA itself is viewed as non-toxic, its production currently requires the use of several highly toxic solvents.

A number of toy makers – like Mattel, Hasbro, and Lego – are already switching to other plastics from PVC. To help, manufacturers could put pressure on plastics manufacturers to further innovate in PVC substitutes that match PVC's performance attributes. For example, Lego has recently announced a $150 million program to support R&D into sustainable plastics. Chemists can work on improving the environmental performance of biobased plastics by tinkering with reaction pathways, replacing toxic inputs and solvents, and making reactions solvent-free.

5. Integrating Ethics into Green Chemistry Decision-Making

Green chemistry offers diverse ways to help overcome technological lock-ins more generally. As we will see, it is more challenging to loosen up the historically rigid PVC production chain than for many other chemicals. Nonetheless, taking a green chemistry lens can enable managers, chemists, and designers to think differently about seemingly long-stabilized technologies, because green chemistry offers concrete solutions for their environmental problems while potentially improving performance.

When companies, chemists, and engineers consider if they should pursue green chemistry solutions, they frequently conduct analysis – whether informal or formal – within a number of dimensions. These include the degree of change to existing production technologies and supply chains; perceived risk to the company vis-à-vis the magnitude and impacts of ecological and human health risks; and the economics and commercial benefits of installing a solution. Even if the precautionary principle is at the core of green chemistry, there can be utilitarian and deontological flavors, leading to more malleable or more absolute outcomes. The technical and economic considerations may be construed rather differently, according to the ethical stance that an actor takes.

We can glimpse how ethical arguments might be made through considering the green chemistry choices that four types of companies and their various employees must make when deciding whether and how to act on PVC risks. Chlorine producers face the most challenging choice of all. Chlorine is clearly a highly harmful substance in both its gas and vinyl monomer forms. Should firms stop making chlorine and move into another chemical sector? In other words, should they abandon their core business? Because many chlorinated substances have been banned (*e.g.*, DDT, PCBs) and chlorine is no longer permitted for use in many applications including refrigerants, solvents, and pulp bleaching, PVC has become an increasingly important market to maintain. As one chlorine producer stated: "Flexibility is no option for our company. The firm is captured in the structure of the molecules" (Mulder & Knot 2001, p. 341). Unless chlorine firms strongly believe in the ethical need to avoid harmful health effects, and are willing to change their business and market altogether, they are unlikely to do anything about chlorine itself. (Conversely, downstream consumer product makers can choose to relinquish chlorine much more readily, which may lead to a dwindling PVC market in years to come.)

Further downstream, companies supplying monomer materials to PVC manufacturers must decide whether they want to use acetylene or ethylene as their input. Acetylene still remains a key route in some countries heavily reliant on cheap coal. Producers still using coal must decide whether to make the acetylene route safer by tweaking its design and technologies. One solution, as we saw, is to replace mercury catalysts with gold nanoparticles. Catalysts are a popular green chemistry solution precisely because they can often be dropped into existing processes and businesses. Business managers may see them as posing relatively low risk to a company in terms of destabilizing production, upsetting customers, and losing market share. Moreover, using catalysts can offer substantial efficiency and financial benefits. Changes that do not markedly alter a product design are more favorably regarded. In contrast, green chemistry solutions calling for extensive change may evoke trepidation. The precautionary principle may help justify more radical changes, or may make otherwise costly and disruptive changes worthwhile.

In this case, the extreme hazards of mercury are robustly known – they are not uncertain or remote, a large number of people are being harmed, and the costs and technological changes are fairly modest. Most policy-makers and analysts, especially in the West, would agree that it is not even necessary to invoke the precautionary principle to justify eliminating mercury outright. Even an utilitarian accounting would likely support this outcome. Nonetheless, the industry calculations may differ somewhat in China and other countries still relying on acetylene, so that strong ethical arguments may be needed to change the status quo. Over the past 35 years, for example, the Chinese government has encouraged extraordinarily rapid industrialization without regard for population health. Hundreds of state-owned coal-fired power stations and many thousands of factories have been built in China, leading to horrendous air pollution and steep rises in cancer and respiratory disease rates. Development has been culturally prioritized ahead of public health. The acetylene route is associated with an industry cost structure based on lowest cost production.

Producers may thus be less willing to modify their technologies because they face the unpredictable risks of deploying gold nanoparticles and possibly partnering with an unfamiliar, foreign catalyst supplier. Chemists at Johnson Matthey (developer of the gold alternative) argue that removing mercury from the acetylene route will enhance worker and resident health (Johnston *et al.* 2015). Yet managers and chemists in the Chinese firms may or may not agree that green chemistry is morally necessary to prevent a clearly damaging risk even if it causes their firm hefty expenses. They may assert that the health benefits of using nanoparticles are too diluted in the larger air pollution problem to matter. Or, they may come to believe that they *do* have the agency to help improve environmental health, perhaps because there is now a broad deontological duty on all manufacturers to alleviate the heavy burden of their people in whatever ways they can. As residents learn more about the costs of the pollution they are being exposed to, they are beginning to mobilize community protests against chemical factories, which may support this ethical view.

In turn, PVC manufacturers increasingly must decide whether to source ethylene from petrochemicals, or from biomass materials. Many companies have long since switched to ethylene from acetylene, largely because of ease of use of oil and natural gas. It would be potentially a green chemistry solution to take up a different route that offers less toxicity, fewer energy and water inputs, and fewer waste by-products. The trouble is that petrochemical-based ethylene is not much better, because of its high, well-known climate change and toxicity impacts. Bio-based ethylene might alleviate these impacts, might release many lock-ins, and is already a proven technology as far as Braskem's green polyethylene goes. Researchers, however, have carried out lifecycle assessments that suggest that biobased PVC could have mixed benefits (Alvarenga *et al.* 2013). While the chemical offers dramatically lower greenhouse gas emissions and consumption of non-renewable resources, it can manifest higher impacts (compared to petrochemical-based PVC) in water use, biodiversity, land use, eco-toxicity, and other areas. Importantly, bio-based chemicals can contain substantial toxicity risks, notably from the petrochemical solvents used in their production (Álvarez-Chávez *et al.* 2012). To some extent, all these adverse effects can be avoided by corporate managers insisting on developing green chemistry designs that foresee them – such as by eliminating the solvents. Because most of these impacts are not yet well-understood and are uncertain, such action would be necessarily precautionary in nature.

Whether or not biobased PVC is actually superior to petrochemical PVC mostly depends on how, and where, biomass is produced. Here, firms have many considerations to evaluate, such as upstream agricultural impacts (Iles & Martin 2013). Corn is a common bio plastic feedstock. Yet corn grown in the US is notorious for its vast environmental effects, including fertilizer run-off that causes an ocean 'dead zone' in the Gulf of Mexico, large pesticide use coupled with GM crop use, and loss of biodiversity due to monoculture farm fields. By contrast, sugarcane grown in Brazil can be less ecologically damaging – unless it causes, directly or indirectly, the conversion of high-carbon containing rainforest and cerrado lands. In both cases, the method of farming matters greatly:

if industrial agriculture technologies are used, there will still be intensive use of fossil fuels for energy and chemical inputs. The next generation of biobased materials – cellulosic biomass such as grasses – is likely to perform better than food crops, but could still be produced on industrial plantations. But if diversified, agroecological farming methods are used to grow sugarcane or, better, grasses, lower environmental impacts may result (Jordan *et al.* 2016). In other words, chemists and managers will need to learn more about previously overlooked areas if they are to appreciate the full implications of their choices. They will need to develop a sense of not only chemical but agricultural ethics in order to better grasp the range of uncertainties that underline biobased feedstock, and thus see whether they must take precautionary action in terms of how this feedstock is produced.

If chemical firms still stick with PVC, their managers must decide whether and how to modify their formulations to address additive risks. Phthalates are only one of the many additives that can be found in PVC. Should PVC makers jettison phathalates altogther in favor of safer alternatives? If so, which alternatives should they consider? Here, the precautionary principle has been intensely debated among firms. While evidence that some phthalates are endocrine disrupters and carcinogens is growing, sizable uncertainty about their effects remains. This is all the more true of most phthalate substitutes: many firms have been developing substitutes on the basis that these could create new markets as phthalates dwindle. Yet little health and ecological data attests to their safety, so that a substitute can be introduced only to be later proven to be harmful. Many firms seem to be adopting an utilitarian precaution: the substitutes still benefit more people in guarding against a known phthalate risk, while the economic and societal costs of eliminating PVC are still too large to accept. Nonetheless, for many substitutes, there is a sliver of scientific evidence of risk. Biobased additives also raise similar moral questions as for biobased ethylene. In response, some chemists may argue that the precautionary principle should be implemented at its most stringent. Even though a substance has not yet been proven to be injurious, enough reliable scientific evidence exists to justify discarding a chemical. Only those substances that have passed exhaustive testing should be accepted onto the market. There appears to be widening agreement among chemists regarding phthalates – but not on which alternatives are tolerable.

Finally, both toy manufacturers and retailers must confront their own ethical quandaries – but ones reflecting the consumer's perspective. These firms may have the greatest capacity to change from PVC and its associated chemicals, in that they do not depend on PVC or chlorine for their business success. They are preoccupied with product functions, not with specific materials. Retailers and toy companies are directly exposed to consumer demands and buying choices. As seen in a recent NGO report ranking leading retailers on their performance in dealing with chemical risks (Safer Chemicals Campaign 2016), retailers are much more visible than most chemical firms. Thus, these firms may have a rather different calculus regarding whether and how to adopt green chemistry, because they are much more exposed to societal values regarding chemical risks. Toy companies can share the views of PVC makers in being reluctant to abandon PVC as an uniquely valuable plastic used in countless toys. But growing consumer *and* scientific concern about PVC additives can lead to managers and product developers making precautionary choices if they are not yet firm believers in endocrine disrupting risks. A large number of firms across many consumer product and medical equipment sectors are simply choosing to 'go PVC-free', preferring safer plastics. In their eyes, it is just too much trouble to try to 'repair' PVC, especially if it has high overall toxicity (due to chlorine and vinyl monomers). Likewise, retailers can tell their suppliers they will not stock PVC toys – sending a powerful market signal upstream to toy and PVC manufacturers. Toy makers can also request safer additives and truly sustainable bio-based ethylene to make PVC less damaging, if they are willing to tolerate its upstream production impacts. The choices that companies make here will be driven by how far they believe they must adopt the precautionary principle, and how far they want to reach back into the production chain.

6. Conclusion

By studying the PVC case, you can learn several important lessons for use in your future careers. PVC underscores the profound ethical challenges of trying to change a plastic manufacturing system that has become structurally inert over many decades. In this case, how should companies deal with the contradictions of using basic chemicals to make numerous intermediate and final products, while also causing numerous environmental, social, and health problems?

As we have seen, a growing number of firms and scientists are experimenting with green chemistry for business, regulatory, and ethical reasons (Iles 2007). In offering powerful new capabilities to create alternative pathways, green chemistry profoundly challenges 'brown chemistry', or the petrochemical-based processes that generate vast quantities of waste and pollution along with poisonous chemicals (Woodhouse & Breyman 2005). Many academic green chemists say their profession helped build a world heavily polluted with toxics; they now want to heal this planet with safer, more sustainable chemicals. Because of its precautionary roots, green chemistry is itself an ethical vision of what chemicals could be. In Europe, California, and many other locations around the world, some policy-makers, industrial chemists, and business executives have joined green chemists in seeking innovations such as discovering safe solvents and making reactions more efficient (Mulvihill *et al.* 2011).

Yet PVC illustrates why and how applications of the principles can be gnarly in practice. The complex PVC production chain means that a single solution is unlikely to succeed fully: multiple solutions may have to be imperfectly patched together. Some solutions may worsen toxicity risks while achieving greater atom efficiency, creating ethical dilemmas that may only be decided by prioritizing some impacts over others. Moreover, numerous companies – petrochemical firms, catalyst makers, specialty chemical manufacturers, plastic processors, consumer product companies, and retailers – are involved in making or using PVC. The ethical responsibility for acting may, then, be diffused across many industry actors that diverge markedly in their philosophies and capabilities. Most crucially, PVC represents an extreme example of technological and business lock-ins that inevitably color company decisions as to what can (or should) be done. Namely, without PVC, affordable chlorine could not be made; without chlorine, PVC could not be made.

As a result, applying green chemistry in practice may not be straightforward. Companies can face difficult decisions regarding whether to use green chemistry solutions, how, and to what degree. As seen in the PVC case, green chemistry practice can have utilitarian and deontological flavors. Scientists and managers are now aware that PVC and its many constituents (from feedstock to additives) cause grave environmental and health problems. Some of these are well understood (like mercury) while others are still ill-known. There are strong reasons – both precautionary *and* proven – to intervene to change PVC production in various ways. But companies may argue that they cannot use green chemistry because it would perturb their existing operations, or because they are locked into particular technologies. Other firms may say that they will adopt green chemistry solutions only to the extent that these can readily be dropped in without financial cost or market disruption. Still other companies may be willing to make fundamental changes on the ground they are morally responsible for the resulting ecological impacts. They may also perceive that customers are demanding safer chemicals or that new markets could materialize.

PVC is only one of many chemicals locked into production chains with major environmental and health impacts resulting. Once a chemical is put on the market, companies have historically had little incentive to review how it was designed and produced. Countless 'older' chemicals persist from the 1930s to 1960s period (often seen as the golden era of chemical innovation). Some examples are polystyrene, flame retardant chemicals, and perfluorinated chemicals, which have permeated consumer products, and whose designs reflect decades of congealed assumptions, negligible toxicological testing, and lax regulation. Green chemistry gives chemists another chance to re-think these legacy chemicals, this time for health and sustainability reasons. Through green chemistry, industry lock-ins can be dismantled. And in developing new materials, further lock-ins can be avoided with greater anticipation, using the precautionary principle. To do so, chemists will need to put ethical questions at the center of their work – and to resist the allure of succumbing to the inertia of 'this is how things have always been done'.

Further Reading

For further investigation of green chemistry ideas, the following reference books are helpful: Lancaster 2016, Matlack 2010, and Clark & Macquarie 2008. These are leading textbooks or handbooks on green chemistry that are accessible at the undergraduate level.

References

Ackerman, F.: 2008, Poisoned for Pennies: The Economics of Toxics and Precaution, Washington DC: Island Press.

Ackley, D.: 2000, 'Effects of Di-isononyl Phthalate, Di-2-ethylhexyl Phthalate, and Clofibrate in Cynomologus Monkeys', *Toxicological Sciences*, **56**, 181-188.

Alvarenga, R.; Dewulf, J.; De Meester, S.; Wathelet, A.; Villers, J.; Thommeret, R. & Hruska, Z.: 2013, 'Life cycle assessment of bioethanol-based PVC', *Biofuels, Bioproducts and Biorefining*, **7** (4), 386-395.

Álvarez-Chávez, C.; Edwards, S.; Moure-Eraso, R. & Geiser, K.: 2012, 'Sustainability of bio-based plastics: general comparative analysis and recommendations for improvement', *Journal of Cleaner Production*, **23** (1), 47-56.

Anastas, P.T. & Warner, J.C.: 1998, *Green Chemistry: Theory and Practice*, New York: Oxford University Press.

Anastas, P.T. & Williamson, T.C.: 1996, *Green Chemistry: Designing Chemistry for the Environment*, Washington, DC: American Chemical Society.

Ashford, N.: 2007, 'The legacy of the precautionary principle in US law: The rise of cost-benefit analysis and risk assessment as undermining factors in health, safety and environmental protection', in: N. de Sadeleer (ed.), *Implementing the Precautionary Principle: Approaches from the Nordic Countries, the EU and the United States*, London: Earthscan.

Chandler, A.: 2009, Shaping the Industrial Century: The Remarkable Story of the Evolution of the Modern Chemical and Pharmaceutical Industries, Cambridge, MA: Harvard University Press.

Clark, J. & Macquarrie, D.: 2008, *Handbook of Green Chemistry and Technology*, New York: John Wiley & Sons.

Collins, T.: 2001, 'Toward sustainable chemistry', *Science*, **291** (5501), 48-49.

European Environmental Agency: 2001, *Late lessons from early warnings: the precautionary principle 1896-2000* [available online at: http://www.eea.europa.eu/publications/environmental_issue_report_2001_22, accessed 3 February 2017.

Faulkner, D.; Shen, L.; Vanessa, Y.; Johnson, D.; Hemingway, R.; Williams, R.; Arnold, J. & Vulpe, C.: 2017, 'Tools for Green Molecular Design to Reduce Toxicological Risk', in: Richardson, R, & Johnson, D. (eds.), *Computational Systems Pharma-*

cology and Toxicology, London: Royal Society of Chemistry, pp. 36-59.

Fratoddi, I.; Venditti, I.; Cametti, C. & Russo, M.: 2015, 'How toxic are gold nanoparticles? The state-of-the-art', Nano Research, **8** (6), 1771-1799.

Frickel, S.; Gibbon, S.; Howard, J.; Kempner, J.; Ottinger, G. & Hess, D.: 2010, 'Undone science: social movement challenges to dominant scientific practice', *Science, Technology, and Human Values*, **35** (4), 444-473.

Geiser, K.: 2015, Chemicals without harm: policies for a sustainable world, Cambridge, MA: MIT Press.

Global Mercury Partnership: 2012, *Conversion from Mercury to Alternative Technology in the Chlor-Alkali Industry*, Nairobi, Kenya: United Nations Environmental Programme.

Grassie N.: 1966, 'Degradation', in: *Encyclopedia of Polymer Science and Technology*, New York: John Wiley, vol. 4.

Grushkin, D.: 2011, 'Breaking the mold', *Nature biotechnology*, **29** (1), 16.

ICIS Chemical Business: 2012; 'China invents and reinvents coal to chemicals' [available online at: http://www.icis.com/resources/news/2012/04/16/ 9549986/china-invents-and-reinvents-coal-to-chemicals/, accessed on 3 February 2017].

Hylander, L. & Goodsite, M.: 2006, 'Environmental costs of mercury pollution', *Science of the Total Environment*, **368** (1), 352-370.

Iles, A.: 2007, 'Identifying environmental health risks in consumer products: non-governmental organizations and civic epistemologies', *Public understanding of science*, **16** (4), 371-391.

Iles, A.: 2008, 'Shifting to green chemistry: the need for innovations in sustainability marketing', *Business Strategy and the Environment*, **17** (8), 524-535.

Iles, A. & Martin, A.: 2013, 'Expanding bioplastics production: sustainable business innovation in the chemical industry', *Journal of Cleaner Production*, **45**, 38-49.

Knot, J.; Van den Ende, J. & Vergragt, P.: 2001, 'Flexibility strategies for sustainable technology development', *Technovation*, **21** (6), 335-343.

Johnston, P.; Carthey, N. & Hutchings, G.: 2015, 'Discovery, development, and commercialization of gold catalysts for acetylene hydrochlorination', *Journal of the American Chemical Society*, **137** (46), 14548-14557.

Jordan, N.R.; Dorn, K.; Runck, B.; Ewing, P.; Williams, A. & Anderson, K.A.; 2016, 'Sustainable commercialization of new crops for the agricultural bioeconomy', *Elementa: Science of the Anthropocene*, **4**, 81.

Liu, X.; He, L.; Liu, Y. & Cao, Y.: 2013, 'Supported gold catalysis: from small molecule activation to green chemical synthesis', *Accounts of chemical research*, **47** (3), 793-804.

Lancaster, M.: 2016, *Green Chemistry: An Introductory Text*, 3rd edition, London: Royal Society of Chemistry.

Lithner, D.; Larsson, Å. & Dave, G.: 2011, 'Environmental and health hazard ranking and assessment of plastic polymers based on chemical composition', *Science of the Total Environment*, **409** (18), 3309-3324.

Lord, M.: 2004, Forever Barbie: The unauthorized biography of a real doll, New York: Bloomsbury.

Markarian, J.: 2007, 'PVC additives: What lies ahead?', *Plastics, Additives and Compounding*, **9** (6), 22-25.

Markowitz, G. & Rosner, D.: 2002, *Deceit and denial: The deadly politics of industrial pollution*, Berkeley: University of California Press.

Martin, A.; Iles, A. & Rosen, C.: 2016, 'Applying Utilitarianism and Deontology in Managing Bisphenol-A Risks in the United States', *Hyle: International Journal for Philosophy of Chemistry*, **22**, 79-103 [also in this volume, chapter 10].

Matlack, A.: 2010, *Introduction to green chemistry*, Boca Raton, FL: CRC Press.

Meikle, J.: 1995, *American plastic: a cultural history*, New Brunswick, NJ: Rutgers University Press.

Mulder, K. & Knot, M.: 2001, 'PVC plastic: a history of systems development and entrenchment', *Technology in Society*, **23** (2), 265-286.

Mulvihill, M.; Beach, E.; Zimmerman, J. & Anastas, P.: 2011, 'Green chemistry and green engineering: a framework for sustainable technology development', *Annual review of environment and resources*, **36**, 271-293.

Perks, B.: 2010, 'Gold Fever', *Chemical World*, September, 48-50.

Raffensperger, C. & Tickner. J.: 1999, Protecting public health and the environment: implementing the precautionary principle, Washington DC: Island Press.

Raloff, J.: 2000, 'Girls May Face Risks from Phthalates', *Science News*, September 9, 165.

Soffritti, M.; Sass, J.; Castleman, B. & Gee, D.: 2013, 'Vinyl chloride: a saga of secrecy', in: EEA (eds.), *Late Lessons from Early Warnings: The Precautionary Principle*, 2nd edition, Copenhagen: European Environmental Agency, pp. 179-202.

Spitz, P.H.: 1988, Petrochemicals: The Rise of an Industry, New York: Wiley.

Thornton, J.: 2000, Pandora's Poison: Chlorine, Health and a New Environmental Strategy, Cambridge, MA, MIT Press.

Tickner, J.; 1999; A Review of the Availability of Plastic Substitutes for Soft PVC in Toys, Greenpeace International.

Tickner, J.; Raffensperger, C. & Myers, N.: 1999, *The Precautionary Principle in Action: A Handbook*, Windsor, ND: Science and Environmental Health Network.

Tickner, J. & Geiser, K.: 2004, 'The precautionary principle stimulus for solutions-and alternatives-based environmental policy', *Environmental Impact Assessment Review*, **24** (7), 801-824.

Tickner, J.: 2011, 'Phthalates and Their Alternatives: Health and Environmental Concerns', Lowell Center for Sustainable Production [available online at: https://www.sustainableproduction.org/downloads/PhthalateAlternatives-January2011.pdf, accessed 3 February 2017].

Wilkes, C.; Daniels, C. & Summers, J.: 2005, *PVC Handbook*, Cincinnati: Hanser.

Wilson, M. & Schwarzman, M.; 2009, 'Toward a new US chemicals policy: rebuilding the foundation to advance new science, green chemistry, and environmental health', *Environmental health perspectives*, **117** (8), 1202.

Woodhouse, E. & Breyman, S.: 2005, 'Green chemistry as social movement?', *Science, Technology, & Human Values*, **30** (2), 199-222.

Vink, E.; Rabago, K.; Glassner, D. & Gruber, P.: 2003, 'Applications of life cycle assessment to NatureWorks polylactide (PLA) production', *Polymer Degradation and stability*, **80** (3), 403-419.

Zhang, J.; Liu, N.; Li, W. & Dai, B.: 2011, 'Progress on cleaner production of vinyl chloride monomers over non-mercury catalysts', *Frontiers of Chemical Science and Engineering*, **5** (4), 514-520.

Zion Market Research: 2016, 'Global PVC Market Set for Rapid Growth, To Reach Around USD 78.90 Billion by 2021' [available online at https://www.zionmarketresearch.com/news/global-pvc-market, accessed February 3, 2017.

Chapter 12

The Ethics of Rare Earth Elements Over Time and Space

Abigail Martin & Alastair Iles

Abstract: Rare earths are a critical resource for contemporary societies. Among their diverse uses, they are key components of sustainability technologies such as wind turbines and electric vehicles. While rare earths can help societies transition away from fossil fuels to renewable energy and conserve energy, their extraction, processing, and use creates serious environmental and social effects around the world, especially in China. We argue that environmental justice and intergenerational justice concepts can provide an ethical framework for navigating this green energy bargain. We survey the environmental and social effects that rare earth production causes and the changing geography of production that means these effects are being distributed worldwide, both in and beyond China. Finally, we consider several strategies that miners, manufacturers, designers, and users can use to achieve greater environmental justice and intergenerational justice, now and for the future.

1. Introduction

The multi-billion-dollar rare earth industry supplies a critical resource for contemporary societies. Rare earth elements (REEs), also known as rare earth metals, are a group of 15 lanthanide metals on the periodic table plus scandium and yttrium, as defined by the International Union of Pure and Applied Chemistry. REEs have become extremely important for manufacturing a wide range of products in consumer, industrial, military, and medical markets. They are also key components in electric vehicles, wind turbines, energy-efficient light bulbs, and other sustainability technologies, making them important for mitigating climate change by helping societies transition away from fossil fuels to renewable energy and conserve energy.

Mining for REEs creates serious threats to the environment and the well-being of communities adjacent to mines. As with mining other raw materials, the extraction of rare earth ore from the Earth has major ecological and public health impacts. Removing large amounts of earth can scar the landscape for generations and limit alternative land uses such as agriculture. Extraction and processing of rare earth materials generates large amounts of waste and releases toxins into the air, water, and soil. People living and working near mining sites face an increased risk of developing cancers and other serious illnesses due to contaminated drinking water and food. Mining can also introduce and exacerbate social conflicts over how resources are controlled, accessed, and profited from.

For chemists, engineers, product designers, and business managers who want to contribute to products for society that can help transition away from fossil fuels, REEs are extremely important. A number of REEs are required to manufacture wind turbines and electric vehicles, but obtaining these elements can have a considerable negative impact on the people and communities involved in REE production chains, creating what may be considered a 'green energy bargain' (Phadke 2018). In addition to the environmental and social costs borne by the communities who host rare earth mining operations, the consumption of REE resources is accelerating at rates that may be unsustainable. In order to meet growing demand for sustainability technologies like wind turbines and electric vehicles, current generations are expected to use more and more of these critical resources. At some point, future generations may lose out if they cannot readily obtain these resources due to reserves being exhausted. The climate-saving benefits of REEs could be distributed unevenly across time, with current and near-future generations benefiting the most. On the other hand, using REEs today could help ensure future generations avoid severe climate impacts. In this chapter, we argue that downstream decision-makers must consider the ethics of REE use from the standpoint of environmental justice and intergenerational justice, which together offer guidance in navigating this green energy bargain.

We begin this case with an overview of the various end uses for REEs. We then review the environmental and social impacts associated with mining and processing REEs, before turning to discuss the ethics of extracting and using REEs. Here, we argue that environmental justice and intergenerational equity direct our attention to specific ethical requirements for those who produce or use REEs. The next section applies the concepts of environmental justice and intergenerational justice to understand where moral hazards concentrate in the changing geography of rare earth mining and processing. Finally, we discuss strategies for reducing moral hazard to advance environmental justice and intergenerational justice.

2. Uses of Rare Earths

The 17 elements commonly referred to as REEs have many end-uses. Their exceptional magnetic, phosphorescent, and catalytic properties make them a valuable component of materials used in manufacturing high-technology consumer products (Balaram 2019). For example, the rare earths neodymium and dysprosium are often alloyed to form magnets that resist demagnetization at high temperatures, which is essential for any heat-generating electronic device such as laptop computers, televisions, flat screens, cell phones, portable DVD players and more. REEs also have end uses in the medical and defense industries, from devices like x-rays and MRI machines, to the military's jet fighter engines, missile guidance and antimissile defense systems, space-based satellites, and communication systems (Machacek & Fold 2014).

Although metals and alloys used to make rare earth magnets are one of the largest rare earth products, the REEs are transformed into a diverse array of materials for many different consumer, industrial, and military applications. REEs lend unique properties to steel and aluminum and have enabled innovations in glass. For instance, the realization of rare-earth-doped glasses has enabled numerous innovations in the chemical formulation of both commercial glasses and new functional glasses. These glasses include filters and lenses, light-sensitive and photochromic glasses, coloring and decoloring agents, X-ray and gamma-ray absorbing glasses, glass with luminescence and fluorescence effects, and communication fibres (Locardi & Guadagnino, 1992). Glass materials containing rare earth elements have been central to dramatic light-based innovations like semiconductor lasers, also called laser diodes (Tanabe 2015).

In addition, the unique catalytic performance of the REEs make them a critical resource for industrial applications requiring catalysts. Catalysts comprise a large share of REE end uses, which includes petroleum refining, the catalytic combustion of fossil fuels, automotive engine emissions control, and the purification of industrial waste, air, and solids. In particular, lanthanum and cerium are used in petroleum refining to make gasoline, which constitutes the largest end-use of REEs in the United States (US DOE 2011). These REEs ultimately make the fluid catalytic cracking process more efficient, which increases the gasoline yield per unit of catalyst (Sadeghbeigi 2012).

More recently, REEs have received increased attention for their use in sustainability technologies that can help wean societies from dependence on fossil fuels. For example, REEs are used to create red (Eu and Y), and blue (Eu) phosphors for energy-efficient light emitting diodes (LEDs), which can provide better energy savings for buildings than incandescent and fluorescent lights. The quantities of REE used to make LEDs are one to two orders of magnitude lower than that required for these other lighting technologies. Thus, expanding the markets of LEDs is not likely to increase REE demand (Ku *et al.* 2015).

Wind turbines and electric vehicles, however, are expected to have a large impact on demand for rare earths, despite their relatively small share of total REE end uses. Electric vehicles rely heavily on dysprosium and neodymium. The hard, magnetic alloy neodymium iron boron (NdFeB) has enabled the development of compact, torque- and power-dense electric traction motors, which has led to greater deployment of hybrid electric vehicles such as the Toyota Prius, and of battery electric vehicles like the Nissan Leaf (Widmer *et al.* 2015). The addition of dysprosium, a heavy rare earth element, to the NdFeB alloy increases its ability to withstand de-magnetization in high temperatures, making it possible for the alloy to perform despite heat-generating vehicle traction. The nickel metal hydride (NiMH) batteries in many electric vehicles also contain cerium and lanthanum. However, some industry analysts expect electric vehicle manufacturers will soon transition to lithium-ion batteries, which would reduce REE demand from electric vehicles because lithium ion batteries do not require REEs (DOE 2011).

Certain kinds of wind turbines also use dysprosium and neodymium. Unlike smaller, onshore wind turbines that use rotating gearboxes, larger offshore turbines use direct drive technology. This involves a generator composed of a ring of NdFeB. By some estimates, a wind turbine that generates 3.5 megawatts of electricity contains about 600 kilograms, or 1,300 pounds, of rare-earth metals (Alonso *et al.* 2012).

Expanding offshore wind capacity and growing electric vehicles sales could therefore lead to greater demand for REEs. Some analysts emphasize that the magnet sector, generally, will become the leading user of REEs based on mass (US Geological Survey 2018). Others predict electric vehicles will be the biggest driver (Alonso *et al.* 2012). As the market for electric vehicles grows, the demand for dysprosium and neodymium could increase by more than 700% and 2600%, respectively, assuming a decarbonization path of electrifying 80% of automobile sales (*i.e.* hybrid electric vehicles, plug-in hybrids, and battery electric vehicles) by 2035 in line with the goal of limiting average global temperature to 2 °C (*ibid.*). Whereas conventional fossil-fuel powered cars may use about one pound of REEs for small motorized components like windshield wipers, the various motors and batteries of an electric vehicle can require nearly 10 times more REE materials than conventional cars (*ibid.*). Using more electricity from wind power will also add to the demand for Nd and Dy. Recent estimates for deploying 80 gigawatts of offshore wind power by 2050 in the US predict the industry's REE demand will increase from about 1,200 tons a year in 2020, to nearly 3,000 tons per year by mid-century (Fishman & Graedel 2019).

These trends pose the question of whether the rate at which REEs are being used now and in the foreseeable future might deplete reserves. Rare earths are not geologically rare, but they are hard to find at levels that make it economically viable to mine (Chakhmouradian & Wall 2012). For example, cerium and yttrium are the 25th and 30th most abundant elements by mass – far exceeding tin, molybdenum, and gold. But other rare earths are less abundant by mass and more geographically dispersed. Usually, REEs are mined as a by-product of extracting another valuable mineral. REE-only mines are very rare. Although there are over

200 mineral ores containing individual rare earths, only 20 of these have been commercially mined, suggesting the economic and technical difficulties.

Many government and scientific agency reports have viewed rare earths as a 'critical' resource in relation to anxieties about China's strategically powerful control over REE supplies. Nasser *et al.* (2015) project that all individual rare earths would take a century to deplete at the rates of use of known reserves as of 2008. If the use rates of a few REEs do rise dramatically, depletion time could speed up by a few decades. Yet rare earths are arguably not yet an issue for most future generations, as there are many other resources for which depletion is on the horizon within decades, such as copper.

The key question is whether actual *supplies* of rare earths are accurately known and whether these supplies can be accessed. The answer is not clear for the long term. New reserves can be found, while existing reserves may be made more accessible through technological advances. For example, known reserves of rare earth ores grew from 88 million tons in 2008 to 130 million in 2014, indicating the importance of exploration work for estimating supplies accurately (Zepf 2016). In addition, REEs escape from supply chains in large quantities because their recovery for recycling is so low (Darcy *et al.* 2013). Most rare earths eventually end up in landfills or elsewhere once products reach the end of their lifetime. Wind turbines, for example, are hardly recycled at present. If rare earths can be recovered at high levels, this will stretch actual supplies more and change the extent to which growing demand for REEs will be met by more mining.

3. Environmental and Social Impacts of REE Production

The environmental and social impacts of REE mining begin during exploration and continue after a mine's closure. During the initial prospecting phase, mining companies must obtain access rights for local land and resources in order to secure project finance. Companies may negotiate access rights for local land and resources with government officials but fail to seek local people's permission (Handelsman 2002). In complex systems of land ownership, what is considered legal property is only one approach to determining who has rights to access land and other resources. When governments and mining companies do not recognize the rules of traditional land tenure systems, there may be lasting conflict. For instance, on Bougainville Island in Papua New Guinea, villagers refused to sell their land for the giant Panguna copper mine. Instead, property negotiations were made according to Australian law and the government expropriated villagers' land to the mining company, Rio Tinto (Denoon 2000). Villagers continued to protest the mining operation over land rights, pollution, and the lack of economic benefit for local people. The conflict escalated into a civil war in the 1990s, in which 15,000 to 20,000 people died (*ibid.*). Rio Tinto was forced to close the mine, and in 2018 the Bougainville government imposed an indefinite moratorium on renewing the company's license due to fears that violent civil conflict would erupt again (Davidson 2018).

After exploration, once a site has secure investment and access to land and resources, mining begins. Different methods can be used to excavate ore deposits from the landscape: open pit mining (removing earth from the landscape), underground mining (involving digging tunnels), or in-situ leach mining (using strong acids to dissolving the ore in the ground so it may be pumped out). REEs typically come from open pit mines, since most REEs are byproducts from other mining operations, as is the case at the most prominent operating mines producing REEs, including the Bayan Obo in Inner Mongolia (the largest REE source in China), the Mount Weld mine in Australia, and the Mountain Pass mine in the US (before it halted production).[1] The significant amount of soil, rock, and other debris dug out may be dumped back into the open pit once the mine is closed, but often it is dumped elsewhere, forming human-made mountains that alter a landscape's sense of place, topography, geology, and ecology (Francaviglia 1992, pp. 137-142).

---

[1] One significant exception is ion adsorption clays in southern China, where most of Chinese dysprosium originates, which use the in-situ leach mining approach (see note 2).

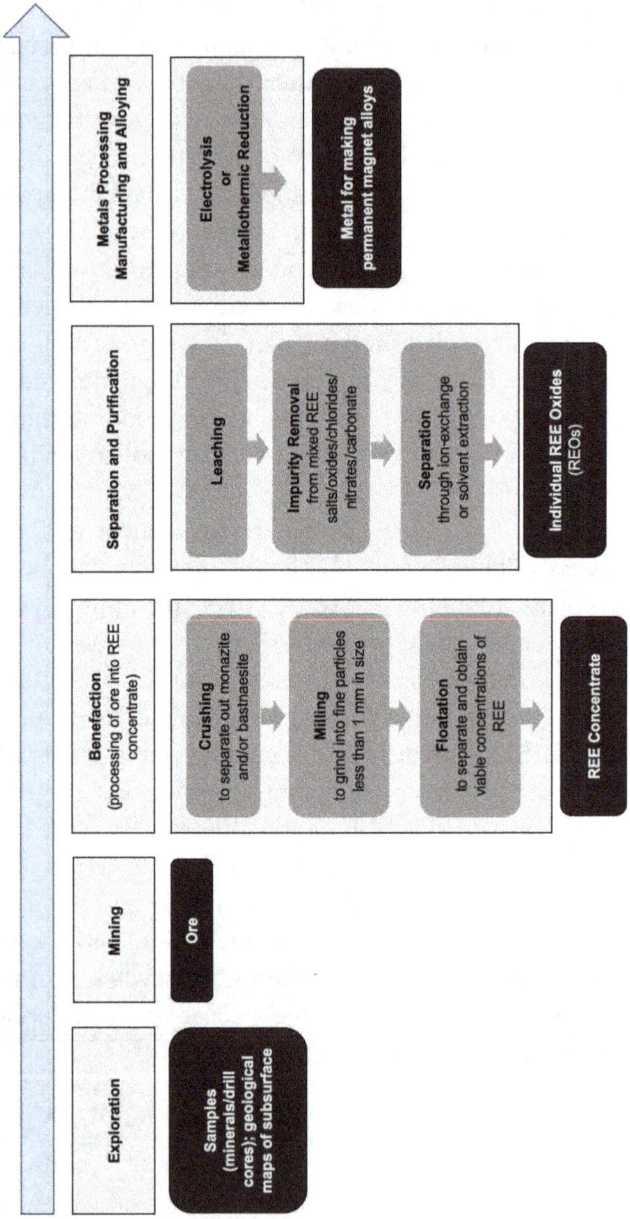

Figure 1. Rare earth production stages and outputs (adapted from Department of Energy 2017).

Removing earth is extremely water-, chemical-, and energy-intensive, resulting in large amounts of waste. Water, electricity, and diesel are needed to run the equipment for blasting, drilling, dredging, and pumping material out of the deposit. Chemical blasting agents and lubricating oil are also used to help move and loosen the earth. In the process, metallic and radioactive dusts, asbestos-like minerals, and exhaust from machinery or generators and a variety of chemicals get released into the air and soil. Excessive amounts of ammonia and nitrogen compounds may leak into the groundwater during the mining-process. Cadmium and lead may also be released into the environment.

Rare earths are not found in their isolated elemental form in nature, and thus their separation and purification make processing challenging, expensive, and polluting. After mining, the REEs must be separated and purified into rare earth oxides through a series of water-, chemical-, and energy-intensive processes (Figure 1). First, the ores are crushed, milled, and separated through froth floatation into dissolved concentrates. To purify these concentrates, there is a second round of processing via complex chemical reactions: dissolved concentrate passes through hundreds of liquid-containing chambers designed to pull out desirable elements or compounds using extraction agents (hydrochloric acid) and precipitating agents (ammonium bicarbonate $(NH_4)HCO_3$ or NaOH precipitation), followed by solvent extraction (*e.g.*, $(C_{16}H_{35}O_3P)$ and HCl) and precipitation steps using ammonium bicarbonate and oxalic acid $(C_2H_2O_4)$.[2] The precipitate oxalates are filtered out and roasted to form a concentrate of rare earth oxides (REO). For most industrial applications, the rare-earth material is supplied as oxides (*e.g.* for automotive catalysis) or in the form of a material obtained from merging oxides with two or more elements to make metals and alloys. However, the final form in which REEs are sold depends on the application. For instance, in the 1970s and 1980s most REE exports from China were mineral concentrates and then mixed REE chemicals (British Geological Survey 2011). Separated REE oxides

---

[2] This general process is not used in the in-situ method in southern China, where an ammonium sulfate solution is pumped into the clay deposit to leach out the REEs into a solution, which is then pumped back to the surface and the REE salts are precipitated from the solution by addition of ammonium carbonate (British Geological Survey 2011).

and metals were in greater demand in the 1990s. Since the 2000s, REE magnets, phosphors and polishing powders have dominated REE trade. The diversity of final forms poses challenges for REE recovery and recycling as discussed later in this chapter.

One of the greatest sources of harm to environmental and human health comes from the production of mine tailings. Tailings are finely ground residual liquid wastes created by separating out the undesired material from the ores, which can include toxic metals, fluorine, and radionuclides, as well as leftover processing chemicals. These dangerous byproducts require careful storage and disposal. Typically, tailings ponds are constructed to hold this toxic wastewater and prevent it from seeping underground. All tailings at the REE mining operation in Bayan Obo, in the Inner Mongolia Province of China, collect in a tailing pond that is over fifty-years-old with a 20-meter thick sludge layer composed of raw ore, iron, niobium, and other substances (Schreiber *et al.* 2016). Left in open air, these ponds emit solvent vapors like sulphuric acid. Moreover, if the tailing ponds leak, their drainage goes on to contaminate other watersheds and soil. Local communities exposed to tailings by drinking water and by eating locally grown food are at greater risk of developing certain illness and premature death.

Water and land contamination from tailings can last hundreds of years, damaging land-based livelihoods for generations. Baotou, China, is a city known as the 'rare-earth capital of the world' because it lies 120 kilometers south of Bayan Obo and is the main processing site. There, leakage from tailings ponds has displaced productive farmland and contaminated crops. By the 1990s, crops in nearby villages were failing, causing farmers to accept that crops would no longer grow and animals could not survive in the area (Bontron 2012). Ten years later, the area population had dropped from 2,000 to 300 people (*ibid.*). According to a New York Times journalist, "Whole villages between the city of Baotou and the Yellow River in Inner Mongolia have been evacuated and resettled to apartment towers elsewhere after reports of high cancer rates and other health problems associated with the numerous rare earth refineries there" (Bradsher 2013). If contaminated groundwater reaches the Yellow River, as many as 150 million people may also be exposed to its risks.

A number of factors will affect the degree of pollution a local community experiences, including fuel sources, the makeup of an area's business cluster, and local permitting and waste regulations. The solvent extraction processes itself varies little from place to place in terms of material efficiencies and chemical usage (Schreiber *et al.* 2016). However, some energy mixes emit more harmful emissions than others, with emissions from coal fired power plants producing the greatest amount of hazardous emissions. In addition, a mining site may attract other polluting industries to the area. For instance, in Baotou, an investigation into mining pollution in the surrounding area by the municipal environmental protection agency found that REE mining and processing facilities had caused the pollution originally, but the problem had been exacerbated by the dozens of factories and industrial services that had been built near the REE processing facilities and the fossil-fuel power station (Bontron 2012). By contrast, this kind of business clustering would be more difficult to achieve in places with strict environmental permitting.

The existence and scope of hazardous waste regulations also determines the impacts of mining pollution. For instance, in Sweden, a mining operation's tailings must be treated and stored in a special facility, with any radioactive waste stored separately. In contrast, China has relatively weak environmental standards when it comes to rare earth mining. Due to China's complex government structure in which local, province, and central authorities often compete against each other for control, enforcing environmental law is a struggle (Packney & Kingsnorth 2016). Bayan Obo's radioactive sludge is also stored separately, but in an open facility (Schreiber *et al.* 2016). This means there may still be exposure to the sludge's radioactive elements, such as thorium, which can cause cancers of the pancreas and lungs and leukemia.

Finally, human rights abuses can be substantial across all stages of a mining operation. Human rights are a set of civil, political, economic, social, and cultural rights articulated in the 1948 Universal Declaration of Rights that have been accepted by most governments as well as by the International Labor Organisation (ILO) (Handelsman 2002). Mining companies have often been criticized for their complicity in human rights abuses related to their treatment of local and indigenous people in mining operations, including negotiation of land access and resettlement of local people. Labor abuses are prevalent as well, including child labor, modern slavery, violating worker rights to organize, and racial and sexual discrimination. The use of security forces to protect mining operations is also a major contributor to human rights abuses and in many cases has fueled conflict in areas already affected by unrest, economic deprivation, and weak governance. Mining companies often turn to police, mercenaries, and private companies for security. When these forces are empowered to commit human rights violations for control over resources, and the host government lacks the means or will to intervene, violent conflict may grow and persist.

4. Ethics in the REE Value Chain

Given the environmental and social impacts outlined in the previous section, what are the ethics of REE production and use? Two concepts of justice are useful here: environmental justice and intergenerational justice. Below, we define both concepts and the moral obligations they raise for those involved in REE value chains, before turning to a discussion of potential strategies for developing REE use in the next section.

The concept of environmental justice emerged from activists and researchers in the US who have shown that the negative environmental effects of industrial activity (*e.g.* air pollution, water and soil contamination from hazardous waste) tend to concentrate in disadvantaged communities (according to race and/or income) (Szasz & Meuner 1997, Schlosberg 2013). In ethical terms, environmental justice invokes the principle of distributive justice by calling for greater fairness in the distribution of risks and benefits of industrial activity – not by redistributing harm to other groups, but by reducing harm in the communities that endure a greater share of harm than other groups. Examining environmental justice in global value chains raises a number of questions about where risk is produced in the global economy (Iles 2004). This requires recognizing those who have been injured (*e.g.* a community or certain groups of workers) versus those who are responsible (*e.g.* corporations or states), and determining what actions or remedies are necessary to address the injustice as well as who has the power to carry this out (*e.g.* government, industry, civil society, or international organizations) (Schlosberg 2013). In other words, who is generating harm, and who is exposed to this harm? Who should take responsibility for the harm, and what should be done for those who have been harmed?

Environmental justice suggests that chemists and others scientists, product designers, and business managers all have a moral obligation to understand the geography of their REE supplies and act to source REEs responsibly. This requires understanding the changing geography of RRE mining and processing. REEs can be found worldwide, but as with most mineral resources, geology is not the primary factor shaping where mining occurs. Both REE supply and demand has been concentrated in China for many decades. China is not only a major producer of REEs, but also a major industrial consumer of REEs. This means that many of the environmental injustices associated with how REEs are extracted and used in the manufacturing sector concentrate in the Chinese communities which host mining, waste disposal, and manufacturing plants. In recent years, the Chinese government has been developing a practical guidance to facilitate the development of 'Green Mines', which focuses on increasing financial support to the different levels of government interested in implementing environmental and efficiency performance standards for upgrading existing mines and building new mines (Dolega & Schüler 2018). However, the Green Mines standard is a management standard applied during mine construction and retrofit, and lacks the influence of strict environmental regulation and enforcement. Moreover, its application is concentrated in Eastern China (Lei *et al.* 2016), whereas most REE mining is in China's northern Inner Mongolia region.

However, as REE mining projects have been expanding outside of China, so too has the map of environmental injustices associated with REE production and industrial use. New REE mining projects have sprung up around the world to capitalize on rapidly increasing REE prices, as the Chinese government has reduced the availability of REE supplies in other countries first through a series of export taxes and export quotas, choosing to promote state-owned mining companies that specialize in REE extraction and processing (Phadke 2018). Investors and governments have argued that diversification of REE mining will reduce dependency on Chinese-sourced REEs (Worstall 2010). Outside of China, notable projects emerged in Australia, the United States, and

Malaysia soon after China restricted supplies (Haque *et al.* 2014), with a growing number of projects coming on line in recent years in Burma/Myanmar, Vietnam, Brazil, Russia, and India (Figure 2). In 2007, Lynas Corporation began mining its Mount Weld deposit in Western Australia, the richest known deposit of rare earths outside of China, and is now the world's only significant rare earths producer outside of China. The US Mountain Pass mine in southern California also temporarily reopened under new management in response to China's trade policies, with a new wastewater system to manage tailings closer to the mine site and avoid the piping of wastewater. Investors are also eying potential REE projects in Vietnam, Brazil, Russia, India, Canada, South Africa, Malawi, Kazakhstan, and other countries, although proposed projects must first be proven to be viable through exploration, and even after that, commercial production will not come online until the late 2020s (US Geological Survey 2018).

After answering the question of who is being harmed by REE production, remedying environmental injustice entails negotiating rights, remedies, and responsibilities. Solutions depend importantly on where the injustice takes places. In countries that have strong environmental and public health laws, remedies to environmental injustice may center on ensuring existing laws are being followed. In countries with insufficient legal protections to offer those communities and the environments impacted by mining, companies will need to either disengage from suppliers associated with serious impacts or increase their involvement in the supplier's operations to ensure REEs are not causing environmental injustices. In either case, companies who are REE consumers have an obligation to avoid creating harm, and the diverse geography of REE extraction and production requires a strong management system that can detect where environmental injustices are taking place.

The demands of intergenerational justice add further complexity to managing REE value chains ethically. Existing REE-based businesses have already caused great damage that will reverberate across future generations. How should future generations be taken into account given REE use is projected to grow? The concept of intergenerational justice extends the timeframe in which we consider the moral claims of actors impacted by REEs, beyond those who are currently alive. Intergenerational justice asks the question, how are future generations impacted by the actions of current generations? Whereas environmental justice focuses our attention on the existing unfair distribution of risk across space and social groups, intergenerational justice focuses on fairness in the future. This idea is becoming prominent in climate change activism. For instance, movements like Extinction Rebellion and Fridays for Future, as well as lawsuits led by school-age children and young adults, argue that their opportunities for their future lives have been imperiled by insufficient government action now and in the past.

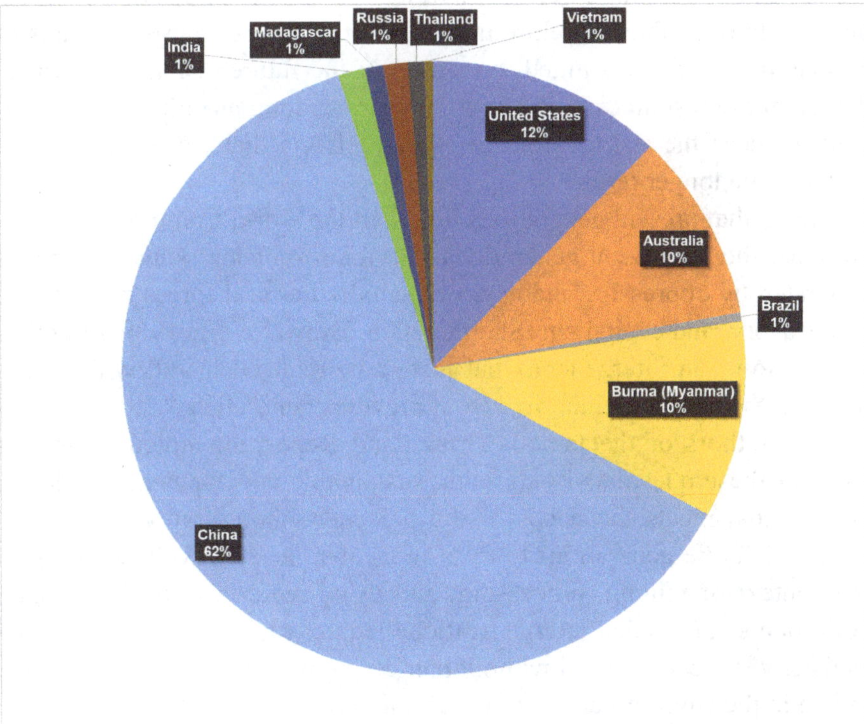

Figure 2. Rare earth production by country in 2019 (based on USGS 2020).

Examining intergenerational justice requires identifying what obligations people today owe to people in future, often referred to as 'intergenerational equity'. One proposal for intergenerational equity is that present generations must preserve the opportunities for future generations to live well (Curren & Metzger 2017). But what standard of well-being should be attained? How far into the future are present generations obliged to consider? Do present generations have obligations to future generations only in their own society/nation/community, or do their obligations extend to future peoples everywhere? Must current generations safeguard the state of the world they inherited, or must they work to make it better (Spjikers 2018)? Is it enough if present generations pass on the capabilities needed to achieve well-being, such as technological innovation or knowledge for repairing or replacing what has been destroyed? For instance, is it ethical for current generations to deplete rare earth resources now, so long as they develop innovations that enable these materials to be manufactured, substituted, or reused in the future? Or is it ethical to deplete these resources in the next 20 years so long as doing so dramatically reduces the risk of climate change effects that cause societies to suffer in the longer term?

Given that present generations inherited the world from previous generations, should present generations also focus on addressing the wrongs generated by oppressive and environmentally harmful forces of colonialism, slavery, and capitalism (Shelton 2008, Brown-Weiss 1989)? In other words, how can intergenerational justice be realized in full without addressing the injustices inherited from previous generations?

The authors of the famous Brundtland Report attempted to answer some of these questions by defining sustainable development as "development that meets the needs of the present without compromising the ability of future generations to meet their own needs" (WCED 1987). In the context of mining, processing, and using materials, the Brundtland definition suggests that intergenerational equity can be achieved if industrial activities can proceed without hampering the ability of future people to live in the environment from which the resource was taken.

In practice, this definition of intergenerational equity means adhering to environmental laws, labor laws, and human rights law; or, in cases where such laws are weak or poorly enforced, a variety of industry best practice standards or codes of conduct can provide guidance. Civil society organizations (*e.g.* Amnesty International), international governance organizations (the Global Compact, the OECD), and industry associations (*e.g.* the International Council on Mining and Minerals) have put forward numerous codes of conduct for the mining industry. Typically these codes require certain impacts to be measured, monitored, and reduced, according to internationally accepted definitions of labor rights and human rights as defined by the 1948 Universal Declaration of Rights, the International Labor Organisation (ILO), and the UN Guiding Principles on Business and Human Rights (UNGP), which is the global authoritative standard on business and human rights.

However, formal laws and codes of conduct may not go far enough to ensure the well-being of future generations. Legal institutions typically struggle to define the rights of local communities and indigenous peoples compared to those for workers because such definitions require cultural-specificity and often entail engaging in ongoing controversies over how nation states have defined the sovereignty of local people and indigenous groups. In other words, formal laws and codes offer little guidance for ensuring intergenerational justice for local communities and indigenous groups in places where there are disputes about how much control these communities have over land, natural resources, and the wealth created by developing the natural resources.

In addition, in situations where a mining company does not adhere to formal environmental laws that prevent or remediate toxic pollution, such pollution may endure for generations. In such situations, some argue that industrialized nations have an obligation to pay for remediation through foreign aid and anticipatory reparations to the communities harmed by destructive industrial practices (Westin 1992). In this view, societies where the consumption of REEs takes place must compensate the REE host societies, and such payments can be used to ensure that current and future generations in those host societies live well. This reasoning is strongest in situations where production costs would be substantially higher (and thus corporate or government profits much lower) if the minerals came from the consumer countries with strong regulations. As one government official in China's Longnan County in Jiangxi Province – the center of REE mining in China – explained:

> [Technology] companies have benefited from using our rare earth resources, they should bear a part of the responsibility and join the pro-

> cess of cleaning up the environment […]. We have made huge sacrifices to extract the resources they need" (Standaert 2019).

Yet realizing intergenerational equity in this case depends importantly on how the host government uses foreign aid payments within its territory. Cleanup of tailings pollution, for example, is extremely difficult and can take upwards of decades to 100 years before local environmental systems recover. In the meantime, local communities suffer from the loss of livelihoods as well as from potential health impacts from exposure to pollution.

Another challenge in enacting intergenerational equity is determining how the removal of a resource from a particular community impacts future generations. Mining permanently pulls resources from underground and relocates those resources elsewhere. For future generations, the inability to access that resource may fundamentally hamper their ability to achieve well-being. For instance, consuming highly desirable REEs now could mean they are not available for future generations who may require them for their livelihood, be it in the form of technologies like wind turbines, electric cars, or technologies we do not yet have that address some not-yet-identified need. This possibility raises additional questions about REE management, such as whether existing and known REE stocks should be managed in some equitable fashion to meet unknown future needs? If so, how should present generations estimate the amount needed for future generations and the proper threshold depletion rate that must be observed in order to prevent scarcity later on? How should present generations balance the need to take extreme climate interventions now (like switching to all-electric cars in a short time) against the possible longer term benefits and losses (such as preventing a 4°C world from happening, in which much of the world may become uninhabitable due to heat, sea level rise, and other effects)? Answering these questions requires giving those whose descendants may be most harmed by REE mining a say in how decision-making about where and how REE extraction takes place.

These challenges suggest the need to take a more expansive view of intergenerational equity – one that goes beyond the definition of the Brundtland Report to include the politically unpopular propositions of redistributing the wealth created by REE mining more equitably and empowering those impacted by REE mining in decision-making about REE activities. The Brundtland definition of sustainable development emphasizes intergenerational equity in terms of preventing or correcting environmental damages in mining communities, without attending to re-distribution and empowerment. Many mining communities oppose mining on the grounds that they are completely left out in decision-making and wealth creation. Typically, the process of licensing and permitting an area for mining is carried out by governments who control access to underground minerals. Governments then require mining companies to pay concessionary fees to prospect minerals and royalties on any mineral earnings. These economic gains may be shared with the community (as described above in the case of the Alaska Permanent Fund) or used to benefit a larger public. For instance, in the case of Botswana, government mining revenue paid for schools, roads, hospitals, and other public infrastructure (Poteete 2009). The Alaska Permanent Fund (APF) is another example of how wealth re-distribution may be addressed in extractive industries (Devarajan *et al.* 2011). The APF is a state-owned corporation that collects revenues generated by Alaska's oil and gas leases and manages these assets through reinvestment. Earnings from these investments are distributed among Alaskan households, and a quarter of the principal (25%) is reserved for future generations. The APF trustees have the authority to distribute dividends accordingly, such that all Alaskans benefit economically from the use of public resources.

But redistribution arrangements alone do not address the need to empower citizens to advance intergenerational justice. Decision-making about how and whether the resource can be used requires open and inclusive decision-making processes in which communities are empowered to determine whether and how resource should be exploited at all, or limited to certain uses (*e.g.* some communities may question the use of raw materials for military applications) instead of indiscriminately in international commodity markets, and how the benefits of exploited resources should be distributed fairly between current and future generations. Of course, such an approach raises further questions of procedural justice, such as who should participate in decision-making about resource access and use. These complex questions will vary by context. Local communities may be united in fundamentally opposing mining activities, especially when those activities threaten to displace local people. But there are numerous cases in which communities look to mining for a combination of economic opportunities and self-determination (Bryceson 2018). For example, as mentioned before, government and mining companies ignored villagers' opposition to the use of land for mining in the Bougainville Island of Papua New Guinea, and violent conflict ensued for decades rendering the mine inoperable (Denoon 2000). Similar conflicts have arisen over mining in many developing countries of the Global South where national governments have deliberately implemented mining regulations and policies for the exclusive interests of large-scale mining companies, while making small-scale mining illegal (Moretti & Garrett 2018). In Bougainville, this conflict is ongoing, but recent developments have empowered local people in mining decisions. The Bougainville Island Mining Act of 2015 contains legal provisions for local elder councils and village assemblies to designate and regulate small-scale gold mining, giving local landowners the power to veto any mining licenses as well as to create 'community mining areas' for smaller-scale mining (O'Faircheallaigh *et al.* 2016).

The important point here is that intergenerational equity requires going beyond the standard of assuring environmental protection. Determining the criteria for extracting REEs will undoubtedly entail conflicts between different beliefs about who should control resources, and whether extraction is justified at all, or for specific markets only – such as for renewable energy technologies – and whether some amount of a resource should be preserved for a community's descendants, who may need them more.

5. Strategies to Advance Environmental and Intergenerational Justice

For chemists, product designers, and business managers working upstream, what strategies are available for extracting non-renewable, depletable REEs in a way that does not create new – or exacerbate existing – environmental injustices, nor impair the opportunities of future generations to live well? Given the lack of existing examples that would satisfy the requirements of environmental justice and intergenerational justice, we discuss some promising strategies and their shortcomings. Calls for 'sustainable mining' have fallen short despite more stringent regulations and industry codes that aim to protect the environment, human rights, and workers. Societies must therefore go beyond sustainable mining protocols and also wrestle honestly with the challenges of figuring out how to meet intergenerational justice claims when it comes to REE use here and now.

5.1 Reducing and replacing REEs

Some large industries that use REEs discovered during the 2010 supply scare that they could do without some of them. When the price of lanthanum soared, oil refinery operators temporarily stopped using this rare earth even though it improves refining efficiency. The glassmaking industry largely abandoned using cerium for polishing. More may be done to find designs that keep REE use to the minimum. However, many REEs needed for high-technology products have no or low potential for adequate substitution with other materials (Graedel *et al.* 2015). For example, dysprosium (used in permanent magnets in computers and wind turbines), europium and yttrium (used in flat panel displays), and thulium and ytterbium (used in laser technologies) do not have straightforward substitutes available. The lack of replacements suggests sharply increasing recycling is one strategy for REE supply chains, which requires implementing a circular flow of REE material through different stages in a product's lifecycle, from design, to end-of-life collection, to separation and recycling.

5.2 Innovations to REE manufacturing and bypass mining

Some industries that rely on REEs are looking for ways to bypass mining entirely by extracting REEs from other materials. For example, the US could someday obtain these elements as byproducts from power plant coal ash and coal mining waste. And the problem of radioactive material mixed in with ores could end up being positive: If thorium-based nuclear plants prove viable, expanded thorium mining would also turn up usable rare earth minerals. However, insofar as such innovations rely on energy production that poses significant risks to local communities, these approaches cannot satisfy the requirements of environmental justice, let alone intergenerational justice.

5.3 Circular economies for REE recovery and recycling

Recovering and recycling rare earth metals is one possible way of avoiding the ongoing environmental and intergenerational injustices of mining. However, only a very small proportion of REEs becomes recycled from products, some estimating less than 1% (Binnemans *et al.* 2013).

One reason is that the amount of rare earth elements that can be recovered from electronics, medical devices, and similar applications is very small, often less than one gram (Bonawandt 2013). Typically, recycling requires that rare earths be separated from metals and alloys created with REEs. For instance, the Japanese mining company Dowa began harvesting circuit boards, hard drives, computer chips and other components for rare earth metals by cutting these components into 2 cm squares, smelting them at 1,400° C, which enables separation of the various components. For every 300 tons of e-waste smelted, the harvestable rare earth material is only about 150 grams. Although REEs are valuable, Dowa would not be profitable were it not for other materials, such as gold, silicon, *etc.* (Tabuchi 2010).

Another issue is that there is no standard method of recycling REEs, and the processes for doing so are considerably costly and environmentally hazardous – some on par with mining. Several efforts are underway to make REE recycling more efficient (Harler 2018). Researchers working under the US Department of Energy's Critical Materials Institute have focused on developing a single-step process to recover REEs from scrap magnets in order to recover the ores from hard drives, magnetic resonance imaging machines, cell phones, and hybrid cars (ORNL 2019). For instance, using membrane solvent extraction, about 3 kilograms of magnets can yield about 1 kilogram of rare earth metals. Other US researchers have been improving an older method of isolating REEs from magnets and scrap metals using molten magnesium (Bonawandt 2013). Researchers in Belgium are using ionic liquids to separate REEs from magnets, a process that uses trihexyl(tetradecyl)phosphonium chloride to transform metals like iron, cobalt, magnesium, and copper into a liquid phase, leaving the rare earths behind in an aqueous state. Researchers at Japanese car manufacturer Honda have found a way to extract rare earths from nickel-metal hydride batteries from hybrid vehicles by using molten salt, and claim as much as 80 percent of REEs being recycled. In addition to these separation challenges, there are also challenges in handling reclaimed REEs due to their air reactivity, which can render them into oxides if left out in the open for too long.

Manufacturing blended REE materials is one alternative to the challenges of purification and the relatively small amounts of pure REE that can be recovered from many products. For instance, scientists and engineers working at Momentum Technologies and the DOE's Critical Materials Institute are producing a blended REE product from recovered hard drives and other technology waste (Harler 2018). After extracting iron and boron, the recovered rare earth metal product includes a mixture of neodymium, dysprosium, and praseodymium. Technology companies and other manufacturers may be willing to take this blended product that combines all three REEs as long as the material meets manufacturing requirements.

One strategy for enhancing the profitability would be to target REE recovery and recycling initiatives in supply chains with much larger REE quantities. For instance, it may be more profitable to work with the REEs in specific supply chains, such as sustainability technologies like wind turbines and electric cars or specific consumer electronics. Some argue that recycling of e-waste will have little impact on REE supplies until there is enough material in the recycling stream to keep up with REE demand. This assumes that manufacturers' only recourse is to wait for a steady flow of recycled REEs to become available for purchase on the world market. However, the recycling of REEs can also be pursued at the firm or industry level through a circular economy approach. The term 'circular economy' refers to 'close the loop' business models that replace the 'take-make-dispose' models, or what some now call the 'linear economy'.

Individual firms could take a product-centric approach to closing the loop for REE reuse as well. A closed-loop system developed internally would keep REEs and other materials in circulation for as long as possible. This would mean that downstream manufacturers, product designers, engineers, and business take control of their upstream REE supply chains, re-circulating REEs rather than purchasing mined REEs or waiting for a sizable market of recycled REEs to develop. Such product-centric design approaches require attention to disassembly: designers and engineers must understand how complex products break down into component parts and how particular materials behave in order to design products for easy separation. For instance, the circuit board of an electronic product may be redesigned so that its metals are easily removed from other plastic, aluminum, and steel parts. Product-centric recycling systems must be designed by those with knowledge of the chemical and physical properties of waste containing REEs, physical separation methods, physical and chemical recycling methods, as well as the thermodynamics of a specific plant's processing to assess material performance with regard to energy efficiency, durability, and manufacturing compatibility, in addition to recyclability (UNEP 2013, Kaya 2016). Liberation modeling is an important tool in this regard because it focuses on defining recyclate grades in a way that allows a common language to develop among engineers, policy specialists, and environmentalists about the trade-offs of different design approaches (UNEP 2013).

Policymaking has an important role to play here. Product-centric design for a circular economy must be undertaken in collaboration with policymakers as well as planning and recycling professionals who can help design collection systems for waste products and discourage informal or illegal disposal. Producer-responsibility laws, recycling targets, and other policy-based incentives can help to incentivize circular economy innovations from specific manufacturers and entire industries. For instance, the 2012 European Parliament law to reduce electronic waste requires member states to collect 45 tons of e-waste for every 100 tons of electronic goods sold in the previous three years, which has pushed companies and governments to develop better collection systems. In 2015, the European Commission launched its Action Plan on the Circular Economy, which aims to go further by pushing companies to re-design products to be durable and made with materials that can be re-used again and again.

In theory, a circular economy would keep harmful material from entering waste streams and reduce the environmental injustices created by e-waste. Instead, companies would assume responsibility and control over the entire lifecycle of all its products materials. Whether this happens in practice remains to be seen. Recycling value chains have created serious global environmental injustices, especially for the discarding and trading of electronics. Collecting and processing of electronic waste like mobile phones, computers, monitors, and televisions is dangerous and expensive to do safely (Amuzu 2018). Countries that accept e-waste from the United States, Europe, Japan, South Korea, and Australia usually lack the means to handle the materials safely (Iles 2004). E-waste workers, including children, are exposed to toxic fumes from smelting electronic parts and using acid baths to recover the valuable components, with little to no protective gear. Processing of scrap that contains lead, phthalates, chlorinated dioxins, and more, creates poor air and water quality for the entire community.

In terms of intergenerational equity, closing the loop on products that use REEs would help preserve the availability of REEs for future generations but it would raise new questions about intergenerational equity for the communities where REEs were originally mined. Thus, even if REE recycling provides a way to avoid new mining in the future, what forms of intergenerational equity are available to communities where lives and livelihoods have become negatively impacted by REE mining or workers who have become dependent upon REE mining despite its pollution, human rights and labor abuses?

6. Conclusion

In conclusion, we have shown that participants in the REE sector have a moral obligation to use these materials in an ethical manner by advancing environmental justice and intergenerational justice. REE value chains are complex, with rare earths feeding into many different end-uses. The increasing demand for REEs has encouraged investors to expand the geography of REE mining to avoid dependency on Chinese imports subject to price spikes. The large-scale mining industry's poor record of environmental destruction and human rights abuses will continue in order to meet REE demand unless more ethical strategies are developed for sourcing REEs. One of the best ways for REE users to advance environmental justice and intergenerational justice is to make REE reclamation a product-centric circular economy. However, consumer products represent only one segment of the REE sector. Where possible, product designers, material scientists, and engineers should fully take into account the risks and limitations of relying on such resources and design new products to reduce the use of REEs.

References

Abraham, D.: 2015, The Elements of Power Gadgets, Guns, and the Struggle for a Sustainable Future in the Rare Metal Age, New Haven: Yale University Press.

Ali, S.: 2014, 'Social and Environmental Impact of the Rare Earth Industries', *Resources*, **3**(1), 123-134.

Amuzu, D.: 2018, 'Environmental injustice of informal e-waste recycling in Agbogbloshie-Accra: urban political ecology perspective', *Local Environment,* **23**(6), 603-618.

Alonso, E.; Sherman, A.M.; Wallington, T.J.; Everson, M.P.; Field, F.R.; Roth, R. & Kirchain, R.E.: 2012, 'Evaluating rare earth element availability: A case with revolutionary demand from clean technologies', *Environmental Science & Technology*, **46**(6), 3406-3414.

Balaram, V.: 2019, 'Rare earth elements: A review of applications, occurrence, exploration, analysis, recycling, and environmental impact,' *Geoscience Frontiers*, **10**(4), 1285-1303.

Binnemans, K.; Jones, P.; Blanpain, B.; Gerven, T.; Yang, Y.; Allan, W. & Buchert, M.: 2013, 'Recycling of rare earths: a critical review', *Journal of Cleaner Production,* **51**, 1–22.

British Geological Survey: 2011, *Rare Earth Elements Mineral Profile* [online available at: https://www.bgs.ac.uk/downloads/start.cfm?id=1638, accessed 17 February 2020].

Bonawandt, C.: 2013, 'Recycling Rare Earth Metals Presents Challenges, Opportunities', *Engineering.com* [online available at: https://www.engineering.com/Blogs/tabid/3207/ArticleID/5693/Recycling-Rare-Earth-Metals-Presents-Challenges-Opportunities.aspx?e_src=relart, accessed 17 February 2020].

Bontron, C.: 2012, 'Rare-earth mining in China comes at a heavy cost for local villages', *The Guardian* [online available at: https://www.theguardian.com/environment/2012/aug/07/china-rare-earth-village-pollution, accessed 17 February 2020].

Bradsher K.: 2013, 'China Tries to Clean Up Toxic Legacy of Its Rare Earth Riches', *New York Times*, 22 October, p. B1.

Brickley, P.: 2016, 'Mountain Pass Rare Earths Mine Gets Financing', *Wall Street Journal* [online available at https://www.wsj.com/articles/mountain-pass-rare-earths-mine-gets-financing-1472674576, accessed 17 February 2020].

Brown-Weiss, E.: 1989, In Fairness to Future Generations: International Law, Common Patrimony and Intergenerational Equity, Tokyo: United Nations University.

Bryceson, D.: 2018, 'Artisanal gold-rush mining and frontier democracy: Juxtaposing experiences in America, Australia, Africa and Asia', in: K. Lahiri-Dutt (ed.), *Between the Plough and the Pick: Informal, Artisanal and Small-Scale Mining in the Contemporary World*, Canberra: ANU Press, p. 31.

Chakhmouradian, A.R. & Wall, F.: 2012, 'Rare earth elements: minerals, mines, magnets (and more)', *Elements*, **8**(5), 333-340.

Curren, R., & Metzger, E.: 2017, Living Well Now and in the Future: Why Sustainability Matters, Cambridge, MA: MIT Press.

Darcy, J.W.; Bandara, H.D.; Mishra, B.; Blanplain, B.; Apelian, D. & Emmert, M.H.: 2013, 'Challenges in Recycling End-of-life Rare Earth Magnets', *JOM*, **65**(11), 381.

Davidson, H.: 2018, 'Bougainville imposes moratorium on Panguna mine over fears of civil unrest', *The Guardian* [online available at: https://www.theguardian.com/world/2018/jan/10/bougainville-imposes-moratorium-on-panguna-mine-over-fears-of-civil-unrest, accessed 17 February 2020].

Denoon, D.: 2000, Getting Under the Skin: The Bougainville Copper Agreement and the Creation of the Panguna Mine, Melbourne: Melbourne University Press.

Devarajan, S.; Raballand, G. & Le, T.M.: 2011, 'Direct redistribution, taxation, and accountability in oil-rich economies: A proposal', *Center for Global Development Working Paper* [online available at: https://papers.ssrn.com/sol3/papers.cfm?-abstract_id=2009385, accessed 17 February 2020].

Dolega, P. & Schüler, D.: 2018, 'China's approach towards responsible sourcing. European Policy Brief, Strategic Dialogue on Sustainable Raw Materials for Europe (STRADE)' [online available at http://stradeproject.eu/fileadmin/user_upload/pdf/STRADE_PB_03_2018_China_responsible_sourcing.pdf, accessed 17 February 2020].

Fishman, T., & Graedel, T. E.: 2019, 'Impact of the Establishment of US Offshore Wind Power on Neodymium Flows', *Nature Sustainability*, **2**(4), 332-338.

Francaviglia, R.: 1991, Hard Places: Reading the Landscape of America's Historic Mining Districts, Iowa City: University of Iowa Press.

Graedel, T.E.; Harper, E.M.; Nassar, N.T. & Reck, B.K: 2015, 'On the Materials Basis of Modern Society', *Proceedings of the National Academy of Sciences*, **112**(20), 6295-6300.

Handelsman, S.D.: 2002, *Human Rights in the Minerals Industry, Mining, Minerals and Sustainable Development*, London: International Institute for Environment and Development.

Haque, N.; Hughes, A.; Lim, S. & Vernon, C.: 2014, 'Rare earth elements: Overview of mining, mineralogy, uses, sustainability and environmental impact', *Resources*, **3**(4), 614-635.

Harler, C.: 2018, 'Rare Opportunity to Recycle Rare Earths', *Recycling Today [*online available at: https://www.recyclingtoday.com/article/rare-earth-metals-recycling/, accessed 17 February 2020].

Hoyle, R.: 2019, 'U.S. Rare Earths Revival Planned Amid Trade Conflict', *The Wall Street Journal* [online available at https://www.wsj.com/articles/joint-venture-plans-to-revive-rare-earths-processing-in-u-s-11558341663, accessed 17 February 2020].

Ichihara, M. & Harding, A.: 1995, 'Human Rights, the Environment and Radioactive Waste: A Study of the Asian Rare Earth Case in Malaysia,' *Review of European Community & International Environmental Law*, **4**(1), 1-14.

Iles, A.: 2004, 'Mapping environmental justice in technology flows: Computer waste impacts in Asia', *Global Environmental Politics*, **4**(4), 76-107.

Kaya, M: 2016, 'Recovery of metals and nonmetals from electronic waste by physical and chemical recycling processes', *Waste Management,* **57**, 64-90.

Keith-Roach, M.; Grundfelt, B.; Höglund, L.O.; Kousa, A.; Pohjolainen, E.; Magistrati, P.; Aggelatou, V.; Olivieri, N. & Ferrari, A.: 2016, 'Environmental Legislation and Best Practice in the Emerging European Rare Earth Element Industry,' in: I. Borges De Lima & W. Leal Filho (eds.), *Rare Earths Industry*, Elsevier, pp. 279-291.

Ku, A.; Setlur, A. & Loudis, J.: 2015, 'Impact of Light Emitting Diode Adoption on Rare Earth Element Use in Lighting Implications for Yttrium, Europium, and Terbium Demand', *Electrochemistry Society Interface*, **24**(4), 45-49

Lei, S.; Hanxiao, K.; Jian, W.; Xu, H. & Ke, H.: 2016, 'The Status and Achievements of Green Mines and Mining Ethics in China', *Journal of Resources and Ecology*, **7**(5), 317-322.

Lipson, D. & Hemingway, P.: 2019, 'Australian mining company Lynas gets permission to dispose of radioactive waste in Malaysia, dividing locals', *ABC News Australia* [online available at: https://www.abc.net.au/news/2019-08-22/malaysians-divided-on-radioactive-waste-from-aussie-miner-lynas/11434122, accessed 17 February 2020].

Locardi, B. & Guadagnino, E.: 1992, 'Rare Earths in Glass Technology', *Materials Chemistry and Physics*, **31**(1-2), 45-49.

Machacek, E. & Fold, N.: 2014, 'Alternative value chains for rare earths: The Anglo-deposit developers', *Resources Policy*, **42**, 53-64.

Nassar, N.T.; Du, X. & Graedel, T.E.: 2015, 'Criticality of the Rare Earth Elements', *Journal of Industrial Ecology*, **19**(6), 1044-1054.

Nieto, A.; Guelly, K. & Kleit, A.: 2013, 'Addressing criticality for rare earth elements in petroleum refining: the key supply factors approach', *Resources Policy*, **38**(4), 496-503.

Oak Ridge National Laboratory (ORNL): 2019, 'From Trash to Treasure: Electronic Waste is Mined for Rare Earth Elements' [online available at: https://www.ornl.gov/news/trash-treasure-electronic-waste-mined-rare-earth-elements, accessed 17 February 2020].

O'Faircheallaigh, C.; Regan, A.; Kikira, D. & Kenema, S.: 2016, *Small-Scale Mining in Bougainville: Impacts and Policy Responses: Interim report on Research Findings*, Griffith University [online available at: ssgm.bellschool.anu.edu.au/sites/default/files/news/related-documents/2016-07/interimresearchfindings_ssm_bougainville_260516.pdf, accessed 17 February 2020].

Off-Earth Mining Working Group: 2019 [online available at: http://www.oemwg.unsw.edu.au/, accessed 17 February 2020].

Packey DJ & Kingsnorth D.: 2016, 'The Impact of Unregulated Ionic Clay Rare Earth Mining in China', *Resource Policy*, **48**, 112-116.

Phadke, R.: 2018, 'Green Energy Futures: Responsible Mining on Minnesota's Iron Range', *Energy Research & Social Science*, **35**, 163-173.

Reisfeld R. & Jørgensen C.: 1977, 'Rare-Earth Lasers', in: Reisfeld R. & Jørgensen C.(eds.), *Lasers and Excited States of Rare Earths*, Berlin: Springer, pp. 64-122.

Roskill: 2019, *Rare Earths Outlook to 2029*, 19th Edition [online available at https://roskill.com/market-report/rare-earths/, accessed 17 February 2020].

Sadeghbeigi, R.: 2012, *Fluid Catalytic Cracking Handbook*, Waltham, MA: Elsevier.

Schreiber, A.; Marx, J.; Zapp, P.; Hake, J.; Voßenkaul, D. & Friedrich, B.: 2016, 'Environmental Impacts of Rare Earth Mining and Separation Based on Eudialyte: A New European Way' *Resources*, **5**, 32.

Schlosberg, D.: 2013, 'Theorising Environmental Justice: The Expanding Sphere of a Discourse', *Environmental Politics,* **22**(1), 37-55.

Shelton, D.: 2010, 'Intergenerational Equity (presentation) and Discussion following the presentation by Dinah Shelton', in: R. Wolfrum & C. Kojima (eds.), *Proceedings of the Solidarity: A Structural Principle of International Law Symposium*, Heidelberg: Springer, pp. 123-168.

Spijkers, O.: 2018, 'Intergenerational Equity and the Sustainable Development Goals', *Sustainability*, **10**(11), 3836.

Standaert, M.: 2019, 'China Wrestles with the Toxic Aftermath of Rare Earth Mining', *Yale e360* [online available at: https://e360.yale.edu/features/china-wrestles-with-the-toxic-aftermath-of-rare-earth-mining, accessed 17 February 2020].

Szasz, A. & Meuser, M.: 1997, 'Environmental Inequalities: Literature Review and Proposals for New Directions in Research and Theory', *Current Sociology,* **45**(3), 99-120.

Tabuchi, H.: 2010, 'Japan Recycles Minerals From Used Electronics', *New York Times* [online available at: https://www.nytimes.com/2010/10/05/business/global/05recycle.html?pagewanted=all&_r=0, accessed 17 February 2020].

Tanabe, S.: 2015, 'Glass and Rare-Earth Elements: A Personal Perspective', *International Journal of Applied Glass Science*, **6**(4), 305-328.

UNEP (United Nations Environment Programme): 2013, Metal Recycling: Opportunities, Limits, Infrastructure, A Report of the Working Group on the Global Metal Flows to the International Resource Panel, Nairobi.

UNEP (United Nations Environment Programme): 2015, Guidance Document: Developing a National Action Plan to Reduce, and Where Feasible, Eliminate Mercury Use in Artisanal and Small-Scale Gold Mining, Nairobi.

US Department of Energy: 2011, 'Critical Materials Strategy' [online available at: energy.gov/sites/prod/files/DOE_CMS2011_FINAL_Full.pdf, accessed 17 February 2020].

US Department of Energy: 2017, 'Report on Rare Earth Elements from Coal and Coal Byproducts' [online available at: https://www.energy.gov/sites/prod/ files/2018/01/f47/EXEC-2014-000442%20-%20for%20Conrad%20Regis%202.2.17.pdf, accessed 17 February 2020].

US Geological Survey: 2018, 'Rare earths' [online available at: https://minerals.usgs.gov/minerals/pubs/commodity/rare_earths/, accessed 17 February 2020].

US Geological Survey: 2020, *Mineral Commodity Summaries 2020* [online available at: https://doi.org/10.3133/mcs2020, accessed 17 February 2020].

Visiongain: 2012, *The Rare Earth Market 2012-2022* [online available at https://www.visiongain.com/Report/843/The-Rare-Earths-Market-2012-2022, accessed 17 February 2020].

WCED (World Commission on Environment and Development): 1987, *Our Common Future*, Oxford: Oxford University Press.

Widmer, J.; Martin, R. & Kimiabeigi, M.: 2015, 'Electric Vehicle Traction Motors Without Rare Earth Magnets', *Sustainable Materials and Technologies*, **3**, 7-13.

Worstall, T.: 2010, 'You Don't Bring a Praseodymium Knife to a Gunfight', *Foreign Policy* [online available at https://foreignpolicy.com/2010/09/29/you-dont-bring-a-praseodymium-knife-to-a-gunfight/, accessed 17 February 2020].

Zepf, V.: 2016. 'An Overview of the Usefulness and Strategic Value of Rare Earth Metals', in: I. Borges De Lima & W. Leal Filho (eds.), *Rare Earths Industry*, Elsevier, pp. , pp. 3-17.

Chapter 13

The Chemical Prediction of Stratospheric Ozone Depletion: A Moral Model of Scientific Hazard Foresight

Joachim Schummer

Abstract: When Mario Molina and Sherwood Rowland in 1974 predicted the depletion of stratospheric ozone through chlorofluorocarbons (CFCs), which posed a threat to almost all terrestrial life, they initiated an unprecedented and still unique political process that led to a global ban of CFCs and other ozone depleting substances. After a brief introduction to atmospheric chemistry and the history of CFCs, the first part of this chapter narrates the history of that prediction, its experimental verification, including the discovery of the Ozone Hole, and the international political consequences. The second part investigates if and to what extent chemists have a moral duty to research and warn us of possible hazards, taking Molina and Rowland as moral role models.

1. Introduction

Chemists have frequently been accused of hubris, *i.e.* the neglect of foresight of the adverse consequences of their own products or processes in an overtly optimistic manner. Prominent cases include DDT (Børsen & Nielsen 2017), thalidomide (Ruthenberg 2016), the Bhopal disaster (Eckerman & Børsen 2018), and many others included in this volume. The advantage of discussing such 'negative' cases from an ethical point of view is that they allow drawing general lessons to avoid similar cases in the future by providing ethical guidance. The downside is that they tend to equate chemistry with the chemical industry and ignore the academic field of chemistry, including ethical issues of understanding rather than issues of making or using things.

However, there are also numerous ‘positive’ cases where chemical research – particularly in analytical, environmental, and physiological chemistry – has allowed the prediction and prevention of hitherto unknown harm. In this chapter, I will call the proactive scientific research in order to foresee and warn of possible harm, which the respective researchers have in no way caused themselves, ‘scientific hazard foresight’. That differs from mere circumspection and foresight to prevent harmful consequences of one’s own actions, for which one is usually responsible. Because ‘scientific hazard foresight’ helps prevent harm, it is of foremost ethical importance, although the disasters that actually occur receive much more public attention than the intricate efforts to avoid them.

To be sure, scientists from other disciplines than chemistry have warned the public of possible and hitherto unknown hazards as well. Prominent historical cases include the marine biologist and freelance writer Rachel Carson pointing to the ecological dangers of insecticides such as DDT. Geologists try to warn of volcano eruptions, earthquakes, and tsunamis; meteorologists of severe weather, including hurricanes, drought, and flooding; physicians of epidemics, malnutrition, and insane lifestyle; astronomers would do so of asteroid impacts; and so on. In addition, climate researchers warn of anthropogenic climate change caused by the emission of carbon dioxide and other infrared radiation absorbing gases.

However, the most prominent and influential case from chemistry, perhaps from all of science, is Mario Molina and Sherwood Rowland’s warning of stratospheric ozone depletion by chlorofluorocarbons in 1974, then widely used as aerosols and refrigerants, posing a threat to almost all terrestrial life. Fortunately the global threat was, based on the chemists’ prediction, fast tackled by a UN convention, a unique achievement in international environmental law up to then. The discovery of the Antarctic ‘Ozone Hole’, a special phenomenon of local and seasonal stratospheric ozone depletion, helped make an unanimous agreement worldwide. Section 2 narrates the history of the prediction of stratospheric ozone depletion (2.3) and its political consequences (2.4) after a brief introduction to atmospheric science (2.1) and chlorofluorocarbons (2.2).

Molina and Rowland are role models, not only in scientific matters for which they earned the Nobel Prize of Chemistry in 1995, but also in moral matters. The ethical lesson to learn from that 'positive' case is that scientists have indeed a moral duty of researching and warning of possible hazards, which Molina and Rowland and many others have perfectly fulfilled, but which many scientists are not always aware of. After a brief introduction, Section 3 provides the ethical justification for that duty (3.2) and discusses its limits (3.3) as well as the question if scientists could delegate the moral duty to an institutionalized form of technology assessment by division of labor (3.4).

2. The Prediction of Stratospheric Ozone Depletion

2.1 The stratospheric ozone layer

In the troposphere, the lowest layer of the Earth's atmosphere with an average thickness of about 15 km, ozone (O_3) is a rare and evanescent gas. It quickly reacts with moisture and many organic substances, dissociates under UV radiation, and is toxic to all living beings, which makes it a standard chemical for water purification because of its short lifetime. Under certain conditions ozone also cleans the air of industrial pollutants, such as carbon monoxide (CO) and hydrocarbons (C_nH_m). However, when nitrogen oxides (NO_x) from high temperature combustion are available and the sun is shining, they catalyze a complex reaction cycle in which ozone is continuously formed, a phenomenon known as photochemical smog. Then the ozone concentration can quickly rise from average tropospheric values of about 20 parts per billion (ppb) to toxic levels of several hundred ppb.

Ozone was not part of the Earth's early atmosphere, neither was free oxygen (O_2) in considerable amounts from which it was originally formed. Instead water vapor (H_2O), carbon dioxide (CO_2), and nitrogen (N_2,) were its main components. Measurable amounts of O_2 in the atmosphere emerged only around 2.45 billion years ago, in the so-called Great Oxygenation Event (Holland 2006, Knoll & Nowak 2017). It was largely created by cyanobacteria in the oceans through photosynthesis.

For the next 1.6 billion years the atmospheric concentration stagnated around 2%, while the oceans became saturated and the minerals in the oceans and on the Earth's surface were oxygenated. In the geologically short period of 850-540 million years ago the concentration quickly rose to its current level of around 20%, while CO_2 was bound in biomass and minerals. The end of that geochemical period, also known as the Neoproterozoic Oxygenation Event, coincides with the emergence of large and complex multicellular organisms, including those that could stand solar radiation on land, such as plants. Note that most organic substances decompose under the UV radiation emitted by the sun, such that terrestrial life as we know it, if not protected in deep sea levels or cavities, would not survive without any UV protection.

The geochemical event that made life on land possible was the formation of the stratosphere, the atmospheric layer at around 15-50 km, in which ozone plays the dominant role. At first glance the layering of the atmosphere appears counter-intuitive. Why, and in what regard, should the gaseous surroundings of a planet be layered? Photochemistry provides the answer.

Imagine an atmosphere without layers in which at increasing height the atmospheric pressure slowly reduces (because of weaker gravitational force), the temperature slowly decreases (because of lower absorption of the infrared radiation from the Earth's surface) and the relative gas concentrations (or relative partial pressures) are stable. Once solar UV radiation acts upon that atmosphere, photochemical reactions take place first in the upper region. In particular, O_2 splits into O atoms by UV absorption (λ<240 nm) (Reaction 1). The O atoms, which are free radicals, quickly react with O_2 to form O_3 (Reaction 2). They can also destroy ozone (Reaction 3) but that is less likely because of the lower concentration of O_3 compared O_2. The main decomposition of ozone occurs by UV absorption (λ<350 nm) and the catalytic help of various other compounds (Reaction 4), which forms O atoms that in turn quickly react according to Reaction 2.

$$O_2 + UV \rightarrow 2\,O \qquad (1)$$

$$O + O_2 \rightarrow O_3 \qquad (2)$$

$$O + O_3 \rightarrow 2\,O_2 \qquad (3)$$

$$O_3 + UV\,(+Cat) \rightarrow O_2 + O \tag{4}$$

The continuous UV-absorbing formation and decay of O_3, according to Reactions 1-4, known as the ozone-oxygen or Chapman cycle since 1930, produces heat and forms an ozone-oxygen equilibrium that depends on concentrations, temperature, and UV radiation, shaping the stratospheric ozone profile (Figure 1). Because the upper region of the atmosphere is more exposed to UV and shields the lower region by UV absorption, the heat producing Chapman cycle becomes less active as we move downwards. Therefore the temperature decreases from upper to lower levels, in the opposite direction compared to the Earth's surface region, the troposphere, where the temperature gradient is governed by the infrared radiation from the Earth. As a result we have a temperature minimum at the level where the Chapman cycle becomes almost inactive, which is called the tropopause, where troposphere and stratosphere meet. On the upper side there is a temperature maximum (the stratopause) because gas concentrations decrease with height, making chemical reactions less likely. The stratosphere is thus that region of the atmosphere where the heat-producing and UV-absorbing Chapman cycle is active, governed by the formation and decay of ozone that protect life from dangerous UV radiation and letting through only some of the much less dangerous UV-A part (λ>315nm). It is a fragile structure because the ozone-oxygen equilibrium also depends on the concentrations of various compounds that catalyze only the decay of ozone (Reaction 4) but not its formation, as we will see later.

The troposphere and stratosphere are not only distinguished by their opposite temperature gradients and different chemical compositions, also the dynamics of their air masses greatly differ from one another. What we call weather – the vaporization of surface water and the condensation of vapor in the form of rain, the buoyancy of warmed air masses on the surfaces, and the resultant temperature and pressure differences that cause the winds – is largely confined to the troposphere. In contrast, the stratosphere is rather calm and protected against tropospheric dynamics by the inversion of the temperature gradient. Heated air masses that move up the tropospheric temperature gradient because of their lower density through buoyancy stop at the tropopause. In the stratosphere, gases usually move rather by diffusion than by bulk movement, such that the exchange between the two spheres is very slow. However, the tropopause is not a fixed level but varies by local turbulences, daytime, seasons, and, to the largest degree, by latitude, overall at altitudes between 6 and 18 km.

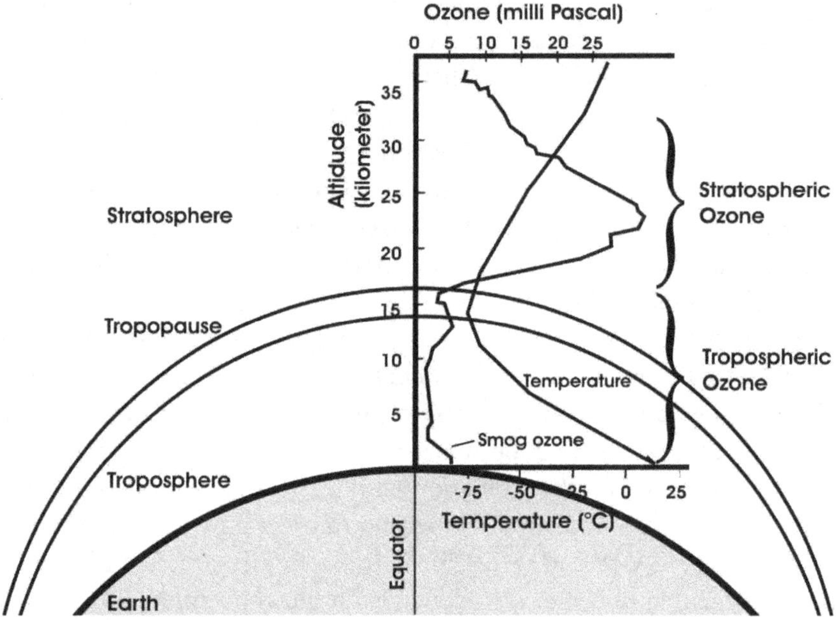

Figure 1: Temperature and ozone concentration in the lower atmosphere (adapted from Parson 2003).

The layering of the atmosphere continues as we move upwards. Like the stratosphere, which is formed by the Chapman cycle, the thermosphere (ca. 80-500 km) is built by UV absorption, but now of higher energy capable of ionizing various compounds resulting again in higher temperature at higher levels, but with ozone being almost absent. Between the stratosphere and the thermosphere is the somewhat ill-defined mesosphere where temperature decreases with height, like in the outermost 'layer', the exosphere, that extends towards the open space. All these layers are not fixed, as textbooks tend to portray them. They greatly vary by various parameters, and the only practicable way to distinguish them is by temperature gradient.

In geological times, the UV-protection shield was built up in steps. First, the thermosphere emerged (then extending to much lower levels) by absorption of the high-energy part of solar UV radiation (ca. 10-122 nm) and the ionization of most substances. Once sufficient amounts of oxygen could survive beneath the ionization radiation shield, it absorbed UV in the range up to around 200 nm wavelength by dissociation into O radicals, which allowed building the stratosphere by the formation of ozone, whose photodissociation absorbs UV up to almost 315 nm, called the Hartley Band or UV-B. Next to the high-energy part, UV-B is particular destructive to organisms, because it affects weaker chemical bonds typical of biomolecules. It is this UV-B that would increase by the emission of chlorofluorocarbons.

2.2 Chlorofluorocarbons

Before we discuss how chlorofluorocarbons (CFCs) were predicted to affect the ozone layer, it is useful to look first at their earlier history. CFCs comprise a large set of compounds, basically hydrocarbons with all hydrogen atoms substituted by various combinations of chlorine and fluorine atoms.

The story of CFCs begins in 1891 when the Belgian chemist Frederic Swarts (1866-1940) tried to produce the first organic fluorine compound by mixing tetrachloromethane (CCl_4) with antimony trifluoride (SbF_3). At first he obtained trichlorofluoromethane (CCl_3F), a colorless liquid that boils at 24° C, and then produced numerous other organic fluorine compounds by the same method (Kaufmann 1955). Swarts seemed to have no direct commercial interest as he did not file any patent; his research aimed at establishing a new substance class and exploring their physical and chemical properties. The strong inertness of the compounds did not suggest any commercial use to him.

That radically changed in the 1930s when a team of chemists at General Motors in the US, led by Thomas Midgley (1889-1944), turned Swarts' synthesis into the large-scale industrial production of CCl_3F, CCl_2F_2, and other chlorofluorocarbons that came to be known as freons under patent protection for a variety of uses. Their low boiling point, chemical inertness, stability, noncombustibility, and nontoxicity made

CFCs ideal refrigerants. Earlier refrigerators, based on closed cycles of dimethyl ether, ammonia, sulfur dioxide, or methyl chloride, all carried severe risks of fire, explosion, or toxicity if the refrigerant leaked out the closed cycle by corrosion or accidents. Perhaps more than any other industrial chemical, CFCs changed people's life style, particularly by the use of refrigerators and air-conditioning, and made living in extremely warm and humid climates comfortable. CFCs soon became also the standard propellants in ordinary sprays (colloquially called 'aerosols'), such as in hair sprays, and replaced the earlier pump spray systems. The chemical industry widely used them as blowing agents for foam rubber and as solvents in chemical processes. Their chemical inertness also made them ideal fire extinguishing substances that would just disappear after their usage without causing any damage.

Based on voluntary reports by the world's leading manufacturers, the commercial production of the two main CFCs (CCl_3F and CCl_2F_2, known as CFC-11 and CFC-12) took off in the mid-1940s, with about 20,000 metric tons in 1945, and rose to more than 800,000 tons in 1974 (see Figure 2, AFEAS 1993, Aufhammer *et al.* 2005). After that it sharply declined in the US, and after about 1987 also in the rest of the world where there was still growth in the early 1980s. At the beginning CFCs were only produced in the US, but in the late 1950s, through licensing and then by fading local patent protection, production started in various other countries and outpaced US manufacturers around 1970.

The first knowledge of the atmospheric distribution of CFCs was produced by the independent British scientist James Lovelock (b. 1919), whom many know only from his Gaia hypothesis: the earth as a self-regulating system in which life, as we have seen in the last Section, shapes its own atmospheric environment. Trained as a chemist and physician, Lovelock became an entrepreneur and writer after he had revolutionized analytical chemistry with his invention of the electron capture detector (ECD) for gas chromatography. When a gas sample is injected into the top of a tube filled with a stationary and permeable material, such as a gel, and carried through the tube by a constant stream of an inert gas, *e.g.* N_2 or helium, the compounds of the sample are emitted one after the other at the end of the tube at characteristic retention times, depending on their different interactions with the stationary material.

Once the apparatus is calibrated with standard samples, you can qualitatively and quantitatively analyze gaseous or volatile mixtures, provided you have a suitable detector at the end of the tube. That technique, gas chromatography, was already developed in the mid-1940s, particularly by the German physical chemist Erica Cremer. However, Lovelock's ECD made gas chromatography useful for environmental analysis, as it increased the sensitivity by several orders of magnitude up to a then incredible level of one part per trillion (ppt) (Morris 2002). The ECD operates with a radioactive element, such as ^{63}Ni, that emits electrons to a positively charged anode, thus closing an open electric circuit and producing a steady electric current. Gas molecules, such as CFCs, that pass through the electron ray and can absorb electrons and are then detected by a temporary decrease of the current.

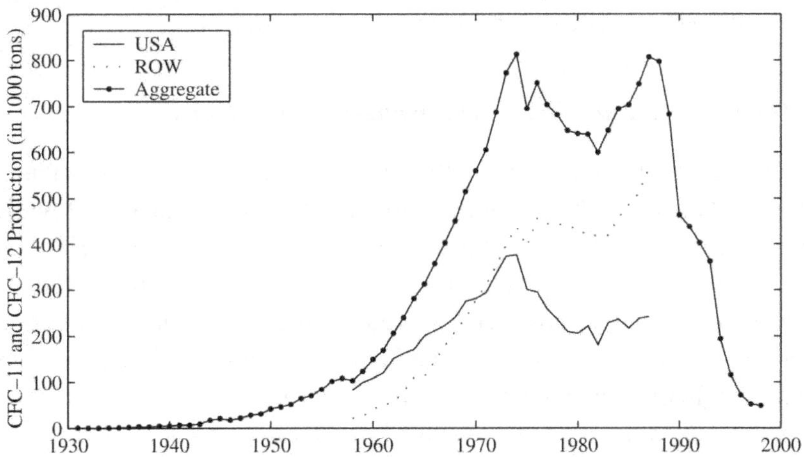

Figure 2: The Production of CFC-11 and CFC-12 in the US and the rest of the world (ROV) (from Auffhammer et al. 2005).

Lovelock, who had already found considerable CFCs concentrations in the air over Ireland, embarked on a ship towards Antarctica in 1971 to measure the latitudinal distribution of CFCs. His original concern was not environmental, as nobody imagined any threat from these inert compounds. Rather he was using CFCs as tracers to better understand tropospheric dynamics, particularly the exchange between the northern hemisphere (where CFCs had been manufactured and used) and the southern hemisphere (where they had not). His original research proposal at a British funding agency was rejected because the expert panel found it impossible that one could measure compounds at the ppt level, particularly not with an instrument that was built "on the kitchen table", as Lovelock loved to portray it (Lovelock 2000, pp. 206ff.). But it worked out, and the results showed tropospheric abundance all the way down to the Antartica, from which Lovelock concluded that CFCs lack a sink and accumulate in the atmosphere (Lovelock *et al.* 1973). In 1974 he also measured for the first time, and by the same technique, the vertical distribution of CFCs in the troposphere and lower part of the stratosphere by taking samples from a military aircraft (Lovelock 1974). This study was now undertaken from an environmental angle, since Rowland and Molina had published shortly before their groundbreaking research on the detrimental effects of CFCs on stratospheric ozone.

2.3 The prediction of CFCs' stratospheric ozone depletion

During the 1960s and early 1970s, when the environmental movement started and raised public concerns about the adverse impacts of various technologies, the stability of the stratosphere became an matter of scientific investigations. Nuclear tests, rockets, and supersonic aircraft all intervened with the stratosphere. Moreover, the Chapman cycle, based on laboratory experiments of the kinetics of the individual reactions, predicted a higher ozone concentration than was actually calculated from the incoming UV radiation on the Earth's surface. Laboratory experiments proved that various possible compounds could selectively catalyze the photodissociation of ozone (Reaction 4), thereby shifting the equilibrium. Thus, if by human activity these catalysts would move into the stratosphere, the ozone layer could easily be damaged.

The first candidate discussed was HO radical formed by photodissociation from the exhaust of water vapor of supersonic aircraft, which flew in the stratosphere. However it turned out that the impact was very low compared to other natural routes of HO formation. A second candidate, first suggested by Dutch atmospheric chemist Paul Crutzen (1970), was NO. It usually oxidizes in the troposphere to NO_2, but it can also newly be formed in the stratosphere from the persistent nitrous oxide (N_2O), or directly exhausted by supersonic aircraft (Johnston 1971). N_2O is naturally produced by soil bacteria and thus is a natural source of ozone depletion, but its effect grows by the increased use of fertilizers (organic or not) in agriculture. The third candidate was the chlorine radical (Cl) of which both natural and human sources at first appeared of minor importance – particular concerns were about the proposed space shuttle exhaust of HCl.

All these scientific hypotheses received tremendous media attention around 1970, particularly in the US. While original concerns about damaging the ozone layer focused on the effect of climate change, fears grew even bigger when physiologists pointed out the effects on life on earth, including rising skin cancer rates of humans. Because of the novel kind of threat posed by humans messing with the stratosphere, numerous experts panels and committees were established, besides the regular meetings of the atmospheric science societies. The debates had a major impact on the formation of a new kind of governmental institution in the US in 1972, the Office of Technology Assessment (Kunkle 1995), later copied in many other countries. The agency should routinely scrutinize the possible adverse effects of new or emerging technologies and provide advice to government.

When the young Mexican chemist Mario J. Molina (1943-2020) joined the physical chemistry group of F. Sherwood Rowland (1927-2012) as postdoc at the University of California, Irvine, in the fall of 1973, Rowland had recently read Lovelock's paper on the global abundance of CFCs. He suggested to his postdoc, among others, to research the photo-kinetics and life cycle of CFCs to see if they could be a source of stratospheric Cl causing ozone depletion, a project for which he had already obtained a grant. Both Molina and Rowland had no particular background in meteorology, they were physical chemists with a focus on kinetics. However, the work was quickly done in four months by literature research because all the required data had been published before. They just had to be composed into a newly developed atmospheric life cycle model of CFCs, predicting stratospheric ozone depletion by human activity, which the authors submitted to *Nature* in January 1974 (Molina

& Rowland 1974) and for which they would share with Paul Crutzen the Nobel Prize of Chemistry in 1995.

Their model consisted of four theses that they quantified as far as possible. First, as Lovelock had argued before, CFCs distribute and accumulate in the troposphere because there is almost no sink (by photolysis, chemical reaction, or rainout) resulting in lifetimes of 40-150 years. Second, the only important sink is the slow diffusion to the stratosphere, where CFCs photodissociate by UV radiation ($\lambda<220$ nm), absent in the troposphere, to form free Cl atoms, *e.g.*:

$$CCl_2F_2 + UV \rightarrow CClF_2 + Cl \tag{5}$$

Third, among all possible stratospheric reactions of Cl the most likely and effective one is the depletion of ozone by formation of clorine monoxide (ClO):

$$Cl + O_3 \rightarrow ClO + O_2 \tag{6}$$

Fourth, and most importantly, the most likely and effective reaction of ClO is with O (produced from the photodissociation of O_2 according to Reaction 1) which again sets free Cl:

$$ClO + O \rightarrow Cl + O_2 \tag{7}$$

Hence, the net reaction of (6) and (7) is the depletion of ozone by O in which Cl is regenerated and thus formally acts like a catalyst:

$$O_3 + O + Cl \rightarrow 2\,O_2 + Cl \tag{8}$$

Because all known competing reactions to (6), which would consume Cl, are slow or reversible, the authors concluded that Cl, once set free in the stratosphere, would deplete ozone for years, at a five times higher rate than NO.[1] And because the amount of CFCs that had already been set free in the troposphere slowly diffuses into the stratosphere, we should expect ozone depletion for a "lengthy period" in the future, even if CFCs production would immediately stop today (*ibid.*).

Molina and Rowland, who elaborated on their model in a series of further papers, were initially not right in every detail. In particular, they

---

1 Replacing Cl with NO in equations (6)-(7) yields the core mechanism for ozone depletion by NO as suggested by Crutzen (1970).

predicted a faster and longer-lasting ozone depletion than was actually measured much later, because they had initially ignored an important sink (the reaction of Cl with N_2O) and used inaccurate reaction rate constants from the literature (particularly for the formation of chlorine nitrate) which they later corrected themselves. Many other chemists helped develop the model further in various details, but the core of the model for predicting global stratosphere depletion has remained largely intact. Various theses were soon confirmed by atmospheric measurements that involved large research teams, including the tropospheric accumulation of CFCs and the presence of CFCs and ClO in the stratosphere. However, because of various difficulties (technical, financial, and natural[2]), it was not before 1988 that global stratospheric ozone depletion could be experimentally verified first by analysis of ground station data (WMO 1991, p. 4)[3] and then combined with satellite data (Stolarski *et al.* 1992). The results showed, for instance, for the northern mid-latitudes an average stratospheric ozone decrease of around 2% per decade in the period 1970-1991 with strong seasonal and regional variations.

In their original model, Molina and Rowland did not yet consider local and seasonal meteorological phenomena and knowingly excluded, for absence of scientific knowledge at the time, the "possible heterogeneous reactions of Cl atoms with particulate matter in the stratosphere" (Molina & Rowland 1974). That turned out to be crucial for what came to be known as the 'Ozone Hole', a temporary strong decrease, up to 50%, of stratospheric ozone in the Antarctic spring and more recently also in the Artic spring. The Antarctic Ozone Hole was first discovered by UV measurements from the British Antarctic ground station (Farman *et al.* 1985). It was only retrospectively confirmed for the previous years by the satellite based Total Ozone Mapping Spectrometer (TOMS), operated by NASA since 1978, who seemed to have mistaken the unusually low ozone values for measurement errors over several years, lacking a routine for pattern recognition in the huge amounts of data produced every year (Christie 2000, pp. 43-52; Conway 2008, p. 73).

---

2 Natural difficulties included accounting for the solar cycle and the impact of volcano eruptions.

3 The main results were already made public in 1988 (Kerr 1988).

Much more so than in the Arctic, a strong circular wind (the polar vortex) surrounds the Antarctic in winter that isolates the polar air masses from the temperate latitudes, resulting in very low temperatures and polar stratospheric clouds (PSCs) composed of solid water and nitric acid (HNO_3). In the absence of light and UV for about 3 months, there are no photochemical reactions forming O and Cl, such that both the Chapman cycle and the Molina-Rowland cycle are frozen in winter. The exact chemical mechanism that leads to the polar Ozone Hole, to the understanding of which both Molina and Rowland considerably contributed, is very complex and perhaps not yet fully understood. In simple terms, stratospheric Cl and ClO is during the polar winter temporarily stored in various compounds (including hydrogen chloride (HCl) and chlorine nitrate ($ClONO_2$)), which on the surface of stratospheric cloud particles react to form molecular chlorine (Cl_2) and hypochlorous acid (HOCl). Once the sun rises in spring, Cl_2 and HOCl are photolyzed to Cl and OH by visual light. ClO can set free further atomic Cl (via dimerization to chlorine oxide dimer (ClOOCl) and photolysis to 2 Cl and O_2). Under these conditions ozone depletion occurs again by equation (6) of the Molina/Rowland model, while atomic chlorine is now recycled from ClO not by equation (7) but via

$$2\,ClO \rightarrow ClOOCl \rightarrow 2\,Cl + O_2 \qquad (9)$$

The crucial point is that in the first weeks of the polar spring, there is already enough visible light for these photodissociations that cause ozone depletion, whereas UV is still too scarce for the photodissociation of O_2 required for ozone formation after the Chapman cycle (Reaction 1). Note that at the small angle of solar radiation in the polar spring, when the sun hardly moves above the horizon, all light towards the polar region enters the atmosphere at much higher latitudes and passes a wide cross-section of the stratosphere. Under these conditions, UV is almost completely absorbed while visual light is only scattered to some degree but reaches the polar region.

In addition to CFCs, there are other ozone depleting substances (ODS). In particular, brominated organic compounds, which were widely used in fire extinguishers, set free bromine radicals in the stratosphere that react in a similar way as Cl. All ODSs came to be regulated and eventually banned by a series of international conventions.

2.4 Political consequences

Molina and Rowland did not only publish scientific papers of their findings, they also talked to fellow scientists, both chemists and meteorologists, at a time when the barrier between both disciplines was still very large. Moreover, after a presentation at an American Chemical Society meeting in September 1974, Rowland gave a press conference that led to a series of alarming articles and reports in national media. Numerous interviews with journalists would follow as well as political consultancy in the form of testimonies at congressional hearings and memberships of expert panels.

When scientific results have political implications, researchers can suddenly find themselves involved in public discourses that greatly differ from that of the scientific community. Emotions like fear and hope, personal rivalry, corporate interests, and indirect political interests might effectively shape the discourse. An important intermediary zone between science and the public are committees, commissions, or panels that are composed by a mixture of scientific expertise, established authority, representation of special interest groups, and political orientation. In the case of the potential ozone depletion by CFCs, several such committees were soon launched in the US, but neither Molina nor Rowland were originally members of any of them because their case had to be assessed 'independently'.

While an interagency task force (IMOS) and two committees by the National Academy of Science largely supported the Molina-Rowland model, as did several other scientists in their publications, and, directly or indirectly, called for legal action, others were more skeptical. The industries involved in the production and use of CFCs launched a Committee on Atmospheric Science to deny the ozone-depleting effect of CFCs, largely by denying all four theses of the Molina-Rowland model and by pointing to volcanos as natural sources of ozone depletion (Oreskes & Conway 2010, pp. 114f.). The debate in the US has frequently been called "The Ozone War", echoing Dotto & Schiff 1978, but that is, compared to other countries, perhaps an exaggeration because most of the counter-arguments were very poor by scientific standards and hardly heard. It is also fair to say that DuPont, the main US manufacturer of CFCs, repeatedly stated since 1974 that they would stop production once credible evidence was provided that CFCs had harmful effects (Parson 2003, p. 33). The intended debate on what counts as 'credible evidence' was certainly a measure to buy time before legal action would be taken (Smith 1998), but many US consumers had already refused to buy spray cans based on CFCs aerosols.

It took hardly more than four years from the first public media report on the Molina-Rowland model to the legal ban of 'nonessential' CFCs aerosols in sprays in the US in December 1978. Canada, Sweden, Denmark, and Norway followed soon, and some other countries required restrictions. Note that at that time stratospheric ozone depletion was still a scientific prediction, not yet confirmed by any measurement.

Two agencies of the United Nations (UN) took the initiative and brought the issue on the international agenda, the World Meteorological Organization (WMO), founded in 1950, and the United Nations Environment Programme (UNEP), established in 1972. In the mid-1970s they both began coordinating international research based on expert panels that defined research needs. Eventually in 1981, UNEP called for an international convention to protect the ozone layer, which would be negotiated for six years (Petsonk 1990). Strong objections to radical measures came from Japan and several West European countries, where manufacturing plants for CFCs had recently been built. As a first step, the Vienna Convention for the Protection of the Ozone Layer was signed on 22 March 1985, two months before the discovery of the Antarctic Ozone Hole was published[4] and years before global stratospheric ozone decrease was confirmed by measurements. The Vienna Convention established only a legal framework for future coordinated research and regulative activities to be defined by 'protocols'. The first one, a true landmark in international environmental law, was the Montreal Protocol on Substances that Deplete the Ozone Layer signed in 1987, now under

[4] The paper (Farman *et al.* 1985) was received 24 December 1984, accepted: 28 March 1985, and published: 16 May 1985.

the impact of the Ozone Hole discovery. It provided a binding schedule for member countries to phase out the production of CFCs and other ODS, varying by the developmental status of the countries and the uses of ODS, the effect of which can be seen in Figure 2. Several amendments that established faster and stricter bans on ODS have followed since, such that by current estimates stratospheric ozone would recover by the end of the 21st century.

The Montreal Protocol is unique in several regards in the entire world history. First, it is the only international agreement ratified by all 196 UN member states, reaching a 100% consensus. Second, it is the first ever international agreement on collectively dealing with an environmental issue, which at the time of the signature was not even sure to affect all latitudes, and which had to be balanced against huge economic interests. The convention and the political process to establish it became a model for many later environmental efforts, particularly on climate change that had been discussed since the 1960s. However, third, the short timeline of 14 years, from the first scientific publication that raised awareness of the issue to international signature, will probably remain unparalleled for a long time. Forth, it is perhaps the strongest impact on international politics that a single chemical research project has ever had, competing only with the discovery of nuclear fission by the chemists Otto Hahn and Fritz Strassmann in 1938.

Although climate change, to which CFCs contribute, has absorbed much public attention and funding of atmospheric science, ozone depletion is still a focus of ongoing scientific research with the aim of preventing harm. For instance, in a recent paper by Stephen A. Montzka *et al.* (2018), the authors find tropospheric increase of CCl_3F since 2012 originating from eastern Asia, which would be a clear violation of the Montreal Protocol.

3. Scientists' Moral Responsibility of Hazard Foresight

3.1. Introduction

The historical case above invites ethical analysis. For instance, it provides the most prominent and historically most influential example of the precautionary principle (PP), before ethicists had discussed it. The PP requires that protective measures should be undertaken if there is considerable evidence, rather than decisive proof, for a hazard, and that the burden of proof is on the side of those who deny the hazard (see also Llored 2017, and chapter 19 of this volume). The international community of responsible politicians did exactly so in 1987, before global ozone depletion was empirically measured. Indeed, the mentioned US task force (IMOS) argued in that direction already in their 1975 report (Parson 2003, p. 35).

A second point for ethical analysis involves the moral responsibility of CFCs manufacturers before and after the Molina & Rowland paper of 1974. That would have to deal with the epistemological view by many academic chemists and physicists, according to which all one needs to know about a compound is its molecular structure. Because the molecular structures of CFCs are very simple and were all well-known by 1974, they created the illusion of perfect knowledge and safety. No chemist would by then imagine that the ozone depletion potential is a property of chemical compounds, which it is of course. The ethical analysis should thereby turn into an epistemological analysis of a dangerous misunderstanding about chemical knowledge, and the uselessness of risk analyses in chemical issues where we do not even know yet where to look for risks.

However, the ethical topic to be discussed in the following, for which there is perhaps no better case than the present one, is this: Do scientists have a special moral duty for researching and warning us of possible hazards? Is there a scientific foresight responsibility justified on ethical grounds? For a start, let us look at four questions.

First, was Molina and Rowland's 1974 publication, in which they warned of stratospheric ozone depletion by CFCs, morally praiseworthy? Everyone threatened by DNA-damaging UV radiation would certainly agree, thereby also agreeing that, apart from the scientific quality, there is an ethical dimension of assessing research.

Second, should Molina and Rowland have been morally blamed if they had not undertaken their life cycle research of CFCs? Here I assume all scientists and most others would strongly disagree. Praise and blame are not symmetrical: we usually do not blame somebody if he has not done something praiseworthy; and we usually do not praise somebody if she has omitted to do something blameworthy, say, a robbery. Instead, we blame someone for failure of the morally expected, and we praise only for doing more than is morally expected. However, what is expected may change over time and depends on whether we look at it in the present or retrospectively.

Third, let us modify the last question further and counterfactually assume that other scientists, X&Y, first discovered the threat posed by CFCs, but much later: Should Molina and Rowland have been morally blamed if they had not undertaken their life cycle research of CFCs earlier, although they had all the required resources and capacities to do what X&Y actually did later? Many scientists, who tend to look at this as a case of research competition only, might disagree. But others would perhaps wonder about the inactivity, the lack of concern for public welfare.

Fourth, let us, more counterfactually assume, an inverted timeline. At first physicians noticed an increased rate of skin cancer, which scientists after a while attributed to higher UV-B radiation, the cause of which is after long research and debates found in the stratospheric ozone depletion by the emission of CFCs. The rapid spread of skin cancer expedited the political process of negotiating a worldwide ban on CFCs. Then people would ask: 'Why didn't scientists investigate the threat of CFCs before it was commercially produced and emitted?' In that case I assume both the manufacturers and the scientists would be morally blamed by most people. Note that the inverted timeline has been the rule in the past for many cases, with notable exceptions such as the prediction by Molina and Rowland.

The three counterfactual examples illustrate the space of morally assessing the omission of foresight research, from being morally neutral to questionable and blameworthy. Before exploring the limits of foresight responsibility in Section 3.3, let us first look at its ethical justification.

3.2 Ethical justification of the scientists' special duty to hazard foresight

In order to investigate if scientists bear special ethical responsibility for researching and warning us of possible hazards, let us first look at the moral duty to rescue and then discuss if the former is a special case of the latter.

The moral duty to rescue is nicely illustrated by an example by ethicist Peter Singer (1997): you notice a person has fallen into a lake and appears to drown. It would be easy for you to rescue the person at the expense of some dirty clothes and a short while of your time. Do you have a moral duty to do so? Most people would certainly agree; in dozens of countries the omission to help would even be a crime. Moreover, the duty to rescue can be justified by all major ethical theories that judge actions.

In utilitarianism, you ought to act so that you maximize happiness and reduce suffering overall. The little inconvenience of your dirty clothes counts close to nothing compared to the drowning of the person to rescue, such that rescuing is imperative here. In deontological ethics, which formulates general moral duties, the general duties of benevolence and to avoid harm are typically on top of the priority list and both imply the specific duty to rescue. Both the Golden Rule and the Kantian categorical imperative provide further justifications. If you were in the case of drowning in the lake, you would morally expect from anyone passing by to help you. And given the risk that you, like anyone else, could by sudden physical misfortune come into such a situation, you would reasonably want that the duty to rescue becomes a general moral obligation for anyone who is able to do so.

The ethical justification of the duty to rescue is based on three important conditions. First, you must be intellectually capable to foresee the risk, here, to distinguish between playful swimming and the danger of drowning. Second, you must be physically and intellectually capable to take useful measures for preventing the risk, here, being able to swim.

Third, your own costs and risks of rescuing should be reasonable, here, for instance, nobody expects you to rescue somebody at the risk of your own life or serious damage to your health.

Taking the first two conditions into account, not everyone has the same moral duty to rescue. A child or disabled person who does not recognize the danger or is unable to swim can in retrospect not be held morally responsible for letting someone drown. On the other hand, people with special capacities to foresee a danger and take counteractive measures do have a special moral duty to act according to their capacities. Because scientists have particular intellectual capacities to foresee risks by knowledge and research, which no one else has, they bear a special duty to warn of possible risk that is justified on general ethical grounds. And in so far as they have particular intellectual and practical capacities for taking or inventing preventive measures against such risks, they also bear a special moral duty to do so.

The general moral duty to rescue thus depends on one's own intellectual capacity with important implications for scientists to research and warn us of possible hazards. While the notion that scientific knowledge implies special responsibilities has frequently been claimed, it has rarely been justified by ethicists and explicitly acknowledged by scientists. Fortunately, however, it has implicitly been followed by many scientists, not the least by Molina and Rowland.

3.3 Limits of scientific foresight responsibility

Some chemists, in an effort to point out the importance of their discipline, tend to say that everything is chemistry. They are likely unaware of the ethical implications of that claim. If everything were chemistry, then every possible harm that happens in the world would have chemical causes and thus could potentially have been foreseen and perhaps even prevented by chemists. Chemists would then always have to be blamed for their omissions.

However, it follows from the ethical justification above that foresight responsibility is restricted to one's actual, rather than pretended, intellectual capacity. That includes, but is not limited to, one's areas of specialization and competence. To the intellectual capacity of scientists also belongs the ability to work themselves into new subject matters, even if that transcends their disciplinary boundary. Recall that Molina and Rowland were physical chemists, with little knowledge about meteorology at first, but could develop an atmospheric life cycle model within a few months. It further includes the ability to identify and contact specialists in different fields, which is how much of today's interdisciplinary research is initiated. Thus, chemists who suspect a potential threat that they do not immediately understand, *e.g.* the uses of a combination of chemicals in a factory that might turn poisonous or explosive, are morally obliged to do some research on their own or contact fellow scientists with respective expertise.

A second limit of responsibility derives from the creativity of research. A potential hazard might not be noticeable by way of conventional thinking, but once you have spent some creative thinking on the issue – usually by looking at it from entirely different angles and questioning the received assumptions – it become obvious. Retrospectively it might even appear so obvious that people wonder why nobody had seen it before, as in the Molina-Rowland case, which can mislead moral assessments afterwards. The intellectual capacity of creative thinking greatly varies among people, including scientists. However, the profession of science consists in producing novel knowledge that no one has ever thought before, for which creativity is essential and expected. Thus, although nobody is expected to do Nobel Prize-winning research, scientists are morally obliged to use their capacities for creative thinking in identifying potential hazards.

Should scientists actively search for potential hazards or just research those cases where they intuitively suspect a threat? Imagine possible but undefinable hazards looming in a laboratory where researchers work and that you are responsible for their health. A responsible chemist would take all safety measures, including the employment of safety devices against unknown risks, and undertake research to identify potential threats. Because the active search for potential hazards belongs to the actual habit and capacity of experimental chemists, unlike for instance of theoretical physicists, it would be expected from them also in cases outside the lab.

However, there are several limits to the obligation of active research. First, research into potential hazards from chemical interactions is indefinite because one would have to investigate all possible chemical combinations under all possible conditions. Hence, it is inevitable to focus at first on cases where hazards of greater harm are suspected. Second, hazard research would consume all time, leaving no more room for other research, and it would lead to vast multiple research in parallel as long as the usual procedures of science are not established, the division of labor by subdiscipline-building and the documentation of research by publications. However, the division of labor creates new problems as we will see in the subsequent Section.

Finally, if chemists actually identify a threat, how far should they go to make their findings public? Of course a scientific publication, in order to receive the honor of being the first one who discovered the hazard, is not what is morally expected. The publication only serves to make the claim sound by scientific standards and communicates the issue to fellow scientists. One has to raise one's voice louder and inform journalists, governmental agencies, and NGOs on the issue. However, much more so than in the times of Molina and Rowland, alarmism has become part of the standard rhetoric of science PR to attract or justify funding. Against the background of that noise, serious issues might remain unheard which is a severe fault of the science-society relationship. Scientists are morally obliged to raise their voices, but society, in particular science policy makers in cooperation with scientific societies, are responsible for keeping effective channels for serious warnings.

3.4 The dilemma of institutionalized Technology Assessment

In order to avoid parallel research and to professionalize scientific hazard foresight, the institutionalization of technology assessment (TA), either as governmental research agencies or as a scientific discipline, appears to be the ideal solution. Indeed, as was argued above, the awareness of novel technological threats to the stratosphere were crucial to the establishment of the US Office of Technology Assessment in 1972, which led to many TA offices worldwide. However, institutionalization has its downsides.

First, possible technological hazards rarely match the disciplinary division of science, as the Rowland-Molina case illustrates, which required a new combination of physical chemistry and meteorology. Although that helped establish the discipline of atmospheric chemistry, other possible hazards are clearly beyond that disciplinary scope. Any effective form of institutionalization would have to be composed of many different, ideally all, disciplines to allow for the fast setup of interdisciplinary project teams tailored to specific problems. However institutionalization is a social process that establishes fixed networks, clear divisions of responsibilities, and strict conventions that tend to assume a life of their own, making it the opposite of a flexible organization that can quickly respond to new challenges. That is also true of scientific disciplines that develop their own ways of identifying and dealing with problems, thereby tending to ignore unconventional approaches. It is even more true of governmental institutes with strict divisions of labor and hierarchical orders. Who would seriously expect Nobel-Prize-winning research, such as the Molina-Rowland prediction, from a governmental officer working in a hierarchical environment? Hence, if the hazard is truly unexpected and cannot be predicted by conventional methods, it would be better to rely in addition on the broad community of scientists that include individuals who dare to use unconventional methods.

The second difficulty, which is more important in the present context, is the erosion of moral responsibility by institutionalization. With any division of labor comes a division of occupational responsibilities, such that everyone has their clearly defined occupational or professional duties. With institutionalized TA, scientists might argue it is not their business to research and warn of possible hazards because other people are responsible for doing exactly that, such that they may exclusively focus on their specific research projects. However, the argument is based on the widespread confusion between occupational and moral responsibility. It is one thing to do what your employer or peer expects you to do, and quite another one to follow ethical guidelines, and sometimes both conflict with each other.

Recall Singer's example above and assume that there is usually a lifeguard at the lake where the person is about to drown, but today the lifeguard is absent or unable to do his job for whatever reason. It is of course the occupational duty of the lifeguard to rescue drowning people, and he might even be sued for his failure. But that is no moral excuse for your own omission, instead it is your moral duty to rescue. Similarly, institutionalized Technology Assessment is no moral excuse for scientists to not care about possible hazards in the world that their intellectual capacity would possibly allow them to foresee.

Because of the first downside of institutionalization, its ineffectiveness, institutionalized TA was in all countries either abandoned or its original goal, scientific hazard foresight, was replaced by various social goals. These goals include awareness rising and consensus building on technological issues, mediation between stakeholders, and the implementation of societal needs and values in technological design. Thus, there is no more institutionalized scientific technology assessment, in the original sense, as the name misleadingly suggest, to which scientist might want to point as an excuse. However, even if that existed, scientists are morally obliged to research and warn of possible hazards, each according to their own intellectual capacities.

4. Conclusion

Most chemists, I assume, are guided in their research by some moral ideas of doing good, of improving our world, if only in tiny parts, by chemical means. The most common way in chemistry is by making new substances that can be employed for some improvements of our living conditions. However, what counts as improvements heavily depends on cultural values and life styles, such that not every change is welcomed by everyone. In addition, material improvements frequently come with downsides, advert consequences that were neither foreseen nor intended. The naive good will of doing good, without considering both possible advert consequences and cultural values, is destined to moral failure. There is no moral duty of improving by all means because that is wrong by all ethical theories.

Another way of doing good has been highlighted in this chapter: doing good by foreseeing and warning of possible harm, for which a moral duty can indeed be derived from the duty to rescue. What counts as harm, in particular harm to health, is less controversial across cultures and life styles than improvements. Moreover, the prediction and warning of possible harm leaves it up to those affected by the possible harm to choose their way of dealing with the issue. The options to choose from frequently include changes of habits and changes of technology, or a combination of both. In the case of CFC aerosols, many people first avoided spray cans and returned to pump systems, *i.e.*, they changed their habits even before the ban. Shortly later, various substitutes of CFCs were introduced as propellants, *i.e.*, a change of technology, but some of the substitutes turned out to have some ozone depletion potential as well. If people are again informed about the possible risks of alternative options, they can make their own responsible choice.

Doing good by material improvements and doing good by researching possible hazards imply different research styles. The traditional focus in chemistry has been on studying how to make things in the laboratory. In contrast, researching possible hazards requires the study of open systems that usually transcends disciplinary boundaries and is thus intellectually more demanding. Although a growing number of chemists have engaged in that kind of research since the 1960s, they are still a minor part, and marginal in the dominant self-image of chemistry, despite the extraordinary societal impact of works such as that by Molina and Rowland. The overall lesson from this chapter is thus: chemist who want to do good by any ethical standards should focus on, or should at least keep in mind at any time, scientific hazard foresight, because that is what they are morally required to do, before making new things.

In the ideal moral world, those who make things and those who research possible hazards work hand in hand from the beginning. Together they can do good by making improvements from which most risks have been eliminated long before the products come to the market.

Further Reading

The story of CFCs and stratospheric ozone depletion has been told from various angles, including epistemological (Christie 2000), political (Parson 2003), and environmentalist (Dotto & Schiff 1978, Roan 1989).

References

Alternative Fluorocarbons Environmental Acceptability Study (AFEAS): 1993, *Production, Sales and Atmospheric Release of Fluorocarbons Through 1992. Technical report*, AFEAS, Washington, DC. [available online: http://www.ciesin.org/docs/011-423/toc.html, accessed 24 October 2019].

Andrew H.; Knoll, A.H. & Nowak, M.A.: 2017, 'The timetable of evolution', *Sciences Advances*, **3**(5): e1603076 [available online: https://www.ncbi.nlm.nih.gov/pmc/articles/PMC5435417/, accessed 24 October 2019].

Auffhammer, M.; Morzuch, B.J. & Stranlund, J.K.: 2005, 'Production of chlorofluorocarbons in anticipation of the Montreal Protocol', *Environmental and Resource Economics*, **30**(4), 377-391 [available online: http://scholarworks. umass.edu resec_faculty_pubs/188, accessed 24 October 2019].

Børsen, T. & Nielsen, S.N.: 2017, 'Applying an Ethical Judgment Model to the Case of DDT', *Hyle: International Journal for Philosophy of Chemistry*, **23**(1), 5-27 [also in this volume, chapter 9].

Christie, M.: 2000, *The Ozone Layer. A Philosophy of Science Perspective*, Cambridge: Cambridge University Press.

Conway, E.M.: 2008, *Atmospheric Science at NASA: A History*, Baltimore: John Hopkins UP.

Crutzen, P.: 1970, 'The influence of nitrogen oxides on the atmospheric content', *Quarterly Journal of the Royal Meteorological Society*, **96**, 320-325.

Dotto, L. & Schiff, H.: 1978, *The Ozone War*, New York: Doubleday.

Eckerman, I. & Børsen, T.: 2018, 'Corporate and Governmental Responsibilities for Preventing Chemical Disasters: Lessons from Bhopal', *Hyle: International Journal for Philosophy of Chemistry*, **24**(1), 29-53 [also in this volume, chapter 5].

Farman, J.C.; Gardiner, B.G. & Shanklin, J.D.: 1985, 'Large losses of total ozone in Antarctica reveal seasonal Cl0x/NOx interaction', *Nature*, **315**, 207-210.

Holland, H.D.: 2006, 'The oxygenation of the atmosphere and oceans', *Philosophical Transactions of the Royal Society B*, **361**(1470), 903-915. [available online: https://www.ncbi.nlm.nih.gov/pmc/articles/PMC1578726/, accessed 24 October 2019].

Johnston, H.: 1971, 'Reduction of stratospheric ozone by nitrogen oxide catalysts from supersonic transport exhaust', *Science*, **173**, 517-522.

Kauffman, G.B.: 1955, 'Frederic Swarts: Pioneer in organic fluorine chemistry', *Journal of Chemical Education*, **32**, 301-305.

Kerr, R.A.: 1988, 'Stratospheric Ozone is Decreasing', *Science*, **239**, 1489-1491.

Kunkle, G.C.: 1995, 'New Challenge or the Past Revisited? The Office of Technology Assessment in Historical Context', *Technology in Society*, **17**(2), 175-196. [available online: https://ota.fas.org/technology_assessment_and_congress/ kunkle/, accessed 24 October 2019].

Llored, J.-P.: 2017, 'Ethics and Chemical Regulation: The Case of REACH', *Hyle: International Journal for Philosophy of Chemistry*, **23**(1), 81-104 [also in this volume, chapter 18].

Lovelock, J.E.: 1974, 'Atmospheric halocarbons and stratospheric ozone', *Nature*, **252**, 292-294.

Lovelock, J.E.: 2000, Homage to Gaia: The Life Of An Independent Scientis, Oxford: Oxford UP.

Lovelock, J.E.; Maggs, R.J. & Wade, R.J.: 1973, 'Halogenated Hydrocarbons in and over the Atlantic', *Nature*, **241**, 194-196.

Molina, M.J. & Rowland, F.S: 1974, 'Stratospheric sink for chlorofluoromethanes: chlorine atom-catalysed destruction of ozone', *Nature*, **249**, 810-812.

Montzka, S.A. *et al.*: 2018, 'An unexpected and persistent increase in global emissions of ozone-depleting CFC-11', *Nature,* **557**, 413-417.

Morris, P.J.T.: 2002, '"Parts per Trillion is a Fairy Tale": The Development of the Electron Capture Detector and its Impact on the Monitoring of DDT', in: P.J.T. Morris (ed.): *From Classical to Modern Chemistry: The Instrumental Revolution*, Cambridge: Royal Society of Chemistry, pp. 259-284.

Oreskes, Naomi & Conway, E.M.: 2010, Merchants of Doubt: How a Handful of Scientists Obscured the Truth on Issues from Tobacco Smoke to Global Warming. New York et al.: Bloomsbury.

Parson, E.A.: 2003, Protecting the Ozone Layer: Science and Strategy, Oxford UP.

Petsonk, C.A.: 1990, 'The Role of the United Nations Environment Programme (UNEP) in the Development of International Environmental Law', *American University International Law Review*, **5**(2), 351-391.

Roan, Sharon: 1989: Ozone Crisis: The 15-Year Evolution of a Sudden Global Emergency, New York et al: Wiley.

Ruthenberg, K.: 2016, 'About the Futile Dream of an Entirely Riskless and Fully Effective Remedy: Thalidomide', *Hyle: International Journal for Philosophy of Chemistry*, **22**(1), 55-77 [also in this volume, chapter 6].

Singer, P.: (1997): 'The Drowning Child and the Expanding Circle', *New Internationalist*, 5 April [available online: https://newint.org/features/1997/04/05/peter-singer-drowning-child-new-internationalist, accessed 24 October 2019].

Smith, B.: 1998, 'Ethics of Du Pont's CFC Strategy 1975-1995, *Journal of Business Ethics*, **17**(5), 557-568.

Stolarski, R. *et al.*: 1992, 'Measured Trends in Stratospheric Ozone', *Science*, **256**(5055), 342-349.

World Meteorological Organization (WMO): 1990, 'Report of the International Ozone Trends Panel, 1988', Geneva [available online: https://www.esrl.noaa.gov/csd/assessments/ozone/1988/report.html, accessed 24 October 2019].

Chapter 14

Ethics of Climate Engineering: Chemical Capture of Carbon Dioxide from Air

Dane Scott

Abstract: In the coming decades, scientists will be increasingly confronted with opportunities to pursue research with implications for one or more climate engineering proposals. Now is the time for chemists to critically reflect on the controversial possibility of managing the climate and its ethical implications. The ultimate goal of this case study on ethics and climate engineering is to promote critical reflection and discussion on the ethics of climate engineering research. To fulfill its goal, this chapter will investigate four questions: (1) Why should scientists and engineers consider climate engineering research? (2) What is climate engineering? (3) What is the substance of a common ethical objection to climate engineering, the moral hazard objection? (4) How might we begin to address crucial ethical concerns with climate engineering?

1. Introduction

As the deadline for implementing effective international political action to prevent dangerous climate change draws closer, interest in climate engineering is growing. In coming years, opportunities for scientists and engineers to conduct research directly, or indirectly, related to climate engineering will increase. But should scientists and engineers contribute to this research? Chemistry research, for example, stands to make important contributions to climate engineering, but would it be right? While modern chemistry has made enormous positive contributions to humanity, some research has caused much harm, particularly environmental pollution. As a result, professional organizations have established ethical guidelines for research. Would climate engineering research be consistent with commonly stated responsibilities to public safety, welfare, and the environment (AIChE 2015)? Would this research contribute to "environmental sustainability" and "protect the environment for future generations" (ACS 2016)? These kinds of questions require careful examination when faced with the controversial idea of controlling the climate system.

Climate engineering raises many ethical issues, but certain issues are central. The Royal Society's report on climate engineering observes that moral hazard arguments are "one of the main ethical objections to geoengineering" (Royal Society 2009, p. 39). Moral hazard objections argue that climate engineering research will undermine efforts "in mitigation and/or adaptation because of a premature conviction that climate engineering has provided 'insurance' against climate change" (*ibid.* p. 37). However, the report states that, "the moral hazard argument requires further investigation to establish how important an issue this should be for decision makers" (*ibid.* p. 45). This chapter will ultimately focus on this important, but confusing, ethical objection to climate engineering research. The moral hazard objection opens the door to other ethical issues that researchers should consider when asking, is climate engineering research the right thing to do? To set the stage for considering this question, this chapter will explore three related issues: Section two will explore why scientists and engineers should consider climate engineering research. The next section will examine the nature and varieties of climate engineering schemes. The fourth will analyze the substance of the common moral hazard objection to climate engineering. The chapter will conclude by identifying principles capable of addressing the deep ethical issues associated with climate engineering research.

2. Why Consider Climate Engineering?

2.1 Climate-intensified disasters: famine, refugees, and war

There is a two-part answer to the question, why scientists and engineers should consider climate-engineering research. The first part points to building evidence that the consequences of climate change could be catastrophic. The second part points to the slow pace of political efforts to prevent this possible catastrophe. The climate crisis is sometimes framed as a transition from the Holocene to the Anthropocene (Rockström & Klum 2015). The Holocene epoch (the past 10,000-plus years) has been exceptionally favorable for humanity. During this short period, the human population exploded from a few million hunter-gatherers to an emerging global technological civilization of over 7.5 billion people. Thousands of years of favorable climatic conditions allowed for the development of agriculture and permanent settlements, which evolved into civilizations. The calm of the Holocene is not the norm for the Earth's 4.5 billion-year history, and we can no longer take this calm for granted (*ibid.*). The impacts of 7.5 billion people on the Earth is ushering in a new epoch, the Anthropocene (Steffen *et al.* 2007). If the planet's climate system moves from the current epoch of predictability to one of volatility, the results could be tragic.

The Anthropocene, the 'human age', ironically threatens to be a tempestuous period, unwelcoming to human civilization. We can already feel the leading edge of these changes in more frequent and powerful tropical storms, longer and more extreme droughts and heat waves, sea-level rise and flooding, and massive wildfires. The American climate activist and writer Bill McKibben remarks that "in almost every corner of the Earth," climate "chaos" is inducing "an endless chain of disasters that will turn civilization into a never-ending emergency response drill" (Mann 2014). There are signs that McKibben's characterization of "climate chaos" is not hyperbole but an emerging reality for many.

In 1980, the United States' National Oceanic and Atmospheric Administration (NOAA) began keeping track of meteorological disasters costing over one billion US dollars. While there are fluctuations from year to year, the trend-line of billion-dollar disasters is clear: the number is on the rise, with 2017 being the most costly year on record. In 2017, hurricanes, floods, and wildfires caused over $300 billion in damage in the United States (Smith 2018). These dollar amounts do not take into account the immeasurable harms of human suffering and loss of life. NOAA identifies climate change as a major factor in increasing the intensity and frequency of all these disasters. The trend of more frequent and severe weather-related disasters is similar in Europe: a 2017 European Commission report warned, that if "not curbed, climate change-related disasters, with heat waves being the greatest concern, could expose some 350 million Europeans to harmful climate extremes every year" (European Commission 2017). The report highlights the need to halt climate change and adapt to its unavoidable consequences.

The countries of North America and Europe are among the largest emitters of greenhouse gases (GHG). These countries have the power to slow climate change and the capacity to adapt to many of its consequences. However, many countries in Africa, and the Southern Hemisphere in general, can do neither of these things. These countries are historically minor emitters of GHG and they lack the governmental, financial, and technological capacities needed for resilience and adaptation. Even under climate change of 2 °C to 3 °C, the countries of Africa will likely experience severe impacts.

In 2016, the United Nations' Environment Programme (UNEP) issued a report that painted a dire picture of Africa's future in the Anthropocene. The report predicts that in coming decades as the climate warms, African countries will increasingly experience more severe and frequent weather-related disasters, which will lead to famine, military conflicts, and political instability. The report warns: "Even with a warming scenario of under 2° C, Africa's undernourished would increase 25-90%. Crop production would be reduced across much of the continent as optimal growing conditions are exceeded. The capacity of African communities to cope will be significantly challenged" (UNEP 2016). If the international community is unsuccessful in reducing GHG emissions, 'business as usual' scenarios foresee an average global temperature rise of 3-4 °C by the end of the century. The consequences for Africa would be truly catastrophic.

The perfect storm bearing down on Africa has serious implications for Europe. Africa's population is predicted to double by 2050 to 2.5 billion. The combination of explosive population growth, climate-intensified disasters, and inadequate adaptive capacity could result in a refugee crisis for Europe. Stefano Torelli writes: "The combination of poverty, dependence on agriculture, environmental degradation, and population growth [...] can be expected to translate into increasing forced migration" (Torelli 2017). Torelli warns that Europe is unprepared to deal with the flood of climate refugees. The forced migration of millions of African would further stress an already-stressed Europe, which will be experiencing climate-related disasters like heat waves. Similar problems could be repeated around the globe.

For instance, South Asia is profoundly vulnerable to climate-intensified disasters due to the region's large population and extreme poverty (Bhatiya 2014). Karachi, Pakistan has a population of over 20 million. It is the economic center of a country where nearly 50% of the population lives below the United Nations' poverty rate of less than one US dollar a day (UNDP 2013). Karachi is vulnerable to climate-intensified disasters such as heat waves and flooding (Nazar 2016): in 2015, the city suffered a heat wave that killed over 1,500 people (*ibid.*). Because of the city's lack of infrastructure and lowland geography, it is prone to flooding. These factors, and others, make Karachi, and Pakistan, acutely vulnerable to climate change. Climate change could displace 40 million Pakistanis (*ibid.*). Another country in the region, Bangladesh, is likely to suffer even more than Pakistan. A 2013 report by the World Bank reports that "Bangladesh will be among the most affected countries in South Asia by an expected 2 °C rise in the world's average temperatures in the next decades, with rising sea levels and more extreme heat and more intense cyclones threatening food production, livelihoods, and infrastructure as well as slowing the reduction on poverty" (World Bank 2013). Climate change could displace 20 to 50 million Bangladeshi (Glennon 2017).

Climate change is often characterized as a 'threat multiplier' for military conflict. Competition for increasingly scarce water, food and energy resources could trigger conflicts in this region with several nuclear powers. (Bhatiya 2014). To further complicate the power dynamic of the region, neighboring China is a now an economic and military superpower that could be drawn into a regional conflict (Lone 2015). The United Nations Environment Programme and the European Union, recently labeled climate change a global security threat (UNEP 2018). Further, many military experts see climate change as threat "to international security and the future existence of modern civilization" (Causevic 2017).

This brief discussion points to a few pieces of evidence that climate chaos could lead to social and political chaos. A group of scientists recently published an article that framed the challenges humanity faces in coming decades in terms of two trajectories, Stabilized Earth and Hothouse Earth (Steffen *et al.* 2018). The scientists used systems theory to identify possible tipping points that could push the Earth System in to a new, irreversible state, Hothouse Earth. This analysis serves a dire warning. "[Hothouse Earth] pose severe risks for health, economies, political stability (especially for the most climate vulnerable), and ultimately, the habitability of the planet for humans" (*ibid.*). However, the door is not shut, but "rapidly closing" on the Stabilized Earth pathway (*ibid.*). The Stabilized Earth trajectory will require more "deliberate management of humanity's relationship with the rest of the Earth System" (*ibid.*). The authors "suggest that a deep transformation based on a fundamental reorientation of human values, equity, behavior, institutions, economies, and technologies is required" (*ibid.*). The question is, can these ethical, political, and behavioral transformations happen in time?

The philosopher Christopher Preston observes that something about the climate change problem has the "temptation of procrastination built in" (Preston 2018). In his article, 'The Perfect Moral Storm', Stephen Gardiner identifies several reasons why this unprecedented, global problem contains the temptation of procrastination (Gardiner 2006). Perhaps the foremost reason is that the lack of immediacy creates a lack of urgency. People living today must start taking action now to avert future catastrophe, but the tragic consequences of inaction are temporally and spatially distant. They seem unreal. Another factor, which will be discussed later, is the daunting task of creating international intuitions capable of building cooperation between nations, which is vital to solve this global problem. Finally, there are powerful vested interests in the current energy system that are resistant to change. For these reasons, and others, we are postponing the difficult Stabilized Earth pathway and are drifting toward Hothouse Earth. This situation is leading some to take a more serious look at climate engineering.

2.2 The slow pace of climate change politics and climate engineering

The United Nations initiated its Framework Convention on Climate Change (UNFCCC) at the Earth Summit in Rio de Janeiro in 1992. After nearly three decades, these efforts have a weak record of reducing GHG emissions. The UNFCCC's two landmark achievements are the 1997 Kyoto Protocol and the 2015 Paris Agreement. The Kyoto Protocol failed to slow rising rates of GHG emissions; the world's economies are no less dependent on fossil fuels today, and GHG emissions continue to rise. The Paris Agreement replaced the Kyoto Protocol and was designed to correct its flaws. In broad outlines, the 175 parties (174 nations plus the European Union) who signed the Paris Agreement committed to collectively reducing GHG emissions with the goal of limiting temperature rise to less than 2° C and strong efforts to limit temperature rise to 1.5° C. Unlike the Kyoto Protocol, the Paris Agreement requires every nation, developed and developing, to submit GHG reduction targets along with a plan called an Intended Nationally Determined Contribution (INDC), to achieve those targets. The agreement also includes the Green Climate Fund, which assists developing countries as they implement adaptation and mitigation plans. The Paris Agreement is a clear advance over the Kyoto Accord. However, until countries develop a sustained track record of fulfilling their commitments, the negotiated INDCs and commitments to the Green Climate Fund are merely good intentions. Paris provides reason for hope, but it is difficult to be too hopeful given the record of past efforts. Increasing doubt about political efforts is leading to a change of attitudes about climate engineering (Boettcher & Schäfer 2017).

Until recently, the last 15-20 years, most climate scientists saw climate engineering as scientifically dubious, ethically suspect, and a dangerous distraction. However, seeing dangerous climate change quickening on the horizon, some scientists and decision-makers began to view climate engineering in a new light. One of those scientists was the eminent, Nobel Prize-winning Dutch atmospheric chemist, Paul Crutzen. In 2006, he published an influential article that broke the taboo on climate engineering. The article asserts that it is time for serious scientific discussion of climate engineering. He writes: "given the grossly disappointing international political response to the required greenhouse gas emissions [...] research on the feasibility and environmental consequences of climate engineering […] should not be tabooed" (Crutzen 2006, p. 214). The article proved to be a watershed for climate engineering, taking it from fringe to mainstream.

In the years before Crutzen's article, only a handful of publications had been devoted to climate engineering. But just one year later in 2007, publications began to surge. A recent article tracks the growth of publications devoted to climate engineering from 1971 to 2013. In 2007, there were 21 publications. In 2008 that number had grown to 73 and by 2013 that number more than doubled to 153 (Oldham *et al.* 2014). Another significant indicator of climate engineering's increasing respectability is its inclusion for the first time in the Intergovernmental Panel on Climate Change (IPCC)'s 5th Assessment Report in 2014 (IPCC 2014). It seems from this evidence that many scientists are accepting the idea of a need for climate engineering research.

By way of summary, the above begins to answer the question, why scientists and engineers should consider climate engineering research. Evidence is accumulating that climate chaos will lead to social, political, and environmental chaos; and it appears that political efforts to avert climate chaos might fall short. If the climate system continues down the Hothouse Earth pathway, the consequences for human civilization could be catastrophic. Further, it seems increasingly unlikely that political efforts alone will be able to put civilization on the Stabilized Earth pathway. It seems right to *consider* climate-engineering research.

However, a technological 'cure' should not be worse than the 'disease'. How can scientists be sure that climate engineering research will do more good than harm? As noted in the introduction, one of the most common ethical concerns is that climate engineering creates a moral hazard. For climate engineering research to be ethically responsible, this concern must be understood and addressed. A first step toward understanding the moral hazard objection is to investigate the implication of climate engineering as a technological fix.

3. What Is Climate Engineering?

3.1 Climate engineering is a technological fix

Climate engineering is often characterized as a technological fix. While the idea of engineering the climate is unprecedented, climate engineering schemes are the product of the commonly applied technological fix strategy. The idea of a technological fix is simple: it is a problem-solving strategy that reframes intractable sociopolitical problems as engineering puzzles that emit technical solutions (Weinberg 1967). Possible solutions to multifarious and capricious sociopolitical problems are more easily identified when these problems are reframed in the clear and predictable terms of physics, chemistry, and engineering.

Modern societies often see technological progress as the quickest and surest path toward economic and social progress; there is a propensity to believe that technological solutions are easier and less painful than behavioral, social, or political solutions (Volti 2014). Governments and industries spend tens of billions each year on research in agriculture, medicine, energy, transportation, the environment, and more, to develop technologies to address problems with sociopolitical roots. It should come as no surprise that climate engineering research would eventually enter into high-level climate change discussions. However, technological fixes are instinctively criticized for being superficial solutions that fail to address the roots of problems. Nonetheless, this problem-solving strategy has several benefits. It offers decision makers additional options for addressing difficult problems. Technological fixes can buy time until problems can be dealt with on a deeper level (*ibid.*). Finally, a technological fix may simply be the best available option all things considered. This might be particularly true for problems with firm deadlines.

Climate change would seem to be a perfect candidate for the technological fix strategy. It is an intractable sociopolitical problem that can be readily reframed as an engineering puzzle. Once the problem is reframed in the terms of physics and chemistry, it presents scientists and engineers with a clear task: develop technologies that will stabilize the Earth's solar energy balance. There are two general approaches to this task: solar radiation management (SRM) and carbon dioxide removal (CDR). SRM research focuses on techniques that increase the Earth's albedo (reflective capacity). CDR research, as the name indicates, focuses on techniques that remove CO_2 from the atmosphere.

3.2 Solar radiation management (SRM)

Some SRM proposals are expensive and futuristic, like installing an array of mirrors in orbit around the Earth. Other plans are inexpensive and less high-tech, like using long hoses suspended by high-altitude balloons to spray sulfate particles into the stratosphere. Crutzen's watershed article focused on this approach. Large volcanic eruptions are known to cool the planet in the same way: in 1991 the cataclysmic eruption of Mount Pinatubo in the Philippines blasted approximately 17 megatons of sulfur dioxide into the stratosphere (Self *et al.* 1996). Microscopic sulfuric acid aerosols formed and circled the planet, reflecting incoming solar radiation back into space. Some estimates suggest that the Mount Pinatubo eruption decreased the global average temperature by 0.4 Celsius for two years (*ibid.*). Stratospheric sulfur injection is an attractive technological fix because it is a quick and cost-effective way to lower the Earth's average global temperature (Moreno-Cruz & Keith 2013). However, the full range of consequences would be difficult or impossible to anticipate prior to full-scale implementation, and the stakes would be extremely high. While the effects of an SRM on global average temperature might be predictable, its effects on regional weather patterns are much more difficult to foresee. Stratospheric sulfur injections and other SRM plans do nothing to remove GHGs from the atmosphere, which would continue to increase (*ibid.*). SRM only masks the warming effects of GHGs. While the following discussions will focus on CDR techniques, the ethical analysis applies also to SRM. I will use the SRM example in the final section to illustrate ideas for developing ethical guidelines for responsible climate engineering research.

3.3 Carbon dioxide removal (CDR)

There is a wide array of possible CDR techniques. To illustrate their diversity, I will mention three: ocean fertilization, enhanced weathering, and bioenergy with carbon capture and storage (BECCS).

Large areas of the oceans do not have adequate concentrations of iron to support phytoplankton blooms. Ocean fertilization schemes take advantage of this fact by proposing to spread powdered iron across vast areas of the oceans to produce huge phytoplankton blooms, which would capture CO_2 during photosynthesis. Once the organisms expire, the captured carbon would sink to the bottom of the ocean where it would be trapped and stored by water pressure (Powell 2008).

Enhanced weathering schemes seek to harness global biogeochemical cycles. The idea is to accelerate chemical weathering processes that capture atmospheric carbon and store it in soils and the ocean. The technique mines calcium- and magnesium-bearing silicate rocks and crushes it to maximize the reactive surface area. The rock debris is then added to soils where it chemically breaks down to release base cations and generate bicarbonate from atmospheric CO_2. The bicarbonate is stored in the soils or it eventually flows into the oceans, leading to carbonate precipitation on the seafloor. Enhanced weathering projects would likely need to be located in the warm and wet tropics where chemical weathering's reaction rates are high enough to be effective (Beerling 2017).

A final example is industry-scale bioenergy with carbon capture and storage (BECCS). BECCS schemes propose to transform energy sectors to burning carbon-neutral biomass. The biomass fuel captures CO_2 from the atmosphere during photosynthesis. Rather than releasing the CO_2 produced during combustion as a pollutant, the gas is captured and stored in underground geologic formations.

CDR techniques can be implemented at scales that would not qualify as climate engineering. Ocean fertilization, enhanced weathering, and BECCS projects would have to be immense to qualify as climate engineering. At these scales, they would very likely have significant social, political, and environmental impacts, which would raise ethical concerns. For example, the quantities of powdered iron fertilizer and the huge phytoplankton blooms could have serious unintended consequences for the world's oceans (Powell 2008). Similarly, mining and crushing rock at climate engineering scales would surely have serious social and environmental impacts. Moreover, it would be very difficult to create the international, democratic institutions needed to equitably distribute the burdens from these impacts (Lawford-Smith & Currie 2017). This could lead to environmental justice concerns; that is, concerns that vulnerable populations would bear a disproportionate burden of any social and environmental impacts.

With this overview in mind, it will be helpful to use a specific case to focus the examination of the moral hazard concerns. The case involves direct air capture (DAC) of carbon with chemicals and it will focus on a company, Carbon Engineering. Carbon Engineering recently reported a major chemical engineering breakthrough in DAC that could have far-reaching implications for climate engineering.

3.4 The case of Carbon Engineering

In 2018, a team of scientists and engineers from the Canadian company Carbon Engineering published the results of a promising technique for directly capturing CO_2 from the air with chemicals. The DAC technique can be used to convert the captured carbon into synthetic fuels or to store it in geologic formations. (Keith *et al.* 2018). David Keith, a Harvard University physicist and leading expert on climate engineering, is Carbon Engineering's co-founder. Significantly, the company's primary financial backer is Bill Gates, who was the co-founder of Microsoft and is one the world's wealthiest people. Carbon Engineering appears to be well on the way to solving two of the major obstacles associated with BECCS schemes: competition for land and high costs. Perhaps an insurmountable obstacle to the wide-scale application of biofuels is that they compete with food production for arable land. Since Carbon Engineering's technique uses industrial chemical processes it does not compete with agriculture. Just as significant, their technique greatly lowers the costs of DAC.

DAC of carbon and storage (DACCS) seemed to many a tantalizing technological fix for climate change, but it proved to be far too expensive. Prior to the publication of Carbon Engineering's 2018 results, the definitive study of the costs of industrial-scale DAC estimated the price to be $1000 US dollars per metric ton of CO_2 (House *et al.* 2011). To put this number in perspective, it would cost approximately $1.2 trillion to capture the CO_2 emitted by coal-fired power plants in the United States during 2017, approximately 1.2 billion metric tons (US Energy Information Administration 2018). A central problem DAC must overcome is the 'net carbon problem' (*ibid.*). The concentration of CO_2 in ambient air is extremely small, approximately 400 parts per million (Burrows 2018). Consequently, large industrial machines must move massive amounts of air through the process to capture enough CO_2 for the technique to work. These industrial machines, and other parts of the operation, require much energy. Energy, of course costs money, and depending on the source of energy, the operation will add more or less CO_2 to the atmosphere. An accurate assessment of this technique for climate engineering would require a full accounting of the CO_2 added to the atmosphere during the operation and for the full lifecycle of the plant. These numbers will not be available until experiments are run at larger scales for longer times.

Carbon Engineering estimates that their technique would lower costs from the 2011 study from $1000 to the range of $94 to $232 per metric ton of CO_2 (Keith *et al.* 2018). They were able to accomplish this by developing a new chemical process and by repurposing existing industrial technologies to run it. Their approach uses arrays of large fans to move massive amounts of air over a chemical solution to capture CO_2. They describe the chemistry as involving two connected loops: "The first loop captures CO_2 from the atmosphere using an aqueous solution with ionic concentrations of roughly 1.0 M OH^-, 0.5 M CO_3^{2-}, and 2.0 M K^+. In the second loop, CO_3^{2-} is precipitated by reaction with Ca^{2+} to form $CaCO_3$ while the Ca^{2+} is replenished by dissolution of $Ca(OH)_2$. The $CaCO_3$ is calcined to liberate CO_2 producing CaO, which is hydrated or 'slaked' to produce $Ca(OH)_2$" (*ibid.*). They produce synthetic fuel by a conventional process commonly used in the oil industry, which reacts CO_2 with H_2 to produce fuel. Carbon Engineering is currently seeking funding to test their chemical processes and technologies at larger scales.

It is important to note that the company is not currently pursuing plans to capture carbon and store it, for example, in geologic formations. Their research focuses on producing 'carbon-neutral' synthetic fuels, which is only possible if the electricity used to drive the process is generated from a non-carbon producing source, such as a hydroelectric plant. Further, it should also be noted that the current process uses some natural gas, which researchers hope to replace with electricity and make the process carbon neutral. That said, David Keith notes that the company could adapt their technique for producing 'carbon neutral' fuels to be used as a negative emissions technology. However, Keith remarks that, "[carbon storage] wouldn't give Carbon Engineering any product to sell, and there are no buyers stepping up to front the effort, for now" (Meyer 2018). For Climate Engineering's technique to realize any potential for climate engineering, a market for removing carbon from the atmosphere and 'permanently' storing it would have to exist. However, this might be an instance where a proof of concept technology could help create a market for the service it could someday provide. Once negative emissions technologies are seen as a possible way to make vast fortunes, while also providing a vital social benefit, it is easy to imagine that the political will to create a market for CCS will somehow emerge – especially with powerful advocates like Bill Gates leading the way.

Carbon Engineering is pioneering a potentially multibillion-dollar industry that might someday serve as a technological fix for the climate crisis. One journalist notes that "[Carbon Engineering] could [...] make Harvard superstar physicist David Keith, Microsoft co-founder Bill Gates, and oil sands magnate Norman Murray Edwards [another powerful financial backer] more money than they could ever dream of" (Vidal 2018). If Carbon Engineering continues to attract wealthy and politically influential backers like Bill Gates and Norman Murray Edwards, the creation of a CCS industry could transform the landscape of climate change politics. This could be a mixed blessing: the promise alone of cost-effective CCS could create a moral hazard, which, as will be explained, could lead to an ethical dilemma for scientists considering CCS research.

4. What is the Substance of the Moral Hazard Objection?

4.1 The temptation of procrastination

As mentioned, moral hazard arguments are "one of the main ethical objections to geoengineering" (Royal Society 2009, p. 39). One problem with these objections is the notion of a moral hazard is not a traditional ethical concept, but originated in the insurance industry. It describes a perplexing problem for insurers: when workers are provided with hazard insurance there is a corresponding increase in risky behaviors (*ibid.*, p. 37). Those who raise moral hazard objections to climate engineering worry that it will be seen as insurance against climate change, which will lead to greater risk-taking behaviors. Stated differently, the possibility of a technological fix will tempt people to delay or avoid taking the difficult steps to put civilization on a Stabilized Pathway, and destine humanity instead to the harsh realities of Hothouse Earth.

This is a sensible argument, but there are difficulties with using the moral hazard objection as an ethical argument. The moral hazard phenomenon identifies a correlation between insurance and risk-taking behaviors, but these behaviors by themselves may not necessarily be unethical, or unreasonable. Moral hazard objections to climate engineering are better seen as premises in a larger argument that includes, as will be seen in Section 4.3, premises grounded in ethical concepts such as justice and fairness. The argument would also need to include premises about the "temptation of procrastination" (Preston 2018). The Carbon Engineering case can serve to illustrate this later point.

Ken Caldeira, a leading expert on climate engineering, points to the allure of an easy technological fix, commenting that, "If [Carbon Engineering's] costs are real, it is an important result [...]. This opens up the possibility that we could stabilize the climate for affordable amounts of money without changing the entire energy system or changing everyone's behavior" (Meyer 2018). More generally, climate scientists Anderson and Peters observe that, "The allure [...] of negative-emission technologies stems from their promise of much-reduced political and economic challenges today, compensated by anticipated technological advances tomorrow" (Anderson & Peters 2016). It is easy to imagine that the promise of a negative emissions technological fix could become a temptation to procrastinate on the daunting task of transforming the whole energy system and changing entrenched energy consumption habits. Further, business leaders and politicians who are heavily vested in the fossil fuel industry could use promising CDR or SRM technologies as temptation for further delays in transitioning from fossil fuels to alternative energy systems. The substance of the moral hazard objection is that promising breakthroughs with potential for climate engineering intensify the temptation to procrastinate, which will lead to delays on the hard sociopolitical tasks climate change requires. Further, it is far from certain that negative emissions technologies, like Carbon Engineering's, will fulfill their promise at scale. Climate engineering is a risky insurance policy for a high-risk scenario.

However, an underlying assumption of the moral hazard-procrastination argument is that the right thing to do is to trust sociopolitical fixes over technological fixes. However, are sociopolitical efforts trustworthy? Given the brief discussion in Section 2.2, should we continue to trust that sociopolitical solutions can be implemented in time? To examine this question, let us look closer at the nature of the political task.

4.2 Climate politics and the dilemma of climate engineering

The political task of achieving an effective international climate change agreement is unprecedented in human history. In the first few decades of the 21st century, nations must transition from the 20th century's ethos of conflict, defined by two world wars, the Cold War, and the War on Terror, to an ethos of trust and cooperation. At least within the sphere of climate change negotiations, this shift requires nations to stop viewing each other as suspicious competitors long enough to cooperate on solving this common problem, but with differentiated responsibilities.

Climate change politics got off to promising start at the 1992 Rio Earth Summit, where 154 nations cooperated in the creation the United Nations' Framework Convention on Climate Change (UNFCCC). The UNFCCC set the right tone with commitments to equity and justice, which it extended to future generations. It states that the Conference of Parties agrees to protect "the climate system for the benefit of present and future generations of humankind, on the basis of equity and in accordance with their common but differentiated responsibilities and respective capabilities" (United Nations 1992, p. 4). However, it has proven difficult for countries to move from acting on self-interest to *consistently* acting on these ethical commitments. The problem of inconsistency is glaring for the United States, the world's second-largest emitter of CO_2 and largest per capita emitter.

The United States is divided politically on climate change. This is reflected in the striking swings in its behaviors toward UNFCCC agreements. For example, in 2015, under President Obama's leadership, the United States helped lead the way toward the landmark Paris Agreement. One year later, after Donald Trump's shocking victory in the presidential election, he promptly took actions to start withdrawing the United States from the Agreement. In one year, the United States went from a vital leader to a major hindrance to the success of the breakthrough agreement. This political divide in the United States renders this key nation an unreliable partner, which threatens to undermine trust in the entire effort.

The United States is singled out here because of the remarkable reversal with the improbable election of Donald Trump, but there are trust issues with other nations. China is the world's largest CO_2-emitting nation and its cooperation is essential for a successful agreement. Several news sources reported in 2015 that based on data from independent monitoring agencies, China had been underreporting its coal consumption by fifteen percent (Buckley 2015). This reinforces suspicions that the Chinese government manipulates carbon emission data for economic and political purposes (Liu 2015, p. 14). It is unlikely that China's behavior is unique and it no doubt added to an atmosphere of mistrust and suspicion about the negotiations and ultimate agreement.

The Paris Agreement does not include effective enforcement mechanisms to discourage countries for reneging or cheating on their commitments. Strong enforcement mechanisms can be added to future agreements. However, until these are added, critics argue that the UNFCCC's efforts will fail. For example, during the Paris conference, the eminent climate scientist James Hansen was quoted in the press as saying the Agreement is a "fraud [...]. It is just worthless words. There is no action, just promises" (Milman 2015). Hansen's words may be too harsh, but the high level of trust between untrustworthy parties required by the Paris Agreement is reason to be skeptical.

The capriciousness and inscrutability of real-world, international politics vanishes when the problem is reframed as a technological fix. But the lure of an easy climate-engineering fix creates a temptation to despair of achieving more long-term sociopolitical change. However, to be realistic is not to give up hope. The benefits of a successful international political process are too great to succumb to despair. Climate engineering is no substitute for transformative social and political change. Also, effective policies to reduce CO_2 emissions would be far safer than climate engineering. There are no guarantees that a climate engineering scheme would be successful at scale or that it would not create more problems than it solves. Fortunately, there is still time for consistent and honest leadership to emerge from key nations. It is also possible for future agreements to include more aggressive, legally binding national commitments to reduce CO_2 emissions and strong enforcement mechanisms. The appropriate attitude might be a hopeful realism, which would require a sustained commitment to sociopolitical efforts while pursuing climate-engineering research. Unfortunately, the moral hazard-procrastination argument points to a dilemma for this middle approach.

Scientists and engineers considering the ethics of climate engineering research must face a dilemma. On the one hand, if we vigorously pursue climate-engineering research we risk undermining political efforts by creating temptations for further procrastination. And, climate engineering is risky and is not a substitute for political efforts (Meyer 2018). On the other hand, if we do not vigorously pursue climate-engineering research and shaky political efforts fail, vetted and tested climate-engineering techniques may not be available and nations could implement climate engineering in desperation. Either choice requires taking risks that could lead to serious consequences. Fortunately, it might be possible to address this dilemma with ethical guidelines for responsible climate engineering research that serve as a foundation for strong, inclusive, international governance.

4.3 Moral hazard, climate engineering, and justice

In an article criticizing an overreliance in the potential of negative emissions technologies, climate scientists Kevin Anderson and Glen Peters relate the moral hazard concern to issues of justice and fairness (Anderson & Peters 2016). They argue that gambling on negative emissions technologies leads to an unjust distribution of risks. If these technologies are pursued but fail to work at scale, the wealthy and resilient communities that are researching and developing them will not be the ones to suffer the most harm. Rather, "communities that are geographically and financially vulnerable to a rapidly changing climate" and future generations (*ibid.*) will face the greatest harms of disasters, famines, and wars. The unfair distribution of risks of climate engineering can be applied to all SRM and CDR proposals.

The moral weight of this argument comes from extending the unjust distribution of risks historically built into the climate problem to the distribution of risks with climate engineering. On the one hand, high-emitting countries owe their prosperity to the combustion of fossil fuels. These countries can do much to address the problem by reducing their CO_2 emissions, and they have greater adaptive capacity because of their prosperity. On the other, historically low-CO_2-emitting countries, who have contributed little to the problem, can do little to address it, and lack the resources to adapt to it. This unfair distribution of responsibilities and risks is magnified with climate engineering. Vulnerable communities are not responsible for research and development of climate engineering technologies and they are at greatest risk if research undermines mitigation efforts or the technology fails.

It is especially unjust to gamble the fate of vulnerable people, and future generations, on risky technological fixes when sociopolitical solutions are still available. Anderson and Peters note that "there are huge opportunities for near-term, rapid, and deep reductions today at little to modest costs, such as improving energy efficiency, encouraging low-carbon behaviors, and continued deployment of renewable energy technologies" (*ibid.*). However, they qualify their argument. Negative emission technologies can "reasonably be the subject of research, development, and potentially deployment" (*ibid.*). But this research must be done with the conviction that if it will not be successful at scale, "failing to do otherwise are a moral hazard par excellence" (*ibid.*).

The above discussion relates the moral hazard-procrastination concerns to justice and fairness concerns. In doing this, it points out that issues of justice and fairness should be paramount in developing ethical guidelines for responsible climate engineering research. Climate engineering research should not intensify and enlarge injustices that are inherent in the climate change problem.

5. Conclusions: How Do We Address These Concerns?

In an article discussing the 2010 report of the United States' Presidential Council on Bioethical Issues, the council's chair, Amy Gutmann, states that the principle of justice and fairness should be broadly applied to powerful emerging technologies (Gutmann 2012). This would certainly include climate engineering. Gutmann writes that the "principle of justice and fairness relates to the distribution of benefits and burdens" across societies and generations (*ibid.*). She elaborates, "a commitment to justice and fairness is a commitment to ensuring that individuals and groups share in the benefits of new technologies and that the unavoidable burdens of technological advances do not fall disproportionately on any particular individual or group" (*ibid.*). A good beginning for efforts to apply the principle of justice and fairness to climate engineering research would start with the following: (1) inclusive and impartial research on the impacts of climate engineering proposals, (2) inclusive international participation in research, and (3) transparency and openness in research funding.

First, the principle of justice and fairness would require scientific knowledge of the possible social and environmental consequences of various climate engineering proposals at multiple levels. Clearly, without detailed knowledge of the possible consequences of a particular climate-engineering proposal, it is impossible to consider a just distribution of risk, harms, and benefits. The majority of research on the most frequently discussed SRM proposals focuses on how the climate system would respond to stratospheric sulfate injections, while the environmental and social impacts at various levels are under-researched and uncertain (Trisos *et al.* 2018). There is a critical need for collaborative research between natural and social scientists to understand the effects on agriculture, community health, and regional and local ecosystems (*ibid.*). For example, some computer simulations of stratospheric sulfate injections indicate that the reflective sulfate aerosols would indeed cool the planet, but with the unintended consequence of reducing the amount of precipitation from the summer monsoons in Asia and Africa (Robock *et al.* 2008). This would have serious consequences for agriculture and billions of people's food supply. It is clear, then, that the principle of justice and fairness would require adequate funding of research aimed at answering the many questions on impacts and whom they would affect in order to prevent injustices.

Second, the principle of justice and fairness would require inclusive international participation in research. Again, it is unjust for resilient, wealthy countries to put vulnerable communities in danger with risky climate engineering schemes when they are not consulted and do not participate in the research. This issue is raised in a recent article in the journal *Nature*, whose title asserts "Developing Countries Must Lead on Solar Geoengineering Research" (Rahman *et al.* 2018). The international team of authors argues that for climate engineering research to avoid unjust governance, underrepresented, developing nations need to take the lead. Currently, North American and European scientists are overrepresented in climate research. This leads to a danger of biases (conscious or unconscious) towards these countries' interests. This situation could ultimately lead to an unfair distribution of the harms, risks, and benefits of a climate engineering project. Rahman and his co-authors argue that since "developing countries have most to gain or lose" from SRM research, these countries need greater representation (*ibid.*). They summarize the situation: "Solar geoengineering is fraught with risks and can never be an alternative to mitigation [...]. It is right, politically and morally, for the global South to have a central role in solar geoengineering research, discussion, and evaluation" (*ibid.*). In an effort to involve developing countries in SRM deliberations, Rahman and his co-authors are engaged in the Solar Radiation Management Governance Initiative (SRMGI). One of SRMGI's objectives is to produce a special Intergovernmental Panel on Climate Change report on the risks and benefits of

SRM (*ibid.*). Their ultimate goal is to create "a coordinated global research initiative [...] to promote collaborative science on this controversial issue" (*ibid.*). By fostering broadly inclusive international representation in research, the SRMGI is an example of the kinds of initiatives needed to implement the principle of justice and fairness.

Third, and finally, the principle of justice would require transparency and openness in research. Transparency and openness in research raise numerous issues, not all of which I can highlight here. But one of the most important is the question of allowing privately funded research and intellectual property rights for novel climate engineering technologies. The profit motive and intellectual property rights are primary sources of funding and incentives for research. Should private funding, international intellectual property rights, and the profit motive be allowed for SRM and CDR? Both publicly and privately funded research come with advantages and disadvantages. On the one hand, publicly funded research could build greater confidence in its focus on public goods, but public financial resources are often limited and centralized and bureaucratic decision-making can limit innovation. On the other hand, privately funded research offers access to additional sources of funding, and with decentralized decision-making can promote greater innovation. However, it risks creating vested interests that could corrupt democratic decision-making. Scientists, investors, or companies who are awarded intellectual property rights for a specific climate engineering technology, for instance DAC technologies, could make a fortune if it is widely adopted. Intellectual property rights would create incentives for people and corporations to promote the adoption of their technology, even if alternative technologies would better serve public interests. This could create a moral hazard, as there would be a financial motive to encourage the adoption of a climate-engineering technology, which could in turn create a temptation to procrastinate on sociopolitical efforts.

This ethical concern is likely more of an issue with CDR technologies than SRM. As was seen with Carbon Engineering, some technologies operate at scales that are too small to be considered climate engineering, but they could be scaled into a CDR project to cool the planet. Private enterprises with the potential to generate fortunes for investors, like Carbon Engineering, raise a difficult problem. It could ultimately serve the global, common good to use the private corporations, the profit motive, and the intellectual property rights system to encourage as much innovation as possible. However, there is the danger that the promise of financial success could shift the political landscape, leading to further procrastination and risking injustices. This type of situation will be a major challenge for responsible climate engineering research.

By way of summary, this chapter started with these questions: Should scientists and engineers contribute to climate engineering research? Is it the right thing to do now? Would climate engineering research be consistent with responsibilities to protect public welfare and safety, contribute to environmental sustainability, and protect the environment for future generations (AIChE 2015, ACS 2016)? The answers provided in this chapter show climate engineering research should be considered. However, because of the unprecedented nature of this technology, it needs clear ethical guidelines and strong governances to provide researchers with the confidence that they are doing the right thing. The moral hazard objection, the temptation of procrastination, and the inherent injustices built into climate change problem lead to serious concerns with this research. The above discussion points to principles of justice and fairness as the right starting point for developing ethical guidelines for responsible climate engineering research. The principle of justice would at least require inclusive and impartial research on impacts and techniques, and transparency and openness on funding sources as starting points for effective governance.

Recommended Readings

Preston (2013) provides a clear and concise overview of the types of climate engineering and the main ethical issues currently being discussed. Keith (2013) develops a detailed argument why societies *must* consider deploying climate engineering. Hulme (2015) is a counterpoint to Keith, he develops arguments why *must not* consider a climate engineering fix.

References

American Chemical Society (ACS): 2016, 'The Global Chemists Code of Ethics' [available online at: https://www.acs.org/content/acs/en/global/international/regional/eventsglobal/global-chemists-code-of-ethics.html, accessed 3 September 2018].

American Institute of Chemical Engineers (AIChE): 2015, 'AIChE Code of Ethics' [available online at: https://www.aiche.org/about/code-ethics, accessed 6 September 2018].

Anderson, K. & Peters, G.: 2016, 'The Trouble with Negative Emissions: Reliance on Negative-emission Concepts Locks in Humankind's Carbon Addiction,' *Science*, **324** (6309), 182-183.

Beerling, D.J.: 2017, 'Enhanced Rock Weathering: Biological Climate Change Mitigation with Co-benefits for Food Security?', *Biology Letters*, 13, 20170149 [available online at: http://dx.doi.org/10.1098/rsbl.2017.0149, accessed 1 September 2018].

Bhatiya, N.: 2014, 'Why South Asia Is So Vulnerable to Climate Change', *Foreign Policy*, April 22 [available online at: http://foreignpolicy.com/2014/04/22/why-south-asia-is-so-vulnerable-to-climate-change/, accessed 26 May 2018].

Boettcher, M. & Schäfer, S.: 2017, 'Reflecting upon 10 years of Geoengineering Research: Introduction to the Crutzen + 10 Special Issue', *Earth's Future*, **5**, 266-277.

Buckley, C.: 2015, 'China Burns Much More Coal Than Reported, Complicating Climate Talks', *New York Times,* 3 November [available online at: https://www.nytimes.com/2015/11/04/world/asia/china-burns-much-more-coal-than-reported-complicating-climate-talks.html, accessed 3 September 2018].

Burrows, L.: 2018, 'Team Provides First Plan for Commercially Viable, Industrial-Scale Carbon Removal Plant', *News and Events,* John A Paulson School of Engineering and Applied Sciences, 7 June, [available online at: https://www.seas.harvard.edu/news/2018/06/team-provides-first-plan-for-commercially-viable-industrial-scale-carbon-removal-plant, accessed 12 September 2018].

Causevic, A.: 2017, 'Facing an Unpredictable Threat: Is NATO Ideally Placed to Manage Climate Change as a Non-Traditional Threat Multiplier?', *Connections: The Quarterly Journal*, **16** (2), 59-80.

Crutzen, P.: 2006, 'Albedo Enhancement by Stratospheric Sulfur Injections: A Contribution to Resolve a Policy Dilemma?', *Climatic Change,* **77**, 211-219.

European Commission: 2017. 'Europe to be Hit Hard by Climate Related Disasters in the Future' [available online at: https://ec.europa.eu/jrc/en/news/europe-be-hit-hard-climate-related-disasters-future, accessed 24 May 2018].

Farquhar, S., Halstead, J.; Cotton-Barratt, O.; Schubert, S.; Belfield, H. & Snyder-Gardiner, S.: 2006, 'A Perfect Moral Storm: Climate Change, Intergenerational Ethics and the Problem of Moral Corruption', *Environmental Values*, **150** (3), 397-413.

Glennon, R.: 2017, 'The Unfolding Tragedy of Climate Change in Bangladesh', *Scientific American*, 12 April [available online at: https://blogs.scientificamerican.com/guest-blog/the-unfolding-tragedy-of-climate-change-in-bangladesh/, accessed 24 May 2018].

Gutmann, A.: 2012, 'The Ethics of Synthetic Biology: Guiding Principles for Emerging Technologies', *Hastings Center Report,* **41** (4), 17-22.

House, K.Z.; Baclig, A.C.; Ranjan, M.; van Nierop, E.A.; Wilcox, J. & Herzog, H.J.: 2011, 'Economic and Energetic Analysis of Capturing CO_2 from Ambient Air', *Proceedings of the National Academy of Sciences*, **108** (51) 20428-20433.

Hulme, D.: 2015, Can Science Fix the Climate? A Case Against Climate Engineering, Cambridge, UK: Polity Press.

Intergovernmental Panel on Climate Change (IPCC): 2014, 'Working Group III' [available online at: http://www.ipcc.ch/pdf/assessment-report/ar5/wg3/ipcc_wg3_ar5_full.pdf, accessed 10 October 2018].

Keith, D.: 2013, *A Case for Climate Engineering*, Cambridge, MA: MIT Press.

Keith, D.; Holmes, G.; St. Angelo, D. & Heidel, K.: 2018. 'A Process for Capturing CO_2 from the Atmosphere', *Joule*, **2**, 1573-1594.

Lawford-Smith, H. & Currie, A.: 2017, 'Accelerating the Carbon Cycle: The Ethics of Enhanced Weathering', *Biology Letters*, **4** (13): 20160859 [available online at: http://dx.doi.org/10.1098/rsbl.2016.0859, accessed 1 September 2018].

Liu, Z.: 2015, 'China's carbon emissions report 2015', Belfer Center for Science and International Affairs/Harvard Kennedy School [available online at: https://www.belfercenter.org/sites/default/files/legacy/files/carbon-emissions-report-2015-final.pdf, accessed 10 October 2018].

Lone, A.K.: 2015, 'How Can Climate Change Trigger Conflict in South Asia?', *Foreign Policy*, 20 November [available online at: http://foreignpolicy.com/2015/11/20/how-can-climate-change-trigger-conflict-in-south-asia/, accessed 25 May 2018].

Mann, C.C.: 2014, 'How to Talk about Climate Change so People will Listen', *The Atlantic*, September [available online at: https://www.theatlantic.com/magazine/archive/2014/09/how-to-talk-about-climate-change-so-people-will-listen/375067/, accessed 24 May 2018].

Meyer, R.: 2018. 'Climate Change Can Be Stopped by Turning Air Into Gasoline', *The Atlantic*, 7 June [available online at: https://www.theatlantic.com/science/archive/2018/06/its-possible-to-reverse-climate-change-suggests-major-new-study/562289/, accessed 16 August 2018].

Milman, O.: 2015, 'James Hansen, Father of Climate Change Awareness, Calls Paris Talks 'A Fraud'', *The Guardian*, 12 December [available online at: https://www.theguardian.com/environment/2015/dec/12/james-hansen-climate-change-paris-talks-fraud, accessed 5 September 2018].

Moreno-Cruz, J.B. & Keith, D.: 2013, 'Climate Policy under Uncertainty: a Case for Solar Geoengineering', *Climatic Change* **121** (3), 431-444.

Nazar, S.: 2016, 'Pakistan's Big Threat Isn't Terrorism – It's Climate Change', *Foreign Policy*, 4 March [available online at: http://foreignpolicy.com/2016/03/04/pakistans-big-threat-isnt-terrorism-its-climate-change/, accessed 24 May 2018].

Oldham, P.; Szerszynski, B.; Stilgoe, J.; Brown, C.; Eacott B. & Yuille A.: 2014, 'Mapping the Landscape of Climate Engineering', *Philosophical Transactions of the Royal Society A*, **372**, 20140065 [available online at: http://rsta.royalsocietypublishing.org/content/roypta/372/2031/20140065.full.pdf, accessed 24 May 2018].

Powell, H.: 2008, 'Fertilizing the Ocean with Iron: Should We Add Iron to the Sea to Help Reduce Greenhouse Gases in the Air?', *Oceanus*, **46** (1) [available online at: http://www.whoi.edu/oceanus/feature/fertilizing-the-ocean-with-iron, accessed 30 August 2018].

Preston, C.J.: 2013, 'Ethics and Geoengineering: Reviewing the Moral Issues Raised by SRM and CDR', *Wiley Interdisciplinary Reviews (Climate Change)*, **4**, 23-37.

Preston, C.J.: 2018, 'Blame and Confusion on a Hothouse Earth', *The Plastocene,* 13 August [available online at: https://plastocene.com/2018/08/13/blame-and-confusion-in-a-hothouse-earth/ (blog post), accessed 4 September 2018].

Rahman, A.A.; Artaxo, P.; Asrat, A. & Parker, A. (2018) 'Developing Countries Must Lead on Solar Geoengineering Research', *Nature*, 556, 22-24.

Robock, A.; Oman, L. & Stenchikov, G.L.: 2008, 'Regional Climate Responses to Geoengineering with Tropical and Arctic SO_2 Injections', *Journal of Geophysical Research*, **113**, 1-15.

Rockström, J. & Klum, M.: 2015, *Big World, Small Planet*, New Haven, CT: Yale University Press.

Royal Society: 2009, 'Geoengineering the climate: Science, governance and uncertainty', Royal Society Policy Document, [available online at: https://royalsociety.org /~/media/Royal_Society_Content/policy/publications/2009/8693.pdf, accessed 10 October 2018].

Self, S.; Zhao, J-X.; Holasek, R.E.; Torres, R.C. & King, A.J.: 1996. 'The Atmospheric Impact of the 1991 Mount Pinatubo Eruption', in: C.G. Newhall & R.S. Punongbayan (eds.), *Fire and Mud: Eruptions and Lahars of Mount Pinatubo, Philippines*, Seattle, WA: University of Washington Press [available online at: https://pubs.usgs.gov/pinatubo/self/, accessed 24 May 2018].

Smith, A.B.: 2018, '2017 U.S. billion-dollar weather and climate disasters: a historic year in context', NOAA Climate.gov [available online at: https://www.climate.gov/news-features/blogs/beyond-data/2017-us-billion-dollar-weather-and-climate-disasters-historic-year, accessed 26 May 2018].

Steffen, W.; Crutzen, P.J. & McNeill, J.R.: 2007, 'The Anthropocene: Are Humans Now Overwhelming the Great Forces of Nature?', *Royal Swedish Academy of Sciences*, **36** (8), 614-21.

Steffen, W.; Rockström, J.; Richardson, K.; Lenton, T.M.; Folke, C.; Liverman, D.; Summerhayes, C.P.; Barnosky, A.D.; Cornell, S.E.; Crucifix, M.; Donges, J.F.; Fetzer, I.; Lade, S.J.; Scheffer, M.; Winkelmann, R. & Schellnhuber, H.J.: 2018, 'Trajectories of the Earth System in the Anthropocene', *Proceedings of the National Academy of Sciences*, **115** (33): 8252-9.

Torelli, S.M.: 2017, 'Climate-driven migration in Africa', European Council on Foreign Relations, 20 December [available online at: https://www.ecfr.eu/article/ commentary_climate_driven_migration_in_africa, accessed 24 May 2018].

Trisos, C.H.; Gabriel, C.; Robock, A. & Xia, L.: 2018, 'Ecological, Agricultural, and Health Impacts of Solar Geoengineering', in: Z. Zommers & K. Alverson (eds.), *Resilience: The Science of Adaptation to Climate Change*, New York, NY: Elsevier, pp. 291-303.

Union of Concerned Scientist: 2013, 'Fossil Fuel Industry Funders of Climate Change Contrarian Groups' [available online at: https://www.ucsusa.org/sites/default/files/legacy/assets/documents/global_warming/Fossil-Fuel-Industry-Funders-of-Climate-Contrarian-Groups-2001-2011.pdf, accessed 28 May 2018.]

United Nations Development Program (UNDP): 2013, 'Human Development Report 2013, The Rise of the South: Human Progress in a Diverse World: Pakistan', UNDP [available online at: http://hdr.undp.org/en/2013-report, accessed 25 May 2018].

United Nations Environment Programme (UNEP): 2016, 'The Adaptation Finance Gap Report 2016', Nairobi, Kenya: UNEP [available online at: http://web.unep.org/adaptationgapreport/sites/unep.org.adaptationgapreport/files/documents/agr2016.pdf, accessed 24 May 2018].

United Nations Environment Programme (UNEP): 2018, 'UN Environment and European Union call for Stronger Action against Climate-Change-Related Security Threat', 22 January [available online at: https://www.unenvironment.org/news-and-stories/story/un-environment-and-european-union-call-stronger-action-against-climate, accessed 5 September 2018].

United Nations: 1992, 'United Nations Framework Convention on Climate Change' [available online at: https://unfccc.int/resource/docs/convkp/conveng.pdf, accessed 24 May 2018.]

United States Energy Information Administration: 2018, 'How much of U.S. Carbon Dioxide Emissions are Associated with Electricity Generation?' [available online at: https://www.eia.gov/tools/faqs/faq.php?id=77&t=3, accessed 28 August 2018].

Vidal, J.: 2018, 'How Bill Gates aims to clean up the planet', *The Guardian*, 4 February [available online at: https://www.theguardian.com/environment/2018/feb/04/carbon-emissions-negative-emissions-technologies-capture-storage-bill-gates, accessed 16 August 2018].

Volti, R.: 2014, *Society and Technological Change*, 7th edition, New York, NY: Worth Publishers.

Weinberg, A.L.: 1967, *Reflections on Big Science*, Cambridge, MA: MIT Press.

World Bank: 2013, 'Turn Down the Heat: Climate Extremes, Regional Impacts, and the Case for Resilience, Executive Summary', Washington, DC: World Bank [available online at: http://documents.worldbank.org/curated/en/843011468325196264/pdf/784220WP0Engli0D0CONF0to0June019090.pdf, accessed 25 May 2018].

Chapter 15

The Ethical Judgment: Chemical Psychotropics

Klavs Birkholm

Abstract: In the case of psychotropic and nootropic substances, evidence is abundant that the pharmaceutical industries are violating elementary ethical norms, implying a serious liability not only for company managements, but also for researchers, laboratory staff, *etc.* Moreover, the rapidly expanding consumption of these substances seems to have radical repercussions on society and cultural norms. This chapter points to three such further consequences: an abnormal spread of diagnostics in human interaction (Section 7.1); a potential suspension of the elementary fight for recognition (Section 7.2); and an ever-present demand for perfection in both working and private lives (Section 7.3). The question of responsibility for these developments is addressed.

1. Introduction: The World-Altering Power of Side Effects

The following analysis applies a three-step model for ethical deliberation and ethical judgment that has not before been presented outside Denmark; therefore I take the liberty of shortly presenting a few basic preconditions for this model (Sections 1, 2, 3) before arriving (in Section 4) at the case proper.

Towards the end of his life, Danish philosopher K.E. Løgstrup (1905-81) makes a famous comment on the unintended side effects of the continuously accelerating technological development. Løgstrup talks of 'the world-altering power of side effects'.

Nowadays, when health authorities and the pharmaceutical industry consider 'side effects', focus is usually on phenomena of the kind also described on the long lists of reservations enclosed with pharmaceutical products: nausea, dizziness, heart fibrillations, 'should not be used during pregnancy' *etc.* – that is to say, medicinal, individually experienced, side effects.

However, Løgstrup thinks along ethical lines, and he does so for two reasons. Firstly, the richest part of the World has left the era in which the purpose of technological progress is to overcome the scourges of poverty and thus can be justified by this raison-d'être alone: "As long as technology served to fight poverty – a battle that has been going on throughout the history of humankind – the purpose was so evidently just and sound that technology did not give rise to further ethical reflection. *This has since changed.*" (Løgstrup 1983, pp. 18-19, my translation and emphasis.)

Secondly, globalization leads to a state where the mutual interconnectedness of humans – by Løgstrup termed 'the interdependency' – has strongly increased, in both close and more distant perspective. We are less than ever before isolated individuals in the sense projected in the pharmaceutical disclaimers mentioned above. On the contrary, we are unceasingly exposed to the consequences of what *other* people do.

Chemistry is both an active subject and a passive object in this development. Subject in the sense that chemical substances – both newly produced substances and chemical emissions and waste products – are to an eminent degree transboundary. Emissions rise up to the atmosphere, wastewater sifts down to the groundwater, *etc.* Only chemical substances, which are deliberately kept sealed in laboratories, may evade this agency. Chemistry is, on the other hand, also an object affected by this transboundary development in the sense that all chemical research and technological development nowadays should be carefully subjected to the strictest ethical protocols.

It is not difficult to explain why the strong side effects often come unnoticed by researchers. In order to be able to produce new knowledge, epistemic sciences like chemistry and physics must almost inevitably focus rather narrowly on a particular, strictly defined object, a cluster of objects, or the interaction of objects. Speaking in metaphors, the natural scientist has to watch the world through a magnifying glass, through a 'microscope'. This is true whether she studies the qualities of specific nanoparticles or the acceleration of the expansion of the universe. Nowadays all disciplines of natural sciences proceed methodologically only by isolating a meticulously selected corner of reality, thus leaving everything else out of sight. They are not able to encompass the wide-angle perspective.

From time to time, scientists and engineers are influenced by side effects while still experimenting. "Already while busy raising the agricultural production by means of insecticides, the unintended destruction of flora and fauna, imposed by the insecticides, sets in", Løgstrup writes (*ibid.*), perhaps with a thought towards DDT. In other cases, the unintended side effects only become apparent much later.

This insight teaches us a simple lesson, which is as important as it is simple: The declared purpose of any given technology should never serve as grounds for the ethical judgment of it. Unfortunately, in institutions that have been officially authorized to make ethical reviews and recommendations one often finds that researchers' own descriptions of their intentions form the basis on which the body concerned estimates the ethical perspectives in the legalizing of some new technology or the performing of an important experiment. I have, however, never met a researcher whose intentions were not of the very noblest kind. Virtually all industries and industrial branches have only the best of intentions when carrying out research!

This is why I always seek to imprint in the minds of my students the classic proverb, usually ascribed to Samuel Johnson, but probably stemming from Bernhard of Clairvaux: 'The road to hell is paved with good intentions.' I advise them to focus instead on

(1) the inherent personal risks (for users and others);
(2) the risks of misuse (possibly malign misuse) and, not least;
(3) the unintended side effects that are not immediately recognizable, but could prove decisive in culture and societies.[1]

[1] Birkholm 2014. I define techno-anthropology with Tom Børsen as studying the interface by which technology changes humans as well as humans changes technology.

2. Epistēmē versus Phronēsis

Apart from Løgstrup's observations, my model for techno-anthropological ethics is inspired by Bent Flyvbjerg. In a seminal intervention in the protracted 'war' between the natural and the social sciences, Flyvbjerg demonstrates that the two branches build on fundamentally different intellectual virtues – and must do so. For hundreds of years this difference has been downplayed, and the reason is obvious: Since the industrial revolution, the natural sciences have gained tremendous prestige in society, because of their success in creating the base for ever new technological achievements, serving to increase the overall wealth and welfare in society. Conversely, the social sciences have been put in the shadows and have, as a result, made the mistake of trying to 'imitate' the methodology of the natural sciences, perhaps in the hope of sharing some of the prestige enjoyed by the latter. However, in doing so the social sciences have compromised themselves, precisely because they cannot hope to prove anything in the sense of the word used among natural scientists. As a matter of fact, this is not their purpose, either (Flyvbjerg 2001).

In applying the term 'prove', I refer to the first of Aristotle's famous five intellectual virtues[2]: *Epistēmē* denotes the ability to point out, through deduction or induction, what is unchangeable and universally true, which is the road to all natural scientific knowledge. The proof may well be concluded by the statement *quod erat demonstrandum.* He who understands epistemic thought is disposed, then, for deftness in the natural sciences. In applying *epistēmē,* the scientist – or the team of scientists – arrives at universal, context-independent truths.

Technē, on the other hand, is the ability to create or shape artful products, artifacts. A clever shipbuilder applies this intellectual virtue when forming the keel of a ship so as to make it cut more effortlessly through the water. Likewise with the deft cither-maker; he builds the resonance chamber in shapes and with opening holes proper for reinforcement of the acoustic waves. The specialized experience of the artisan or craftsman plays an important role here. Through *technē,* he arrives at truths

[2] Aristotle 1982: Book VI, pp. 324-373.

that are pragmatic, context-dependent and variable. Nowadays, the *technē* of the carpenter, the architect or the engineer is often referred to as instrumental rationality.

The third of Aristotle's virtues, the one to which Flyvbjerg primarily refers, is *phronēsis*. It denotes the ability to choose the acts that are required in a specific situation to ensure the good life – 'the common good' within the community (the family, the society, the state). In other words: An individual may be clever in calculating how a number of different chemical substances will react when mixed at certain temperatures (= *epistēmē*); or she may be clever in designing a thermostat with the ability to reduce heat loss in the rooms of a house (= *technē*); but she may also be clever in simply 'being a human' (=*phronēsis*).

Phronēsis is the most valuable of all the intellectual virtues, says Aristotle, because its presence – or absence – defines the ethos of society and the overall condition of the state. Accordingly, those who are chosen for political office, should possess a high degree of *phronēsis*. This virtue is sometimes translated as 'prudence' or 'good sense' and is also somewhat present in the contemporary concept of 'value-rationality'.

In the context of the present chapter, we may ignore Aristotle's two final intellectual virtues (*nous* and *sophia*). What matters is the acknowledgement of how fundamentally different the epistemic and phronetic disciplines are to each other. The latter analyses values – what is good and what is bad in human life. *Phronēsis* is the reasoning that is directed towards practice, action. It examines relations that vary according to context – specific relations, not universal 'first principles'. Hence the result of a phronetic analysis is always temporary and might be changed when circumstances shift. In contrast, the natural scientist is bound to strive for a definite, once and for all solution to the problem at hand, a solution that makes a clear cut between true and false.

Reflecting on ethics is, to an eminent degree, a phronetic exercise, which is why the training of young researchers in making sound ethical estimates must take place during work on specific cases. Context – all that is particular – plays a vital role, and therefore decisions on what to do can rarely claim any definite necessity; circumstances may always change, leading to altered and maybe wholly dissimilar ethical judgments.

3. The Ethical Dilemma

All ethics are, of course, normative; the ethical deliberation refers to certain norms or to conflicts between different norms. Norms are commonly accepted moral truths about the nature of the good life, such as 'You must not kill', 'You must not steal', or 'You must not piss in the village well'.

Certain norms may primarily be characterized as prohibitions; this goes for ancient taboos such as 'You mustn't eat your fellow man', 'You mustn't commit incest', 'You mustn't treat the dead unseemingly', or the killing taboo. Other norms may be characterized, rather, as injunctions, such as 'You must help your fellow man who is suffering'. In the European cultural sphere, this norm is symbolized in the evangelical parable of the Good Samaritan – we refer to this norm as 'mercy', sometimes 'care' – which today has global validity, even though perhaps less so in *e.g.* Hindu cultures.

What ethical norms do have in common, whether prohibitions or injunctions, is the way in which they impose themselves on us as spontaneous incentives (Løgstrup 2007). We are spontaneously prompted to treat a deceased person seemingly (we wouldn't merely dump the body in a garbage container). We are spontaneously prompted to rescue a drowning person shouting for help, by jumping into the water. And, to include yet another norm, we spontaneously meet our fellow humans with trust, when, for example, we make an agreement or accept a promise. Society would not at all function if, instead, we were to meet with mistrust the man who tells us that we may find a supermarket further down the street or the woman who tells us that the bus drivers are on strike, so there's no need to wait for the bus today. Were we to assume, in short, that our fellow humans are probably out to cheat us, society would fall apart.

Other ethical norms of great significance today are justice, fairness, authenticity, autonomy, and the right to self-determination. Contemporary philosophers like, among others, John Rawls, Charles Taylor, and Michael Sandel have made significant contributions to the discussion on these norms. On the whole, however, such discussions within the community of academic philosophy have no direct impact on the training in ethical reflection among natural scientists; they are not at it to become academic philosophers. What matters is their ability to identify *ethical dilemmas* and to mobilize a certain *phronēsis* in the handling of them.

There are, indeed, examples of obviously 'evil' research and development where no ethical dilemmas present themselves. Within chemical science, one might point to the development of poisonous gasses and other chemical weapons, the purpose of which is to kill people in war. Such examples are rather uninteresting here, since participation in such research is unconditionally condemnable, excusable with regard to neither homeland security nor orders from company management.

An ethical dilemma, on the contrary, expresses a normative conflict, the solving of which must rely on prudent estimates. Today, chemical research and industry present us with a long list of such difficult dilemmas. These are only a few examples:

- Should care institutions – nursing homes, hospitals, hospices, rehabilitation centers, *etc.* – ban employment of people who have received cosmetic treatment with Botox? The neurotoxin *Botulinum toxin* has since the 1950s been used for medical treatment of muscular spasms, but has also during recent years been used for cosmetic interventions, mainly face-lifting. Now, recent research suggests that facelift treatments with Botox do not only weaken the affected person's ability to convey emotions through facial expressions, but also weaken their empathy for other people (Neal & Chartrand 2011). This corresponds well with other research showing that learning as well as empathy build on the human ability to mimetic mirroring of others; we 'feel', so to speak, the other person's grief or anger as though it were our own – we 'imitate' it. If further research confirm the findings of Neal and Chartrand, should it then lead to the health care sector taking this into account when employing staff, or maybe even to political regulations on chemical production of the toxin?
- As part of the struggle to prevent the climate of the Earth from collapsing, a range of geoengineering projects are currently being developed. Some of them are about powdering cloud formations with a sulphurized powder, or about spraying salt water into the atmosphere (see among others Alterskjær *et al*. 2013). The aim is to 'whiten' the clouds, in hopes that the rays from the sun will be partly blocked from

access to the lower parts of the atmosphere. The potential chemical side effects are not easily predictable. One possible side effect, which is being discussed, is the salting of soils through precipitation – agricultural land, woodland. Another possible side effect may be a decrease in precipitation, with an especially significant impact on regional monsoon seasons, this being due to decreased sunlight causing decreased evaporation (Ferraro *et. al*. 2013). Both effects, though seemingly opposite, might lead to crop failures and, perhaps, famine. The question, therefore, is: Who could claim the authority to decide whether to implement such a project, the result of which might have crucial impact on life conditions over the entire Earth? On the one hand, we feel a normative obligation to protect life-forms from the consequences of global warming. On the other hand, we hold on to the norm of self-determination, both to individuals and to nations.

Here, then, I have merely mentioned two very different and currently highly relevant examples of how chemical research and development today is intertwined with central techno-ethical dilemmas. Other examples might easily be found by the dozen. However, as the aim of the present essay is to demonstrate how ethical judgments are made, I will – for the remainder of the text – confine myself to one single case: The chemistry of psycho- and nootropics.

4. The Substances and their Spread: Denmark

To estimate the ethical challenges it can sometimes be useful to know about numbers. In 2015, a total of 156,982,000 DDD (daily doses) of anti-depressive drugs were sold in Denmark, making this group of products by far the most common psychotropic, even the most common pharmaceutical drug as such, in that country. The demographic backdrop to this figure is a total population of 5,627,235 people. Of these, 419,062 people – 7.4 percent – redeemed prescriptions for anti-depressive medicine during the year mentioned. In 2011, figures were even higher (8.29 percent of the Danish population). But in 1996, figures were significantly lower, as 'only' 106,476 Danes – 2.0 percent of the population – consumed anti-depressants. When converted into statistics the amount of sold drugs within this category rose, over a period of 15 years (1996-

2011) with 333 percent! This increase is quite evenly distributed across the years.[3]

The largest group is, by far, the SSRIs (selective serotonin reuptake inhibitors).[4] Add to this the anxiolytic benzodiazepine and benzodiazepine-like drugs, which were sold to 550,117 Danes in 1999, but to 'only' 336,514 fifteen years later (2014). A range of lithium-based products were sold in 3,154,400 DDD in 2014 (statistics on the number of people buying these products are not available).

An even more dramatic development may be observed when looking at methylphenidate-products, which are prescribed for the treatment of ADHD (attention deficit hyperactivity disorder). In this case, the number of users has risen from 1,812 in 1999 to 41,612 in 2015 – a total increase in the number of users, within 16 years, of 2,196 percent!

Common to all the drugs mentioned above is the purpose of regulating feelings, moods, memory, the ability to concentrate, *etc.* by influencing the neurotransmissions of the brain. But, contrary to psychedelic drugs on the black market (such as LSD, ecstasy, or cocaine) and 'natural drugs' taken in shamanistic contexts (mescaline, among others), these products are produced and ordained by pharmaceutical companies. SSRI products are ordained for the treatment of depression, anxiety, obsessive compulsive disorder (OCD), shyness, stress, and sometimes posttraumatic stress disorder (PTSD). The slightly different SNRIs (serotonin-norepinephrine reuptake inhibitors) are ordained for the treatment of just about the same disorders. Ritalines may be ordained to children and youths with hyper-kinetic disorders (ADHD) and otherwise only for the treatment of the relatively rare disorder of narcolepsy.

It seems relevant to ask what has caused this enormous increase in the consumption of Central Nervous System stimulants?[5] Has bio-chemical

---

3 All figures have been obtained from the official Danish Registry of Medical Statistics (www.medstat.dk) which is administered by the Danish Health Data Authority. The reason that the increase was brought to a halt in 2011 may be that a report from the Danish Council of Ethics was published in November 2010, initiating a prolonged debate in the Danish media. That, however, is only a hypothesis.

4 Among the SSRI products, figures were: 96,492,000 DDD in 2014, 112,307,000 DDD in 2011, 27,013,000 DDD in 1996. Marketing names in Denmark are, among others, Citalopram, Escitalopram, Fluoxetin, and Paroxetin.

5 A similar development is discernable in a range of countries; see Whitaker 2015, pp. 363f.

research and the pharmaceutical industry succeeded in developing new substances with the ability of fighting widespread diseases, which were before left untreated? Or, are we witnessing epidemic outbreaks of completely new disorders? Is it mostly about the medico-chemical industry's commercial interest in making up new markets and producing new consumers of future products? Or are we seeing such fundamental changes to our social life conditions in modern societies as to make necessary new forms of mental regulation? I will try to answer these questions as part of my ethical judgment.

5. Step One: Inherent Personal Risks to the User

The essential property of the group of chemical substances dealt with in the present essay is that they influence the functions of the brain, altering mood and consciousness. In this they are similar to recreative drugs, some of them almost identical by their chemical composition – *e.g.* methylphenidate is very close to amphetamine (which is used, by the way, in the North American parallels to ritalins: Adderall, Dyanavel, and others).

Since the human brain, by far the most complex organ in nature, has not yet been satisfactorily mapped by science, it would seem obvious that systematic intervention in its neurochemical processes must inevitably imply a certain, smaller or bigger, risk. We do something to the brain – we observe an effect, but we do not know precisely what we are doing. The different substances affect the serotonin, the dopamine, or/and the noradrenaline receptors respectively, thereby amplifying the levels of these transmitters in the brain – but the substances work very differently, depending on their exact composition, the way and the amount by which they are induced, and the receiving human bodies. The whole thing is still confusingly complex and incomprehensible.[6]

In this situation, the application of the precautionary principle is obviously relevant. On European scale, the precautionary principle is

---

[6] The number of neurotransmitters has – until now – been estimated to be more than one hundred, but more will most likely be discovered. Mentioned in this paper are only: Serotonin ($C_{10}H_{12}N_2O$), Norepinephrine ($C_8H_{11}NO_3$), and Dopamine ($C_8H_{11}NO_2$).

detailed in Article 191 of the Treaty on the Functioning of the European Union. But it is also emerging as a factor of growing importance in global bodies like the UN World Commission on the Ethics of Scientific Knowledge and Technology (COMEST 2005) and the World Health Organization (WHO 2016, Martuzzi & Tickner 2004).

However, in this case precaution has mostly been bypassed, and based on the above mentioned, rather incalculable complexity, it was foreseeable that a number of unwanted side effects would occur – also, that some of these would be subject to fierce debate. (1) First, the use of all psychoactive drugs entail a propensity for creating addictions. Already in 2003, an expert WHO committee determined that a substantial number of SSRI-users show signs of abstinences upon ceasing to use the drugs. WHO in fact complains that pharmaceutical companies and the psychiatric establishment seek to create terminological confusion by distinguishing between 'addiction' and 'withdrawal syndrome' (WHO 2003). Other frequent side effects are (2) dizziness (potentially fatal to older people) and (3) sexual disorders such as decreased libido. Providing a systematic overview of known side effects of psychotropic and nootropic medicine is not the aim of the present essay, but I wish to present a selected example: Paroxetine.

In September 2015, *British Medical Journal* (BMJ) published research that seriously incriminates the pharmaceutical giant GlaxoSmithKline (GSK) in particular, and the pharmaceutical industry in general (Le Noury *et al.* 2015). The investigation is the first of a planned series of revisions of earlier pharmaceutical product tests, with new researchers looking over the originally collected data and the published conclusions. In this case, focus is on the anti-depressive drug Paroxetine, which following a 2001 test was claimed by GSK to be safe for both adults and children. Paroxetine is also being sold in Denmark under the name of Seroxat, and 6.0-6.7 million DDD are being prescribed annually.[7] The new revision concludes that the beneficial effects of Paroxetine on children and adolescents are smaller and the harmful (side) effects far

[7] Estimated by the author on the basis of data for 2011, 2012 and 2013 from the Danish Registry of Medical Statistics. If sales figures within the private sector (pharmacies) are added to the handing out of prescriptions at hospitals, the numbers for 2011 are 6.735 million DDD; for 2012 6.403 million DDD; and for 2013 5.977 million DDD.

more serious than claimed by GSK in 2001. One of the seven authors, Professor David Healy of Bangor University, was quoted in the British newspaper *The Guardian* for saying that around 12 out of 93 children risk suicidal thoughts when using the drug, a figure that is clearly discernable from the original data! “This is a very high rate of kids going on to become suicidal. It doesn’t take expertise to find this. It takes extraordinary expertise to avoid finding it.” (Boseley 2015)

In an editorial comment BMJ claims it to be a blemish on medical research that the 2001 report has not been withdrawn and that none of the 22 researchers behind it have wished to modify any of their previous statements (Doshi 2015). The report has long been criticized. Already in 2002, the US Food and Drug Administration claimed that documentation of the alleged beneficial effects of Paroxetine was non-existent and in 2012 GSK were sentenced to a penalty of 3 billion dollars for misguiding and exaggerated marketing of the product.

Here, then, we have an example of unethical circumvention of the precautionary principle. Both the authorities and the public have deliberately been misled in order for GSK to obtain a permit that the products in question would otherwise not have been able to achieve. When reactions have not been stronger – in spite of suicides among adolescents, which might have been prevented – it is no doubt down to the enormous power held by the pharmaceutical industry over research institutions as well as political circles.

In a comment on the publication of the sensational article, the editor in chief of BMJ, Fiona Godlee, remarks that this first revision report clearly demonstrates the degree to which the current regulation of pharmaceutical products has failed. It is absolutely necessary, she claims, to establish independent clinical testing instead of tests that are financed and carried out by the industry itself, the latter practice being unfortunately commonplace today.

In Denmark, patients may seek information on Paroxetine products on the website www.min.medicin.dk, a sort of user’s guide run and financed by the pharmaceutical industry itself. On Seroxat, this ‘user’s guide’ states that it may be used for the treatment of depression, anxiety, OCD, social phobia, *etc.* The side effects are listed in four categories: ‘common’, ‘not so common’, ‘rare’ and ‘extremely rare’. In none of these categories, the risk of suicidal thoughts is mentioned; it is stressed, on the other hand, that most side effects appear only in the early phases of the treatment!

In his very critical book on the subject, *Deadly Psychiatry and Organised Denial*, the leader of the Nordic Cochrane Center, Peter C. Gøtzsche, lists numerous examples of devastating side effects of psychotropic products that are today being generously prescribed by psychiatrists as well as general practitioners (Gøtzsche 2015). Gøtzsche is a professor of clinical research design and the entire book is strictly based on demand for severe evidence. The focus point of the book is that test results are too often interpreted at pleasure by the pharmaceutical industry and – especially worrying – that inconvenient test results are kept quiet.

6. Step Two: Potentials for Misuse

In my model for ethical judgment this step normally refers to finding out whether other persons or other interests might have access or capacity to apply a given technology in ways not originally anticipated by the researchers. SCiO for example, is "a pocket size molecular sensor for everybody".[8] The device, produced by Consumer Physics Inc., is a handheld scanner that uses spectroscopy to analyze the chemical composition of anything it's pointed at. When the company startup was launched on Kickstarter in 2014, the purpose was presented as "identifying foods for diet tracking and checking medications to make sure they're not counterfeit" (Strictland 2016). On this prospect the startup managed to allocate $ 2.7 millions from "enthusiastic backers". Now, investigating the use of the device after it has been brought to the market, Eliza Strictland of the Institute of Electrical and Electronics Engineers found that the two top threads in the "developer forum" on Consumer Physics's website were proposing apps to test the purity of illegal drugs, starting with ecstasy. Ecstasy, she explains, "is often cut with other substances, ranging from inert fillers to dangerous chemicals, but since most labs won't test illegal drugs, users have no way of checking what they're

8 See https://www.consumerphysics.com.

taking. This situation sounds like a problem that the SCiO can solve [...] although it might not be what Consumer Physics wants to be known for" (*ibid.*).

The case of Paroxetine, though, is unusual because it is the actions of the pharmaceutical company itself that qualify to the category of 'misuse'. Putting other people's health and life at risk for the sake of company growth and profits is evidently morally reprehensible.

However, it seems to be a common notion that responsibility in such cases lies with the managers of the company, whereas individual employees have merely carried out work that they were ordered to do. This perspective is convenient but untenable. It denies the existence of certain codes of professional ethics – a factor that perhaps especially ought to demand the attention of everybody whose work is related to the treatment of diseases.

This can clearly be derived from the history of the Third Reich. Not only was the chemical industry a crucial factor in the economy of the Third Reich (BASF, IG Farben, *etc.*), but many of the Nazi crimes against humanity were committed by the medical professions – doctors, clinical assistants, laboratory staff, *etc.* Atrocities ranged from euthanasia-programs for handicapped children, via cooling down prisoners of war (testing what humans can endure), to the infamous genetic experiments of Joseph Mengele.

The legal process against these war crimes forms the historical backdrop to the Helsinki Declaration where the World Medical Association (WMA) recognizes a set of ethical principles for all research and all experiments on human beings, including research that is combined with medical treatment and care.[9] The meaning is unmistakable. The verdict in 1947 of the court in Nürnberg upon the Third Reich doctors also applies to present-day Denmark and other modern states. No one who partakes in forms of medical treatment or development of medicaments that violate the ethical principles of human dignity or informed consent, can possibly evade co-responsibility for their actions by claiming that

9 The declaration was adopted in 1964 and has since been updated several times. In Denmark, it has not least been influential in connection with the formation in 1980 of the Danish Committee System of Research Ethics and later (in a more indirect manner) in relation to the founding of the Danish Council on Ethics.

they were merely following orders and performing the duties of their work, ignorant of the final purpose of their contributions. Even at the cost of losing one's job, one is in such circumstances obliged to 'blow the whistle' and demand a change. This is simply the very essence of what's termed 'professional ethics', also including all staff at Glaxo-SmithKline.

Another group of psychotropic preparations charged with misuse are the ADHD products: Ritalin, Adderall, *etc.* Elsewhere, I have documented how students at US colleges feel pressured towards consuming what is commonly referred to as 'brain enhancers' or 'brain doping', simply because their fellow students do it and because it seems an accepted 'fact' among students that the consumption of such amphetamine or methylphenidate drugs enhances the ability to cope successfully with written assignments, tests, and examinations (Birkholm 2015, pp. 145-150). Already in 2008, a joint group of doctors and philosophers suggested that these prescription substances become regular over-the-counter drugs, so that students in the USA and Europe may freely buy them without running the risk of incrimination (Greely *et al.* 2008). Instead of seeing these students as drug abusers, we should – according to the authors – consider them "early adopters of a trend that is likely to grow", in other words, a kind of pioneers.

"Human ingenuity", they write, "has given us means of enhancing our brains through inventions such as written language, printing and the Internet". And: "The drugs just reviewed, along with newer technologies such as brain stimulation and prosthetic brain chips, should be viewed in the same general category as education, good health habits, and information technology – ways that our uniquely innovative species tries to improve itself." (*Ibid.*)

I fully agree with the authors that enhancement of our species is the real issue at stake here, but I do not acknowledge their assumption that engineering interventions in the human brain and nervous system belong to the same categories as former inventions like printing, libraries, or coffee. Adding all the present enhancement endeavors within different scientific fields (genetics, robotics, brain-computer interfaces, pharmacy *etc.*), the whole thing sums up to a both dangerous and preposterous attempt to intervene in the evolutionary process that was for millions of years driven by hazards, natural selection, and other very wise principles as demonstrated by Charles Darwin.

Since the potential for enhancing humans is tantalizing to these doctors and philosophers,[10] they do not consider excessive consumption of these drugs a kind of misuse. Hence they do not at all touch upon the question of the identity of future well-educated people: Do we want future 'academics' to be persons who perform specific, highly specialized analyses only by the aid of cognition-enhancing drugs? Or do we, rather, prefer 'academics' to go on being persons who by the aid of their naturally and socially endowed cognitive and analytic abilities are capable of acquiring substantial knowledge and complex skills within certain subject areas?

These questions of authenticity are examples pointing to the real ethical issues (some of which I discuss in Section 7). To most of the authors joining the discussion on cognitive enhancement, the ethical task seems to be an estimate of the possible risks for the individual users – a set of problems I would term *ante ethics*: The real ethical deliberations can only begin by assuming this or that new technology to be safe for use.

Greely *et al.* made their contribution in December 2008. A more recent debate, grossly along the same lines, was initiated in July 2013 in the *American Journal for Bioethics*. Here, Serbian philosopher Veljko Dublević – linking up with Greely *et al.* – asks what would more specifically be a "responsible use" of Ritalin and Adderall for enhancement purposes. Dublević (2013) expresses his intention to clarify this by answering the following question: "what exactly should be the moderately liberal public policy" for regulating such drugs? His answers are commented among others by Hall *et al.* (2013), Faulmüller *et al.* (2013) and LaBuzetta (2013).

This is a policy issue, clearly besides the focus of the present chapter.[11] On these pages I am dealing with the industry behind the drugs, not with questions of prohibition or regulations on the market. What are the responsibilities and the ethical challenges facing the chemistry researchers and the producers behind the drugs? That is the question here.

[10] One of the authors, John Harris, is – indeed – professor of bioethics (University of Manchester).

[11] I have, however, dealt with it elsewhere (Det Etiske Råd 2010).

Some interesting points, however, are worth noting from the advocates for cognitive enhancement. Their main concern is to decriminalize the growing number of young people using the drugs, referring explicitly to the ethical principle of autonomy. If, say, person NN is perfectly able to use Adderall in a controlled way, capable of monitoring his own reactions carefully and preventing a slip into addiction, then a prohibition would be a paternalistic violation of his autonomy.

This a somewhat reduced understanding of autonomy, very far from the original concept as formed by Immanuel Kant in the wake of European Enlightenment. If respecting the norm of autonomy sums up to respecting personal preferences, there is really nothing left for ethical reflection. As observed by Wayne Hall *et al.* (2013) this whole philosophical literature is dominated by libertarian views – views that are generally not shared by drug policy analysts.

In Denmark, the National Board of Health (SST) dictates that Ritalin products may be prescribed to children and adolescents between the ages of 6 and 18, who have been diagnosed with ADHD. Nonetheless, half of all prescriptions are currently being handed out to adults, thus failing to meet the prescribed indication area (Danish Council on Ethics 2010, p. 71). Also, the Council found that children and adolescents are in many cases pressured into an ADHD diagnosis, which in Denmark may release funding for extra staff resources in school and offer relief to tormented parents who often receive complaints about their unruly children.

7. Step Three: The Effects on Culture and Society

To investigate eventual users' risks inherent in a new (or old) technology is really to inspect whether any malfunctions were neglected by the inventor (Section 5). Then, to investigate the potentials for misuse is to turn one's view from the microscope-lens of the laboratory to the world outside: Did we overlook some other possible applications of this invention, more or less unpleasant and not intended by the scientists? (Section 6)

But the real challenge when making a prudent ethical judgment is the third step: To address the future, attempting to assess whether this scientific invention, provided it proves effective in the way we want, might have any ethically significant effects on society and culture.

Since most of the drugs we are discussing in this chapter have been accessible for decades, it is easier than usual to give answers on all steps in this model of ethical judgment. Concerning the last step, I will now deal briefly with three questionable phenomena: (1) the expansion of a culture of diagnoses as a consequence of the ample supply of psychotropics; (2) a suspension of the essential human fight for recognition as a possible threat to society's checks and balances; (3) the implacable demand for perfection in both working and private lives.

7.1 Life diagnosed

The transition during the last 25 years in the conception of mental frailties from seeing them as something caused by important life experiences to regarding them, rather, as biologically determined, has been fairly well described (see *e.g.* Lane 2007, Mayes & Horwitz 2005). The subtitle of Lane's book, *How Normal Behavior Became a Sickness*, is characteristic. The crucial paradigm shift took place with the release of DSM-3, the third version of the diagnostic manual of the American Association of Psychiatrists, published in 1980 and completely overturning our former notions on the nature of mental illnesses. Though the DSM manual is prepared by a US professional consortium it has international impact. Not least because it underlies the WHO classification manual ICD-10, which is used in most countries of the world.

Until 1980, a certain friction existed between, on the one hand, the medically trained psychiatrists, used to coping with mental problems by means of pharmaceutical products and, on the other hand, analytically trained psychologists who viewed mental illnesses in a broader context. The latter group was on the advance during the 1970s, and this caused disturbance in the USA where more precise definitions were in demand. At the same time, US insurance companies began covering the expenses of psychotherapy via health insurance schemes, but only in case of 'real' diseases or illnesses – in cases of 'reactions to existential problems' no insurance coverage was offered.

From 1980 onwards procedures were changed, so that when a therapist could mark out a certain number of symptoms from a list this would constitute a diagnosis, on the basis of which treatment – always a medical treatment – might be initiated. Which 265 diagnoses were included in the 1980 catalogue was decided by show of hands at the professional consortium behind the register.

A diagnosis denotes that something is out of course, deviating. But how and when can something be defined as 'deviating'? If you are tagged with a certain diagnosis, you belong to a certain group of people who are different in a very specific way. Such people are, then, the same, and they will often unite in their sameness – this is called 'a patients' association'. But in reality we are all both 'the same' and 'different'. Some people just attach more importance to the differences than do others.

Were we to ask what society stresses most – the similarities or the differences – the answer would be that this changes with time. A hundred years ago, at the time of Sigmund Freud, all women who did not feel at ease in the bourgeois patriarchal family were considered neurotic. Like Madame Bovary, Flaubert's famous fictional character. Today, not a single neurotic person is to be found, simply because this diagnosis has been deleted, struck off the register. Instead, we now have the television series *Desperate Housewives*.

In 2013, the diagnosis register was published in a new version, DSM-5 (the previous, DSM-4, was released in 1994). The debate prior to the publication of the new manual made it clear that the distinction between the sick and the healthy mind is something that is actually determined through negotiations. There were votes, horse trading, and a multitude of economic interests involved. When, for example, shyness may now be diagnosed as *Social Anxiety Disorder (SAD)* and when the intense grief over the death of a closely related human qualifies to the diagnosis of *Major Depressive Disorder (MDD)* – that is, if this state of grief lasts for more than two weeks – this reflects the social norms as well as the economic interests of a certain group of professionals in a certain culture at a certain historical point in time. *Disruptive Mood Dysregulation Disorder* (DMDD) is a diagnosis for children who burst into heavy outbreaks of temper at least three times a week. Yuck! And gluttony is now diagnosed as *Binge Eating Disorder (BED)*. On the other hand, Asperger's Syndrome no longer exists.

Diagnoses come and go, and the diagnoses we make tell us much about the dominant norms of our time and society. They also tell us much about the development of the pharmaceutical industry, though. Diagnoses, as it is, tend to appear in the wake of the development and approval of new medical products. The above-mentioned Paxil, Seroxat, and Nardil products are recommended, for example, to people suffering from shyness; the former two are variants of Paroxetine (described in Section 6).

Parallel with the increased medicalization, large groups within Western populations seem to have become participants in an intense hunt for diagnoses. In many countries, diagnoses have also become the admission ticket to treatments, economic benefits, support services, *etc.* The other side to this, though, is the intensified standardization of human beings. Individual life stories are suspended, thus cutting off sources to the selves. The ethical demand for authenticity is jeopardized.

If, say, a young man is hyper-attentive, having been raised in a family dominated by a violent father, in a home where family members must constantly be vigilant in order to anticipate the father's sudden mood swings, this part of the young man's identity is erased when he is diagnosed and medicated. When an older woman is sad because her husband left her ten years ago for a younger, perhaps more attractive woman, this important life story is easily eliminated when she is merely given a diagnosis and offered anti-depressants.

In such a medicalized culture, diagnoses make sure that individuality is left in the dark – and along disappears the individual's struggle for freedom and autonomy.

7.2 The Fukuyama-Kojève argument

Advocates of human enhancement generally show more interest in the ethical dimensions of cognitive enhancers although mood enhancement poses larger challenges. In my book, *Efter Mennesket (After Humans)*, I have adapted the argument brought forward by Francis Fukuyama in 2002, that psychotropic medicines represent a potentially dangerous intervention in some of the most fundamental, socially balancing mechanisms inherent in all kinds of societies: the mechanisms linked to the fight for recognition (Birkholm 2015).

Inspired by Plato (427-347 B.C.), Hegel (1770-1831) and Kojève (1902-1968), Fukuyama identifies *thymos* as the true origin of man's commitment to both work and politics. *Thymos* is the part of man, which has to do with neither rational thought nor basic instinct, but with emotions such as honor, shame, justice, ambition, self-esteem, and self-respect, emotions that we wish to see recognized by our fellow men.

"What, must I hold a candle to my shames?", Shylock's daughter Jessica cries out in the second act of Shakespeare's *The Merchant of Venice*, when young Lorenzo in the darkness of night abducts her by a ladder and at that point presents her with a flaming torch (Shakespeare 1967, pp. 55f.). By running away, Jessica abandons her Jewish father, even taking with her a small chest containing part of his treasures. It is shameful, and Jessica wants her suitor and his helpers to recognize this feeling.

This same Shakespeare demonstrates, with the tragedy of Richard III, how a born cripple claws his way towards the English throne by cheating and murdering, in hopes that political power will satisfy the longing for recognition that neither beauty nor physical abilities have ever granted him.

Some might argue that these Shakespearean examples are far away from the ethical challenges of chemistry. Quite the contrary. They are both vivid illustrations of some of the fundamental human feelings that pharma-chemical research and industry now tries to modify or eliminate.

The thymotic urge towards recognition is, according to Fukuyama, a vital fuel for all human development and civilization:

Virtually all human progress has been the by-product of the fact that people were never satisfied with the recognition they received; it was through struggle and work alone that people could achieve it. Status in other words, had to be earned, whether by kings and princes, or by your cousin Mel, seeking to rise to the rank of shop foreman. The normal, and morally acceptable, way of overcoming low self-esteem was to struggle with oneself and with others, to work hard, to endure sometimes painful sacrifices, and finally to rise and be seen as having done so." [Fukuyama 2003, p. 46]

As we know today, the achievement of being recognized also comes with a chemical side: A reward inside the individual's brain in shape of high levels of serotonin. Winners of the Olympic gold medal or winners of a prestigious song contest may experience a 'shower' of serotonin in their brains. So – this is the argument of Fukuyama – would Thomas Jefferson ever have achieved to write the American Declaration of Independence, would Winston Churchill for all his speech impediment ever have endeavored to lead Great Britain in World War II, would Thomas Mann ever have written some of the greatest novels of our time, would Bob Dylan ever have written all his ballads, and would Jon Stewart ever have managed to host 16 years of *The Daily News* if all of them had free access to Prozac or Zoloft? These anti-depressants are sold in the USA with the advertised idea that the pills can provide self-esteem (in biochemical language: they delay the reuptake of serotonin in your brain).

The question, then, is not only whether anti-depressants (as discussed in Section 5) may harm the individual consumer. The question is, rather, whether the aggressive marketing of them may, in the long perspective, disturb some of the most delicate checks-and-balances in the fabric of society.

7.3 The Quest for perfection

I see a common denominator between: the dissemination of mood enhancers; the explosive use of 'brain doping'; the promotion of advanced technologies for prenatal diagnosis; the aspirations to do gene-editing in human embryos; the strive to delay human aging at cellular level; experiments with augmented vision; and lots of other emerging biotechnologies – both genetic, pharmaceutical, and cybernetic. And that common denominator is the pursuit of Human Enhancement.

The *Zeitgeist* gives us the impression that humans are not good enough as they are. We have too many shortcomings and imperfections, hence we must take advantage of the ongoing technological revolution to reshape humankind: make us more fit, more intelligent, able to live much longer, equipped with superhuman abilities. This idea has become almost a religious obsession to 'posthumanists' and 'transhumanists' like Nick Bostrom, Julian Savulescu, John Harris, Kevin Warwick, and many others. They are now in the limelight of academic conferences on ethics (or 'applied ethics'), always able to vindicate some new technological progress and explain away all ethical concerns.

And you do not want to be considered a reactionary, do you? A reactionary in the field of science and technology today seems to be a person who does not unconditionally support the next 'inevitable' step in human evolution: A merging of man and intelligent machines to create a whole new species, a kind of *homo technitos* or, as I prefer to frame it: *homo artefact*. After all, this is the golden gate that most of the advanced technologies envision: A leap into the making of some kind of cyborg-creature, which can travel in space, communicate directly with machines and other forms of intelligence, and which are not dilapidated like us.

Most of these ideas still have the character of science fiction, of course. The creation of a memory expansion slot to insert in human brains still has a long way to go. Intelligent prostheses with more advanced abilities than natural arms and legs can be made today, but we are far from ready to apply them on a mass scale. Recent progress in gene-editing techniques (CRISPr) may be encouraging to the posthumanist dream, but also in genetics we are still miles away. Only one branch of science is really able to deliver for the moment: The chemistry of psychotropics and nootropics. So, this science today maintains and nurtures the dream of human perfection.

One could ask: Is it only the new and emerging technologies, including those of pharmacy, that produce the quest for perfection? Or is it – the other way round – our present obsession with competitiveness that produces a demand for such technologies? That may be a question of the-hen-and-the-egg, and a fair answer might be that the technological innovation in its present directions and the culture of neoliberal economy mutually nourish each other.

It is, in fact, not difficult to find cultural explanations of the appetency for perfection. The competition of admission to both academic and non-academic educations is becoming increasingly tough, and the same applies to job admissions. People are obliged nowadays to perform 24/7 – performing, that is, on a strangely impalpable scale, where the job is not to build a brick wall, to mount a valve, or to make drawings for a new building, but to reach certain abstract targets, evaluation criteria, quotas, *etc.* Moreover, while performing excellent you have to look like a complete success: fluent, painless, fulfilling without effort, easily spending, and very happy.

A Danish Professor in the History of Ideas, Lars-Henrik Schmidt, recently studied Job Advertisements and found the overall most demanded competency (sic) to be: 'cheerful', in 'a light mood', 'generally gladsome', *etc.* (Brinkmann 2010, p. 133).Today this is taken as an indication that you are cooperative. Furthermore, add Schmidt and his co-author Claus Holm, the opposite mood is now pathologized. To be sad, melancholic, disconsolate, or apathetic is considered abnormal (*ibid.*). So, if you are a highly skilled chemistry laboratory technician with a penchant for philosophy and melancholy, you do not need to apply – you will not get the job anyway! The employer prefers a less skilled rival who is always happy.

This demand for both effortless performance and an ever-happy face bear witness to a cultural schizophrenia with a lot of implications, one of the most well-documented being stress and depression. Certainly, this state of mind creates a lot of jobs in the businesses of coaching, wellness, fitness, and therapy, but it also nourishes the underlying utopian idea of perfection. Today, Lara Croft, 007, Batman, and Superman are no longer just mythological figures of fantasy, they are real role models. They are no longer presented to us merely for our amusement in the cinema, we have to live up to them.

In this perspective, all sorts of optimizing technologies seem to offer a great relief, not least the chemical ones abundant today. But they also make an important contribution to maintaining a quest, partly of their own creation, that threatens to alienate humans from their authentic selves.

8. Conclusion

The chemical sciences are right at the core of the complex ethical dilemmas in present-day techno-anthropology. Virtually no corner of our society is independent from the findings and the innovations within chemistry, even our supply of energy depends on it. And virtually no spot on the earth, be it land, water, or air, can evade the consequences of chemical production. It is therefore essential that universities and technical high schools educating chemists include mandatory training in making ethical judgments. This can be done by applying the three-step model described in this essay: (1) focus at the inherent personal risks (for users and others); (2) then investigate the risks of misuse (possibly malign misuse); and, not least, (3) finally discuss unintended side effects that could prove decisive in culture and societies.

In the case of psychotropic drugs, chosen here as an example, ethical considerations are shown to be long overdue. Not only is a rampant excess consumption – co-produced by pharmaceutical companies, psychiatrists, and general practitioners – constituting a systematic and truly unethical misuse of these chemical technologies. Management and employees are putting millions of people's health at risk, and thereby fail to honor the ethical responsibilities of their respective professions.

At the same time, though, the production and distribution of these substances are fueling certain deep tendencies in contemporary societies, which at best we ought to analyze and make choices about, at worst they are just 'happening' to a generation of helpless humans: (1) the tendency to circumscribe all individuals within certain digital standards that constitute a diagnosis, but at the same time threaten to eliminate any individuality and freedom to define your own path in life; (2) the danger of substituting the joy of reward for doing something excellent with a chemical surrogate, thus halting a decisive mover of the dynamics of history; (3) the demand for flawless performance, leaving all sorts of imperfection and impairment in society's dustbin.

Perspective: All three tendencies could motivate for a moratorium on further development of the substances concerned, if needed for a couple of years, thus making way for a necessary ethical deliberation in science and society.

References

Alterskjær, K.; Kristjánsson,J.E.; Boucher, O.; Muri, H.; Niemeier, U.; Schmidt, H.; Schulz, M. & Timmreck, C: 2013, 'Sea-Saltinjections into the low-latitude marine boundary layer: The transient response in three Earth system models', *Journal of Geophysical Research: Atmospheres*, **118**, 12,195-12,206.

Aristotle: 1982, *The Nicomachean Ethics*, English trans. by H. Rackham, London: Loeb.

Birkholm, K.: 2014, The Ethical Judgment: Teaching and learning techno-ethics, Aalborg University, unpublished.

Birkholm, K.: 2015, *Efter Mennesket. På vej mod Homo Artefakt*, Copenhagen: Forlaget Samfundslitteratur.

Brinkmann, S. (ed.): 2010, Det diagnosticerede liv: Sygdom uden grænser, Aarhus: Klim.

Boseley, S: 2015, 'Seroxat study under-reported harmful effects on young people, say scientists', *The Guardian*, 16 September.

COMEST (World Commission on the Ethics of Scientific Knowledge and Technology): 2005, *The precautionary principle,* Paris: UNESCO.

Det Etiske Råd: 2010, *Medicinsk Optimering. Etiske overvejelser og anbefalinger.* (Redegørelse udarbejdet af en arbejdsgruppe med Klavs Birkholm som formand), Copenhagen: Det Etiske Råd.

Doshi, P.: 2015, 'No correction, no retraction, no apology, no comment: paroxetine trial reanalysis raises questions about institutional responsibility', *BMJ*, **351**, h4629.

Dubljević, V.: 2013, 'Prohibition or Coffee Shops: Regulation of Amphetamine and Methylphenidate for Enhancement Use by Healthy Adults', *American Journal of Bioethics*, **13** (7), 23-33.

Faulmüller, N.; Maslen, H. & de Sio, F.S.: 2013, 'The indirect psychological costs of cognitive enhancement', *American Journal of Bioethics*, **13**(7), 45-47.

Ferraro, A.J.; Highwood, E.J. & Charlton-Perez, A.J.: 2013, 'Weakened tropical circulation and reduced precipitation in response to geoengineering', *Environmental Research Letters,* **9**, 14001.

Flyvbjerg, B.: 2001, Making Social Science Matter. Why Social inquiry fails and how it can succeed again, Cambridge: Cambridge University Press.

Fukuyama, F.: 2003, Our Posthuman Future. Consequences of the Biotechnology Revolution, London: Profile Books.

Greely, H. *et al.*: 2008 'Towards responsible use of cognitive-enhancing drugs by the healthy', *Nature*, 456 (11 December), 702-705.

Gøtzsche, P.C.: 2015, *Deadly Psychiatry and Organised Denial*, Copenhagen: People's Press (Danish original: *Dødelig psykiatri og organiseret fornægtelse*, Copenhagen: People's Press, 2015).

Hall, W.; Patrigde, B. & Lucke, J.: 2013, 'Constraints on Regulatory Options for Putatively Cognitive Enhancing Drugs', *American Journal of Bioethics*, **13** (7), 35-37.

LaBuzetta, J.N.: 2013, 'Moving Beyond Methylphenidate and Amphetamine: The Ethics of a Better 'Smart Drug', *American Journal of Bioethics*, **13** (7), 43-45.

Lane, C.: 2007, *Shyness: How Normal Behavior Became a Sickness*, New Haven: Yale University Press.

Le Noury, J.; Nardo, J.M.; Heayly, D.; Jureidini, J.; Raven, M.; Tufanaru, C. & Abi-Jaoude. E.: 2015, 'Restoring Study 329: efficacy and harms of paroxetine and imipramine in treatment of major depression in adolescence', *British Medical Journal*, **351**, h4320.

Løgstrup, K.E.: 1983, *System og Symbol*, Copenhagen: Gyldendals.

Løgstrup, K.E.: 2007, *Beyond the Ethical Demand*, Notre Dame: University of Notre Dame Press.

Martuzzi, M & Tickner, J.A. (eds.): 2004, The Precautionary Principle: Protecting Public Health, the Environment and the Future of Our Children, Copenhagen: WHO.

Mayes, R. & Horwitz, A.: 2005, 'DSM-III and the Revolution in the Classification of Mental Illness', *Journal of the History of the Behavioral Sciences*, **41** (3), 249-267.

Neal, D.T. & Chartrand, T.L.: 2011, 'Embodied Emotion Perception Amplifying and Dampening Facial Feedback Modulates Emotion Perception Accuracy', *Social Psychological and Personality Science*, **2**, 673.

Shakespare, W.: 1967 [1600], *The Merchant of Venice*, ed. J. R. Brown, London: Methuen.

Strictland, E.: 2016, 'The Pocket-Sized Lab's Killer App: Analyzing Illegal and Semi-Legal Drugs", *IEEE Spectrum*, 2 May.

Whitaker, R.: 2015, *Anatomy of an Epidemic*, New York: Broadway Books

WHO: 2003, 'WHO Expert Committee on Drug Dependence', *Technical Report Series, No. 915 – Thirty-third Report*, Geneva: WHO.

WHO: 2016, 'Climate Change and Human Health', Report prepared for COP22 in Marrakech, chap. 12.

Chapter 16

'Are You Playing God?': Synthetic Biology and the Chemical Ambition to Create Artificial Life

Joachim Schummer

Abstract: Throughout history, chemists have faced the accusation of 'playing God' or similar devilish associations, overshadowing all moral judgments of chemistry. The chapter provides an ethical analysis of the accusation with focus on Craig Venter's 2010 announcement of having produced the 'first self-replicating cell'. Against the deeper historical background of the ambitious projects that came to be known as 'synthetic biology', I describe Venter's actual research and its international media reception. Then I analyze both the ethical and theological implications of creating living beings in the laboratory. In conclusion I argue that the Venter case, like many others cases from chemistry before, is a case of unfortunate science-public interaction that mislead both ethics and science.

1. Introduction

Chemists working on apparently innocent research projects might be surprised if not puzzled when they suddenly face the public accusation of 'playing God'. They probably think: 'I am just doing good science for the benefit of society. How is that related to ? Why should that be bad?'

Much more than any other science, chemistry carries a heritage of religious presumption and hubris continuing over 2,000 years. Responsible chemists should be particularly aware of their cultural history, cautious about societal provocations and pitfalls, and versed in ethics.

In the Christian tradition, the accusation of 'playing God' was originally applied to the 'Fallen Angels' who would later be called devils under the leadership of Satan. The *Book of Enoch*, an Apocrypha of the

Old Testament written around 300 BC by a Jewish sect in Ethiopia, narrates the story of a conspiracy of a group of angels who disobeyed the commands of God and traveled down to Earth in order to mate with human females and build their own reign, in this sense 'playing God'.[1] The worst crime they committed was revealing to women certain secret crafts, in particular metallurgy and the production of pigments, dyes, and colored stones (glasses). These crafts, which all involve chemical transformations, were secret because they employed knowledge of the 'Primordial Creation'. At the time the text was written, they were actually developed in Alexandria and became the starting point of what was later called alchemy.

The story or myth is remarkable in several regards. First, it explains the origin of human technology through the impact of evil forces. Second, it relates what we would today call chemistry to the divine knowledge of the 'Primordial Creation' and declares it particularly forbidden. Finally, it is the founding myth of the devil, whom Christianity and Islam, unlike Judaism, incorporated in their theology. In some sense, the invention of the devil was the oldest critique of chemistry.

Chemical crafts, alchemy, and eventually chemistry would never get rid of their devilish association in Christian culture and thereby of the accusation of 'playing God' in the sense of changing or imitating the divine creation with the help of demonic forces.[2] There are numerous cases similar to the one discussed in this chapter, such that this theme has overshadowed all public moral judgment of chemistry.

For instance, church father Tertullian in around 200 CE argued that the dying of wool is against God's will and an association with Satan because God had not made colored sheep in the Primordial Creation. Note that he did not argue against weaving because God had not made sheep with woven wool. Instead the general argument, which has ever since been directed against new technologies, applies to essential transformations only. Chemical transformations with their radical changes of

---

[1] On the following see Schummer 2003.

[2] In the Christian tradition the accusation of 'playing God' can have different meanings depending on God's role as creator (messing with the Creation/Nature), moral lawgiver (giving oneself own moral laws), or ruler of destiny (controlling the course of events). This paper refers only the first meaning.

material properties, which transcend ordinary experience and are so difficult to explain, have played a central role. Those who tried to change the Creation, that is to produce something 'unnatural' as it was later called, assumed the role of the Creator God, either alone or with satanic help. In alchemy, every 'unnatural' chemical transformation, be it in metallurgy or for the production of medicines or other uses, was suspected to involve demonic forces. Therefore alchemists, up to the 18th century, felt obliged to explain their successful work as a 'gift of God' or a process helped by good angelic forces (Karpenko 1998).

Nonetheless, from the 14th century onwards, alchemy became the target of critique by many famous European writers (including Petrarch, Chaucer, Erasmus) and artists (including Dürer, Brueghel, Teniers), modifying the demonic myth.[3] Whether in writing or in painting, they all produced variations of the same theme: a man is tempted by a fiendish alchemist into practicing alchemy, becomes obsessed with his work, and eventually ruins his life. In paintings, the obsessed or mad alchemist was usually depicted with worn clothes, disheveled hair, and surrounded by a mass of alchemical laboratory equipment. During the 19th century, writers such as Mary Shelley, Balzac, Dumas, and Hawthorne transformed the 'mad alchemist' into the 'mad scientist', who as a rule was then a chemist or a physician performing chemical experiments. While the 'mad alchemist' was generally unsuccessful in his work, the 'mad scientist' is at first glance successful, but eventually fails either because his work gains uncontrolled and unforeseen momentum or because he has morally perverted ambitions that end in self-destruction. Eventually Hollywood's movie makers were so fond of the topic that they produced hundreds of variations, drawing on the late medieval imagery of the 'mad alchemist' that was first set in motion picture in Fritz Lang's *Metropolis* (1927). Since then, a chemical laboratory with various types of glassware is in public culture firmly associated with demonic work.

In the literary and artistic traditions the fabrication of simple biological organisms played little role, for reasons explained in Section 2.1.[4] Instead, the creation of human or human-like beings figured prominently,

---

[3] On the following, see Schummer 2006.

[4] A rare early exception is Theodore Sturgeon's *Microcosmic God* (1941), see Schummer 2011, pp. 184-186.

from Mary Shelley's *Frankenstein* (featuring a mad chemist) to Wells' *Doctor Moreau* (featuring a mad physician). They drew on earlier literary works or myths, such as the Jewish Golem legend and folktales about the alchemical fabrication of a homunculus (Newman 2004). Among all life forms it was only the creation of human life that was at stake and rose the accusation of playing God.

As will be shown in Section 2, only in the 20th centuary did chemists' scientific fascination and the public's moral excitement about the creation of life begin. That culminated in Craig Venter's announcement of having fabricated the first artificial cell in 2010, and a worldwide media response of 'playing God', which echoed critiques of chemistry for more than two thousand years. Section 3 conducts an ethical analysis of that case and the ambitions of synthetic biologists to create life, before I draw general conclusions about unfortunate science-public interactions that have provoked the accusation of 'playing God'.

2. From Spontaneous Generation to Synthetic Biology

2.1 The spontaneous generation and artful creation of life in early history

For most parts of the history, and probably in all cultures, the spontaneous generation of simple living beings from inanimate matter was considered a banality by most people.[5] Whenever some dirt and moisture was incidentally or intentionally mixed together, mold, worms, flies, lice, *etc.* emerged. Because nobody liked those creatures, they would blame the inadvertency of people who made that vermin, instead of accusing them of 'playing God'.

There was thus ample evidence of spontaneous generation in ordinary life. Also many Holy Scriptures describe it. For instance, in *Exodus* 8 of Judaism and Christianity, frogs and mosquitos are skillfully produced as plagues against the Egyptians. The Hindu *Laws of Manu* (I, 45) divides all living beings into five groups according to their origin, one of which is characterized by their spontaneous emergence out of warmth and

---

5 On the following, see Lippmann 1933, Farley 1977, Schummer 2009, 2011, chap. 3.

moisture. Many ancient Greek natural philosophers expressed their views on which elements are necessary for the creation of life. Aristotle even provided a detailed description and explanation of the generation of *testaceae*.

In the first-century BC textbook on agriculture, the Roman writer Virgil laid down the principles of *bugonia*, the art of creating honey bees. To illustrate how those views flourished among educated churchmen, I translate from the German Conrad von Megenburg's *Book of Nature* (ca. 1350):

> Bees emerge out of the bellies of young cows [...]. You need to cover the bellies with muck to get bees. Bees also generate from the buried skin of oxen; wasps from the skins of donkeys; Worms out of the muck of pigs; frogs out of turnips or chard; horseflies out of bad air or rotten breath [...]. From the body of dead horses you get wasps and hornets, from the body of donkeys you get a kind of flies called glow worms [...]. Note that the bees that generate from cows propagate, and that their offspring are of similar kind as real bees.

The most detailed early-modern technical description of generating plants and animals, of cross-breeding and improving them for human purposes, is the widely read second edition of *Magiae Naturalis* (1589) by the Italian Giambattista della Porta (1535-1615). Unlike the first edition it passed Roman censorship without problems, even though it now included a new part on cross-breeding animals, including the production of human-animal hybrids, and techniques for determining the sex of human offspring. One generation later, the English Lord Chancelor Francis Bacon transformed in his utopia *New Atlantis* (1623) della Porta's material into a grand systematic research project of improving living nature, from basic living beings made from the optimized composition of inanimate matter to perfecting human nature. Influential as Bacon's program still is nowadays – from synthetic biology to transhumanism – it then faced neither ethical nor religious criticism.

Religious objections against the idea of creating life in the laboratory arose only after the acceptance of 19th-century theories of evolution, from Lamarck to Darwin. If humans have ultimately emerged through evolutionary processes from simple living beings, the first creation of those beings would potentially produce humans, and thus the laboratory work would have to be considered comparable to divine creation. That critique was forcefully voiced by the French catholic chemist Louis Pasteur (1822-1895), who experimentally worked hard on refusing the possibility of spontaneous generation, in a famous speech of 1864 before the political and intellectual elite of France:

> What a triumph, gentlemen, it would be for materialism if it could affirm that it rests on the established fact of matter organizing itself, taking on life of itself; matter which has in it all known forces! [...] What good then would it be to resort to the idea of primordial creation, before which mystery it is necessary to bow? Of what use then would be the idea of a Creator-God? [quoted from Geison 1995, p. 111].

Suddenly the question of spontaneous generation (and intentional creation) of life, which had been considered a banality for millennia, became a religious issue.

2.2 Chemical ambitions in the 20th century

Scientists, who used to think that all natural phenomena have a natural cause, continued to believe in spontaneous generation, both at the historical origin of life and in the laboratory under suitable conditions. However by the end of the 19th century all presumed incidences of spontaneous generation had been found to be caused by spores or germs of microorganisms. Moreover, organic chemists had isolated and structurally described a wealth of diverse compounds in biological organisms, many of which were stereoisomers, which proved the chemical complexity of even the simplest life form.

Starting with Nobel laureate Emil Fischer (1852-1919), organic chemistry took a new systematic and visionary approach.[6] Nowadays he is well known for the structure elucidation of sugars, which he achieved without any spectroscopic means by using the controlled re-synthesis of the natural products and their chemical modification. Rather than just focusing on the naturally produced, he developed a classification of all possible sugars as well as a synthetic repertoire to produce them, in order to find sugars with new or more useful properties, for instance sugars

---

6 On the following, see Schummer 2011, chap. 7, on Fischer see also Johnson 2015.

with higher sweetness or as substitutes for diabetics. He applied the same strategy to purines (which includes the nucleobases of DNA and RNA as well as many alkaloids), to amino acids, and small peptides.

In the footsteps of Francis Bacon and always eager to collaborate with biologists, Fischer in 1890 envisioned the chemical modification of organisms far beyond the possibilities of breeding and cross-breeding, what a century later would be called metabolic engineering and eventually synthetic biology. For instance, the chemically modified organisms should be able to digest artificial sugars and to produce out of them, with their modified metabolism, new kinds of fatty acids and proteins with desired properties. In 1915 he first called that approach 'chemically synthetic biology'.[7] Furthermore, in an unpublished lecture delivered to his working group in 1907, he predicted that chemists would soon be able to create life from scratch in the laboratory. With a strange sense of humor, he added that one day chemists would also be able to create humans, including chemistry professors, so that his own profession would eventually become obsolete.

In the following decades many scientists claimed the imminent life creation through public announcements, for instance the physiologist and president of the British Association for the Advancement of Science Edward A. Schaefer (in 1912), the physicists Oliver Lodge (in 1923) and Paul Renno Heyl (in 1930), and the chemistry Nobel laureates Theodor Svedberg (in 1937) and Glenn T. Seaborg (in 1965). The nuclear chemist Seaborg even asserted that it would be achieved within a couple of years.

In his Presidential Address of the American Chemical Society of 1965, Charles C. Price (1913-2001), a physical organic chemist, demanded that "synthesis of life should be a national goal" for chemistry, similar to the prestigious well-funded grand projects of physics during the Cold War (nuclear energy, radar, weapons, and space exploration). Optimistically he proclaimed: "The job can be done – it is merely a matter of time and money". The "economic consequences of such a breakthrough would dwarf those of either atomic energy or the space program. Success could lead to modified plants and algae for synthesis of

---

7 At that time, the term 'synthetic biology' was used to mean a holistic approach in biology, from which Fischer clearly wished to distinguish himself.

foods, fibers, and antibiotics, to improved growth or properties of plants and animals, or even to improved characteristics for man himself." (Price 1965, p. 91)

The most visionary and ambitious 20th-century prophet was the British physical chemist James Frederic Danielli (1911-1984), who after his emigration to the US turned for a while to microbiology. In 1970 he hit the headlines for the first time announcing 'the first synthetic cell' that he had produced by recomposing the nucleus, membrane, and cytoplasm of three different amoebae into one new organism. A few days after that rather trivial experiment he was again cited in the media, claiming that such work is dangerous and should be controlled by the government – thereby inventing moral alarmism as PR for science. Soon he became a spokesman for 'synthetic biology', as he first called the then evolving field of genetic engineering and molecular biology in 1975. He also created the by now standard rhetoric of the field, according to which biology would follow the model of chemistry turning from an analytical period of understanding to an synthetic era of making.

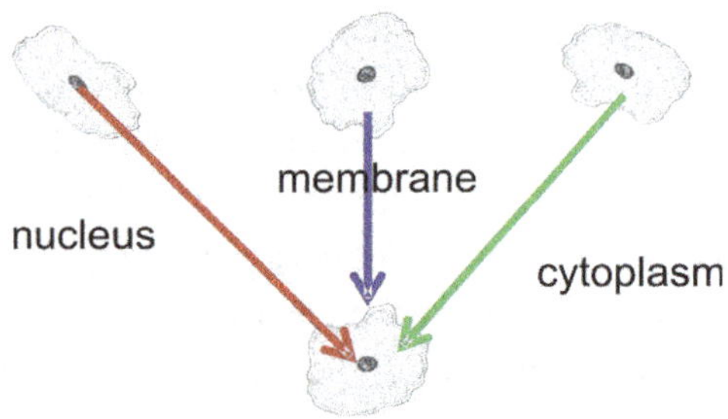

Figure 1. The 'first synthetic cell' by James Frederic Danielli.

Danielli's agenda of Bacon's program includes virtually everything which four decades later the 'new' synthetic biology would again promise to achieve. Relying on genetic determinism he envisioned the complete synthesis of genomes being transferred to host cells, in order to create any desired organism. Once a minimal organism is created, it could arbitrarily be enhanced for food production (*e.g.*, nitrogen fixation, synthesis of essential amino acids or desired proteins), cleaning the environment (*e.g.*, waste water treatment, heavy metal removal, sea water desalination), energy (*e.g.* the production of oil), medicine (*e.g.*, production of hormones, antibodies, human genes) as well as for data storage and processing (modified neurons for biological computing).

Furthermore, anticipating the later religious sect of transhumanism, he developed an eugenic program for the genetic enhancement of humans. A modified genome would produce humans with higher creativity, intelligence, longevity, and less aggression. Even more, if one produces a set of genetically identical humans, one could conduct better sociological experiments on them for the benefit of our knowledge of society.

Throughout the 20th century, scientists, and chemists in particular, have publicly announced that they soon would be able to create life in the laboratory. While this turned out to be wishful thinking, it revealed three important aspects of their views. First, because their predictions all turned out to be crudely false, their exaggerated views of chemistry's potential undermined their scientific credibility. In fact, hardly any worked in the field, many were physical chemists or physicists with apparently only limited knowledge of biology. Second, because the creation of life had in the 19th century turned from a banality to a religious issue, it is likely that they all sought public attention for themselves or their profession through their predictions. This is further supported by the fact that most of them, I assume intentionally, confused the modification of organisms with the creation of life. Third, in a surprisingly naive way they transgressed the border to science fiction as well as moral boundaries by lightheadedly relating the potential fabrication of simple life forms to the creation of humans, either for humoristic reasons (Fischer) or for eagerly attracting public attention to their science (Price and Danielli). In particular, Danielli seemed to have lost any moral sense, not in his scientific work, but in his words that suggested the production of biologically identical humans for human experiments. His unbridled ambition to popularize science through its fantastic power made him look like the protagonist of a mad-scientist movie from Hollywood.

The media reaction was expectable. For instance, the German political magazine *Der Spiegel* (21 December 1970) displayed on its cover page a naked women sitting in an alchemical retort with modern glassware in the foreground and wrote "Biochemistry: The human being is reconstructed". Its cover story was entitled "Biochemistry: Straight to Hell".

2.3 The New Synthetic Biology

At the beginning of the 21st century, two main approaches competed with each other for media attention to become the new synthetic biology: proto-cell research and synthetic genomics. Both drew on many decades of former research.[8]

Proto-cell research has its roots in chemical evolution, the study of prebiotic conditions on earth that allowed for the stepwise formation of life: from simple organic substances and complex biomolecules to organismic structures and eventually life itself. The classical breakthrough experiment was already conducted in 1952. After heating a gaseous mixture of water, hydrogen, methane, and ammonium for a week and treating it with electrical sparks, Stanley Miller (1930-2007) found a broad spectrum of amino acids as well as sugars, lipids, and components of nucleotides. Further experiments during the next decade could reproduce the formation of nucleic acids and proteins under prebiotic conditions. Studies on the spontaneous formation and dynamics of membranes, vesicles, and micelles, which were already pioneered by Nobel laureate Irving Langmuir (1881-1957) in the early 20th century, added the necessary compartments for organismic metabolism. Since RNA had been found to act not only as a carrier of genetic information but also as an enzyme that might control its own self-replication, much of the research on the chemical origin of life has studied vesicles or micelles filled with RNA plus some other ingredients. The idea is to develop a system that shows many or all characteristics associated with life, such as metabolism, self-replication, growth, homeostatis, and chemical exchange with the environment.

Almost unnoticed the field has moved from understanding the historical origin of life under prebiotic conditions to the creation of simple life forms in the laboratory under any condition and by any means. Proto-cell

---

[8] On the following, see Schummer 2011, chap. 8.

research distinguishes itself from chemical evolution by emphasizing creation at the expense of understanding the origin of life.

Synthetic genomics is an ambitious sub-field of genetic engineering. On the one hand, it draws on metabolic engineering, *i.e.* the genetic modification of entire protein systems rather than just single proteins, in order to add desired functions or to knock out undesired ones in organisms. That is largely the vision of Emil Fischer which has become possible since about the 1980s. Instead of making bacteria that produce just one new protein, *e.g.* insulin for diabetes treatment, they insert entirely new functions, for instance, a sensor for a chemical stimulus that triggers a color signal, such that the bacteria turn red whenever the stimulus is present. Encouraged by approaches from software engineering, the success depends on whether such functions are freely combinable in biological organisms, which is of course frequently not the case. On the other hand, synthetic genomics employs the entire repertoire of chemically and biochemically synthesizing and multiplying DNA sequences of maximum length and accuracy, ideally of an entire genome without any errors. Landmark achievements were the syntheses of the genes of the growth hormone somatostatin (1977) and alpha-interferon (1981) by Keiichi Itakura and Michael Edge, respectively. Others include the syntheses of the entire genomes of the poliovirus with 7,000 base pairs (bp) by Eckard Wimmer in 2002, the Spanish influence virus with 13,000 bp by Jeffery Taubenberger in 2005, a mycoplasma bacterium with almost 600,000 bp by Craig Venter's group in 2008. Combining both strands, synthetic genomics aspires to design genomes of organisms with any desired function to be then synthesized in the laboratory and finally 'brought to life'. Again, synthetic genomics distinguishes itself from traditional genetic and metabolic engineering by its emphasis on creating life rather than just modifying existing organisms.

Most synthetic genomicists have, knowingly or not, taken over Danielli's idea of a minimal organism, which includes just enough biological functions (and genetic code) for survival under some standard conditions, and on which then desired functions should be genetically attached. They aspire to reach the minimal organism either by step-wise knocking out of gene sequences of some bacteria or yeast or by designing the minimal genome on their computer. The two approaches are usually called 'top-down' and 'bottom-up', but they rather reflect the different methodologies of biologists and software engineers who are both involved. Next we analyze in some detail the most famous example.

2.4 Craig Venter and the Media

On 20 May 2010 a press release entitled 'First Self-Replicating Synthetic Bacterial Cell' received worldwide media attention.[9] It was issued by the J. Craig Venter Institute, a private non-profit genomic research institute in the US, founded and directed by biochemist J. Craig Venter. The institute is one of the most productive and creative ones worldwide. A forerunner institute, Venter's commercial company Celera, became famous in 2001 for beating the international consortium of the Human Genome Project in the race to decode the human genome by employing unconventional methods.

The three-page press release was comparably rich in scientific detail and referred to a research paper published online in *Science* on the same day (Gibson *et al.* 2010). However its main message was that scientists had achieved the goal of synthesizing life in the laboratory. How did the media respond?

Not surprisingly, journalists ignored the details and focused solely on the main message. Across all levels, from simplistic tabloids to sophisticated newspapers, they all joined a worldwide chorus: 'Craig Venter is playing God'. A detailed international analysis of online media reports during the following week brought about a surprising geographic pattern according to religious predominance. The chorus was dominant in all Catholic, Anglican, and Hindu countries. In Austria the religious tone was even reminiscent of the former Roman inquisition, relating Venter to witchcraft and the devil. In prevailingly protestant areas of Europe, the most popular association was to Frankenstein, which is only a pseudo-secular version of accusing somebody of aspiring to be Creator. Within Europe only Sweden, Denmark, France, and some parts of Switzerland had nuanced reports with little or no religious allusion, as in the USA, where national pride was unmistakable. Most Christian orthodox coun-

---

9 On the following, see Schummer 2011, chap. 9.

tries preferred metaphors from Greek mythology for uttering critique, like Pandora's box, except Russia, where the Frankenstein-motif seems to be popular. Although Judaism and Islam directly or indirectly refer to the same creation myth as Christianity, the playing-God motif was absent in Israel and Arab countries, notwithstanding harsh criticism from Arab commentators. Similarly in China, Japan, and other Buddhist countries, where no comparable creation myth exists.

A follow-up analysis illustrates how uncorrectable public stereotypes are. Through its *L'Osservatore Romano*, the Vatican immediately issued an official statement on the case, acknowledging the work by the Venter team as an interesting scientific contribution and rejecting all associations with divine creation. That notwithstanding, several media, incl. ABC News and Associated Press, titled 'Church warns cell scientists not to play God'.

Let us now have a closer look at what the Venter team actually did and if it was justified to describe the work as producing the first synthetic cell. The paper in question reported only the final step of a series of work on two species of mycoplasma bacteria, let us call them A and B. Mycoplasma bacteria, which cause many serious human infections such as peneumonia, are special bacteria because they have no cell wall and contain a very small genome of only approximately 1 million base pairs. In previous work, the team of 24 scientists from different countries, had, according to a standard procedure known since the 1980s, replaced the genome in species B with that of species A to produce a hybrid bacterium. Luckily the hybrid was able to self-replicate, which is not typically the rule in such hybrid productions. After a complete genome analysis of A they had further performed a complete genome synthesis of A. All of which had previously been published. The final step was to repeat the production of the hybrid bacterium, now with the synthetic genome, which turned out to be successful.

The proper scientific goal of this work consisted not in producing a hybrid with a synthetic genome, 'synthesizing a self-replicating cell' as the press release suggested. Instead, the goal was to prove the accuracy of the complex genome synthesis, for which the team had developed several new techniques to minimize errors. And the proof of the error-free synthesis was the self-replication of the hybrid bacterium. In other words, the creation of a self-replicating cell was not the goal but a means of the scientific work.

In press conferences and various media interviews Venter systematically misrepresented the work of his team. For instance, he argued that the cell was produced out of 'four bottles of chemicals'. As a matter of fact, the team neither produced a cell, nor did they chemically synthesize the genome. They actually bought about 1,000 sequences, each of about 1,000 bp, from a commercial DNA-sequence manufacturer who had produced them from enzymatically linking smaller parts that were chemically synthesized. Living yeast cells then combined them to form longer sequences, first to 100 different sequences with each 10,000 bp, then to 10 with each 100,0000 bp, and eventually to one with about 1,000,000 bp.

Furthermore, the hybrids were not produced by literarily replacing the genomes. Instead they mixed synthetic genomes A with cells of species B whose membrane were treated with a kind of soap, such that external bodies could occasionally slip into the cell (Fig. 2). It very rarely happens then that a cell with two different genomes A and B incorporated turns through cell division into two cells of which one contains only genome A. If genome A has implemented an antibiotic resistance, all B cells and mixed cells can be killed by the antibiotic, such that only the hybrid cells survive. From the unsurprising fact that these cells produced proteins similar to species A, the team concluded that the hybrids would be of species A.

In their research paper the authors initially described the procedure more carefully than in the press release and denied that they had synthetically produced an artificial cell. Indeed they wrote that it only looks 'as if' an artificial cell was produced, because they had actually produced only the genome. But then they blurred that difference with computer metaphors arguing that the 'software' (genome) would 'reprogram' or 'reboot' the 'hardware' (rest of the cell) resulting somehow in an artificial cell. While these metaphors are biologically disputable, it seems clear that they had to hold to some naive form of essentialism, according to which the DNA is the essence of life, in order to infer the synthesis of the cell from the synthesis of the DNA.

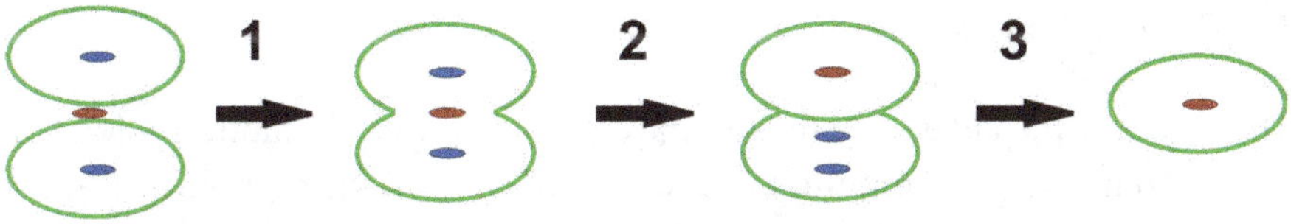

Figure 2. Schematic representation of the experiment by the Venter team. (1) In a mixture of synthetic genomes and surface-treated cells, two cells sometimes merge into one cell that contains two natural and one synthetic genome. (2) Natural cell division of such a cell sometimes results in two cells, one with the synthetic genome alone, the other one with two natural genomes. (3) If the synthetic genome includes an antibiotic resistance, applying the antibiotic kills all the other cells.

The Venter case is an illustrative example of how respectable scientific work (here: the almost error-free synthesis of a huge DNA-sequence due to sophisticated control techniques) can be turned into public media excitement through systematic misrepresentation by the authors themselves: by intentionally mixing up means and ends (self-replicability as a means to prove the error free DNA-synthesis), exaggerations (*e.g.* describing complex biochemical synthesis in living organisms as simple wet chemical synthesis), confusing parts with wholes (DNA and cell), avoiding unwelcome truths (*e.g.* that genome replacement is a standard procedure of hybridization rather than cell creation), and wrapping up everything in dubious metaphors to hide the confusion.

Moreover, in many interviews Venter argued that the experiment would challenge our philosophical understanding of life and show that 'we' (humans) are information machines, without any further explanation. Because the traditional philosophical understanding of life has largely been about human life, it is not clear how the successful hybridization of closely related bacteria can contribute to that. Nor is it clear how the bacteria experiment can revive the old information-machine metaphor that had stimulated the early days of genetic engineering. However, by relating his pretended bacteria creation to human beings through quasi-metaphysical comments, Venter could indirectly point to the idea of creating or modifying humans. Such allusions are sufficient to create media hype and public excitement because they trigger stereotypes that are deeply rooted in Western culture, from the homunculus myth that surrounded alchemy to the Frankenstein myth.

3. Ethical Analysis

This section provides an ethical analysis, not about synthetic biology in general but about its ambition to create artificial life as highlighted in the previous sections.[10] The analysis is confronted with a strange situation that is very unusual in current ethics of emerging technologies. Typically research and development proceeds much faster than the ethical reflection which is lagging behind. In our case, however, the situation seems reverse: Over thousands of years before the 19th century, people had believed that life can easily be made from inanimate matter and that it had frequently been done so (Section 2.1). But nobody had raised any moral or religious objection against that; ethicists (and theologians) had simply been disinterested in the matter. Nowadays we see stimulated media hype and public moral excitement about the artificial creation of life (Section 3.4), although the goal is not yet achieved. The critique is poorly articulated. As if ethics – *i.e.* the moral reflection on values, norms, obligations, judgments, intentions, virtues, and consequences of actions – did not exist, a broad public expresses their moral concern, if not horror, through the accusation of 'playing God'.

The following subsections look at the artificial creation of life from six different ethical angles. The first three approaches are deontological, *i.e.* they judge actions independent of their specific consequences, only according to whether they meet or violate certain general obligations. The three corresponding obligations are (1) not 'playing God', (2) respecting the dignity of life, and (3) reverence or responsibility for life. In contrast, the subsequent three approaches, which philosophers call consequentialist positions, judge actions only according to whether their consequences are desired or undesired. Particularly they look at three kinds of risks (*i.e.*, undesired and unintended potential consequences): artificial organisms could (4) interfere with the ecosphere, (5) be misused as bioweapons, and (6) cause harm to laboratory staff and local environment.

[10] On the following, see Schummer 2011, chaps. 13, 15.

3.1 The accusation of playing God

As we have seen above, the dominant media response in Christian countries to scientists who claim or aspire to create an artificial organism in the laboratory is the accusation of 'playing God'. The objection is deeply religious and has meaning only within a religious context that includes a divine creator of all life. If A accuses B of playing God, A means that B pretends and intends to be like God.

In the Christian tradition that intention came to be the cardinal or worst of all sins, much worse than robbing, raping, child abuse, or murdering. Through the story of the fallen angels, who disobeyed God and tried to establish their own regime on earth, the sin is firmly related to Satan (Section 1). Thus, accusing somebody of playing God is another way of calling him Satan.

The religious prohibition of creating artificial life rests on three assumptions that are all problematic:

(1) Historical assumption: All life on earth originated from divine creation. However, most scientists believe that life originally emerged by chemical evolution.
(2) Metaphysical assumption: Every natural organism owes its existence indirectly to divine creation in a way that would be circumvented by human life synthesis. However, most theologians, including the Vatican statement cited above, believe that all life originates from God, no matter what humans make in their experiments. Matter might originally be imbued with the potentiality of life or with sperms that can turn into real life under certain conditions. Church Father Augustine explained spontaneous generation in this way already in the fifth century.
(3) Psychological assumption: Scientists who try to create life in the laboratory intend to be like God. However, unless they clearly express such an intention, there is little reason to assume that. Note that the sin does not consist in doing something, but in pretending and intending something. If the scientist is an atheist, the assumption is plainly wrong. Somebody who does not believe in the existence and concept of God can hardly intend to be like God.

Since all three assumptions are problematic, it is difficult to understand how the accusation could be meaningfully applied to synthetic biologists by either Christians or atheists. It rather originates from a folk religion of an artisan-like creator God, mixed up with literary myths such as the Frankenstein and homunculus stories. However, as the Venter case illustrates, some scientists subtly play with that folk myth in order to raise media attention for their own research.

3.2 Dignity of Life

A frequent objection against the creation of artificial organisms worries about the dignity of life. If life can be synthesized in the laboratory, it is said that it would lose its dignity, which is considered a bad thing. The objection has been raised throughout the 20th century and recently echoed by many critics of synthetic biology. It is worth considering the argument in detail because it typically comes with hidden assumptions and conceptual confusion.

Note that the argument cannot be applied to individual organisms synthesized in the lab. Because these organisms did not exist and thus could not have any dignity before their synthesis, it is logically impossible that they would lose their dignity by their creation. In order to make sense of the argument, we need to reformulate it so that it applies to *kinds* of organisms: 'The knowledge that organisms of a certain kind can be synthesized in the lab eliminates their dignity, which is bad.'

There are at least two major problems with the argument. First, it is not clear if simple life forms actually bear any dignity and where that would come from. In the religious creationist reading, the dignity is endowed by divine creation, which is missing in human creation. The argument is then just another version of the playing-God objection that was rejected above (van den Belt 2009). In any secular reading, dignity is ascribed by humans through appreciation and moral respect. However, today's wide use of antiseptics and antibiotics for killing bacteria without the slightest moral concern suggests that humans ascribe no dignity to microorganisms. There is also no historical evidence that humans ascribed dignity to basic living beings before the mid-20th century, when philosophers began to develop related ideas to which we will come back in the next section. Starting in the Renaissance, dignity has been ascribed solely to human beings in order to establish respect for humans and natural human rights. For instance in the Kantian tradition, human dignity rests on the moral freedom of humans, opposed to the causal determination of other living and non-living nature.

It seems that the argument relates the life of basic organisms to human life, such that the synthesis of simple life form would affect our understanding of humans and eventually human dignity. Indeed most newspaper commentaries of Venter's hybrid bacterium, as well as Venter himself, tried to establish that relationship – most prominently by the Frankenstein myth. That is however both biological and philosophical nonsense. If a simple organism could be made in the laboratory, that would have no biological implications about how to make a human. Moreover, philosophical traditions derived human dignity from moral and intellectual capacities rather than from biological functions that we might share with other organisms.

Second, even if simple life forms had some kind of dignity, the argument would not stand ethical scrutiny because it discredits knowledge. Imagine somebody ascribing dignity to a bacterium based on the idea that this bacterium cannot artificially be made. Imagine further that scientists find out that the idea was wrong, that the bacterium can actually be made. The only reasonable ethical response would be to recognize one's mistake, that the ascription of dignity was based on a wrong assumption. In contrast, the argument suggests that such knowledge is bad even if it is true. However, such a position that defends dogmas by discrediting knowledge would hardly count as an ethical standpoint.

If simple life forms can generate from inanimate matter, the natural response would be a shift in one's attitude, as the Scottish biologist John Arthur Thomson already wrote in 1922, "it would increase our appreciation of what is often labeled as 'inert' matter" (Thomson 1922, vol. 1, p. 62).

3.3 Biocentric Ethics (1): biological individualism

For most parts of history, Western ethics was anthropocentric, *i.e.*, humans (frequently excluding slaves and women) were the only beings that were considered moral objects in moral deliberations. In accordance with the biblical task that man should rule over Earth (*Genesis* 1, 28), nonhuman nature, both living and nonliving, lacked any intrinsic value and was considered only instrumental to human interests as long as it was not eradicated. Only during the Enlightenment period the scope of morally relevant objects was extended to include higher animals capable of suffering, in particular by Jean-Jacques Rousseau (1712-1778) in France and Jeremy Bentham (1748-1832) in England.

Truly biocentric approaches, which attribute intrinsic values to all living beings and appreciate them for their own sake in moral consideration, were developed not before the 20th century. Most famously the French-German physician and theologian Albert Schweitzer (1875-1965) developed his theory of 'reverence for life' for which he was awarded the Peace Nobel Prize in 1952. He maintained that all living beings – human and nonhuman, including pathogenic bacteria (!) – share a 'will for life' that establishes in humans the moral emotion of respect or reverence as the fundamental basis of ethics. Similarly, in his *The Imperative of Responsibility* (1984), the German philosopher Hans Jonas argued that all living beings have an intrinsic purpose, they defend themselves against death and call for protection, from which he derived the fundamental human responsibility for all life.

Can we infer from these classical biocentric approaches ethical objections against the artificial creation of life? Unlike what one would expect at first glance, the answer is a clear No. If the artificial organism meets all the biological criteria for life, as synthetic biologists aspire to achieve, it would deserve the same reverence and responsibility as a natural organism. Moreover, if living beings unlike inanimate matter have intrinsic value, one could even derive the moral obligation to increase that value by turning inanimate matter into life. Thus classical biocentric approaches, which focus on the intrinsic value of biological individuals, would encourage rather than prohibit the artificial creation of life.

3.4 Biocentric Ethics (2): biological holism

Unlike the previously mentioned approaches of biocentric ethics, which consider biological individuals to have intrinsic values, there is another branch of environmental ethics that appreciates the biosphere as a whole, a dynamic ecosystem in which individuals only play temporary and exchangeable roles. The difference between both positions echoes two extreme views of the human sphere: individualism (society is an aggregation of independent individuals with certain rights) versus socialism (society is a complex holistic system composed of the interdependencies between exchangeable individuals).

Not surprisingly, biological holism, which is sometimes also called 'deep ecology' and which can be traced back to early 19th-century romanticism, comes with various religious and political flavors, each celebrating its own founder. Yet, from a biological point of view it makes perfect sense to look at life not as an arbitrary aggregation of individuals, but as a dynamic system, both at the local level of ecosystems and the global level of the biosphere. If one considers large time scales, the biosphere undergoes evolutionary changes by its own population dynamics as well as by outer impacts such as natural disasters and eventually human interference.

Biocentric holism, as an ethical approach, ascribes intrinsic values to the ecosphere as a dynamic whole and calls for its protection by sustaining various characteristics, such as species diversity and adaptive flexibility. The call is difficult to justify, though, within a biocentric framework that takes humans as part of the global ecosphere, because that would allow any human interference to count as an acceptable factor of biological evolution. Justifications therefore include the Christian appeal of human stewardship for nature and the anthropocentric appeal that human life depends on its biological environment to be sustained for the sake of humans. Whatever the justification may be: can we derive from biocentric holism moral constraints on the human endeavor of producing artificial organisms?

Humans have ever interfered with the biosphere, from agriculture, horticulture, forestry, and the breeding and domestication of animals to the use of antibiotics, antiseptics, pesticides, and genetic engineering. In some cases, biocentric holism provides clear guidance, for instance against large deforestation of rain forests and huge monoculture production of crops upheld by massive employment of pesticides, both resulting in vast extinction of biological species and instable ecosystems. Also genetic engineering can be problematic if the gene-modified species massively and irreversibly change the ecosystem through direct interaction with other species or through gene exchange on an evolutionary scale. Even though genetic engineering has employed from the beginning 'natural' enzymatic processes for gene insertion, replacement, and modification, the extend of biological gene mobility among species has long been underestimated.

If the artificial organism is equipped with normal DNA or RNA, the problems of genetic engineering will largely repeat. If the DNA or RNA is modified, by chemically modified nucleobases or nucleotides, the possible genetic interactions with the biological environment are hardly foreseeable. If, on the other hand, the artificial organism is composed entirely differently than natural organisms, genetic interactions with the environment becomes unlikely, but such organism might be able to drive out other organisms and thus massively change the ecosystem. To prevent such 'run-away' cases, one could genetically incorporate an invulnerability that allows killing the artificial organisms when they might become a biological thread. However, such in-built vulnerability is not a handy solution for all cases, as synthetic biologists sometimes suggest. If the artificial organism is sufficiently similar to natural organism, the vulnerability could easily spread to natural species with undesired effects on the ecosystem.

Despite the shortcomings regarding its ethical foundation, biocentric holism can give us some moral guidance for the artificial production of life. However, at this point there is hardly more than a rule of thumb: particular precaution is due if the artificial organism is neither very similar to nor very different from natural organisms.

3.5 Bioterrorism

A frequently voiced argument against synthetic biology warns of the misuse of dangerous creatures by rogue states or terrorists (*e.g.*, Schmidt 2009). To be sure, the argument points to an important risk. However, one should evaluate the threat with a sense of realism.

It is more than unlikely for several reasons that in the near future somebody would create a new organism from scratch for malign purposes only. First, it is extremely difficult to do, in fact no scientist knows how to do that by now. Second, the research and development would be extraordinarily dangerous (see below), which rules out all amateurish endeavors. Third, there are many malignant micro-organisms in wild life, like the Ebola virus or *Bacillus anthracis* (anthrax), which could be collected and employed in a much easier and less dangerous way based on long term experience with these infectious diseases. Forth, it is much easier to modify existing organisms by standard genetic engineering such that they become malignant. If the presumed rogue state or terrorist had only the slightest sense of proportion for the required effort and the desired effect, they would of course collect or modify natural pathogens.

Nonetheless, bioterrorism remains a threat, not by new organisms from scratch, however, but by classical genetic engineering. Commercial firms offer synthetic gene sequences at rapidly decreasing rates, currently for about 20 US cents per base pair. That makes it cheap and easy to insert into bacteria sequences that code for toxins either as plasmids or as parts of the genome.

Not because of synthetic biology but because of the rapid decay of prices, there is an urgent need for international regulation of gene synthesis companies, their clients, and orders with regard to pathogenic genes. Thus far, there are only some national recommendations and voluntary commitments by companies.

3.6 Risks for laboratory staff and the environment

Last but not least, the creation of new and the modification of existing micro-organisms is a high risk for all laboratory staff including the cleaning personnel and the local environment. Ethicists frequently ignore that when they consider scientific research to be done by the individual researcher who might be responsible for his own harm. However, in current laboratory research projects usually dozens of people are involved with different backgrounds in science and safety instruction.

For instance, the synthesis and modification of pathogenic viruses, such as the poliovirus and influenza virus, may improve our medical knowledge but is at the risk of all people involved, including their social environment. Mycoplasma bacteria, which are favorite test organisms in synthetic biology because of their lack of a cell wall, are frequently pathogenic and cause chronic infectious diseases, probably for the same reason. When handling sequences of DNA and RNA for whatever reasons, one should keep in mind that these biopolymers can interfere with the genetic material of humans and other living beings in hardly predictable ways.

When synthesizing a new, hitherto unknown chemical compound, chemists are used to treat it in their laboratory with all precaution, as if it would be highly toxic. Even if one has perfect structural knowledge of the compound, its effects in complex biological contexts are unpredictable. Much stricter safety measures must be applied to research with microorganisms, modified or artificial, and all biological material that is capable of replication because even the smallest amount could multiply in suitable environments. All experimental settings need to be strictly double-isolated against both the laboratory and the outer environment.

For any such experiments, the research goals should be well balanced against the remaining risks for laboratory workers and the surrounding natural and social environment. One must have good reasons – ones that would be morally acceptable in case of an accident – to justify the research. In would not help much saying, 'I just wanted to try something out'.

4. Conclusion

The history of life creation and the way people looked at it is more than puzzling. For thousands of years it was considered a banality that happened everywhere, whether intentionally or not; nobody cared about it or raised any moral or religious concern. Nowadays it is both an overarching goal for many scientists and an outrageous idea for large parts of the public in Christian countries. The reversal of views could not be more extreme. What made this happen?

As pointed out in Section 2.1, evolution theories moved the focus of scientific and moral attention to simple organisms as the potential beginning of life, from which human beings would eventually have emerged. However that cannot provide a full explanation of the public consternation. Both creationist, who do not believe in evolution, and people in secular societies express a similar kind of dismay with the accusation of 'playing God' or imitating Frankenstein, the 'modern Prometheus' as

Mary Shelley subtitled her influential novel. Moreover, as we have seen in Section 3.1, the accusation of 'playing God' has no ethical or theological basis, neither within nor without the Christian doctrine. One might instead claim a general shift in the evaluation of life forms during the 20th century supported by environmental movements. However, we are talking here mostly about bacteria that are invisible to the naked eye and beyond the horizon of most people, although about 10^{14} bacteria live in every human and approximately $5*10^{30}$ on Earth making up most of the biomass (Whitman *et al.* 1998). Who would actually go as far as Albert Schweitzer and call for reverence to pathogenic microorganisms?

It seems more likely that the public dismay about life creation expresses a diffuse fear and dismay of science in general, and chemistry in particular, that is composed of literary myths epitomized in the mad scientist (Section 1). As we have seen in Section 3, however, the public dismay does not stand up to ethical scrutiny. None of the three deontological approaches (the obligations to not 'play God', to not touch the dignity of life, and to reverence and be responsibility for individual organisms) can be used to derive serious objections. Of course the three consequentialist approaches point to important risks (interference with the ecosphere, misuse as bioweapons, and harm to laboratory staff and local environment). However these risks of the creation of artificial organisms do not essentially differ from, are sometimes even minor to, the risk of modifying existing organisms. In sum, it is difficult to find any particular ethical substance behind the public dismay expressed in phrases like 'playing God'.

Unarticulated as these phrases are, they are extremely powerful in creating public moral excitement, with two negative side effects. On the one hand, they distract the public from proper ethical deliberations and issues. Because they express radical rejection, they undermine any dialogue between science and the public. That might even hinder constructive ethical work in which scientists and ethicists ideally work together. On the other hand, they mislead science by pushing the creation of life as an end in itself, because that is exactly what draws public attention and thereby indirectly research funding. Scientists like Price, Danielli, and Venter have been tempted into alluding to these phrases, for instance by confusing 'modification' with 'creation' of life and microorganisms with humans. Venter, as we have seen in Section 2.4, even went further and systematically misrepresented the research of his team, among others by putting the creation of life as the presumed research goal, whereas their hybrid bacterium was actually only a means.

One might think that this is only PR on the surface and does not touch real research. However, several developments suggest that it has also an impact on scientific values. For instance, proto-cell research neglects or even gives up the earlier scientific quest for understanding the origin of life in favor of making something alive, whatever it is and under whatever conditions (Section 2.3). Thereby the epistemic value of understanding, of knowing how and why something happened, is replaced by the technological value of knowing how to do something. In synthetic genomics it is also not clear, whether the enormous efforts at building an organism from scratch improves our scientific understanding of life and its historical origin. The supposed epistemological principle of knowing from making, which was once so successful in organic chemistry, is difficult to apply to whole biological systems. Moreover, it is far from clear whether building an organism from scratch would have any technological advantage over the targeted genetic modification of existing organisms. Thus the effort might even violate the technological value of keeping a reasonable cost-benefit ratio.

Finally, unfortunate science-public interactions do serious harm to the profession of scientists. As the cases of Danielli and Venter illustrate, the literary myths of the fiendish mad scientist, which somehow began with the Book of Enoch 2300 years ago, are deeply entrenched in popular culture and can be revived at any time. Minimal allusions suffice to reinforce age-old stereotypes. Chemists, due to their special cultural heritage outlined in the Section 1, are particularly affected and should meticulously avoid any allusion. If they ignore that, because of lack of knowledge about the cultural history or out of selfish desire to be in the focus of public attention, they damage the image of science. In the end neither scientists, nor the public, nor ethicists are happy.

Further Reading

A much more detailed presentation with background information and hundreds of references, which are here omitted for space reasons, is Schummer 2011, albeit not yet translated into English. More details on Section 2 can be found in Schummer 2003, 2006, 2009. There are numerous anthologies on the societal and ethical aspects of synthetic biology, from earlier ones such as Bedau & Parke 2009, Schmidt *et al.* 2009 to Boldt 2016, Hagen *et al.* 2016.

References

Bedau M.A. & E.C. Parke (eds.): 2009, The Ethics of Protocells: Moral and Social Implications of Creating Life in the Laboratory, Cambridge, MA: MIT-Press.

Boldt, J. (ed.): 2016, Synthetic Biology: Metaphors, Worldviews, Ethics, and Law, Wiesbaden: Springer.

Farley, J.: 1977, The spontaneous generation controversy from Descartes to Oparin, Baltimore: John Hopkins UP.

Geison, G.L: 1995, *The Private Science of Louis Pasteur*, Princeton: Princeton UP.

Gibson, D. *et al.*: 2010, 'Creation of a bacterial cell controlled by a chemically synthesized genome', *Science*, **329**, 52-56.

Hagen, K.; Engelhard, M. & Toepfer, G. (eds.): 2016, Ambivalences of Creating Life: Societal and Philosophical Dimensions of Synthetic Biology, Basel: Springer.

Johnson, J.A.: 2015, 'From Bio-organic Chemistry to Molecular and Synthetic Biology: Fulfilling Emil Fischer's Dream', *Proceedings of the International Workshop 'Transformation of Chemistry from the 1920s to the 1960s'*, 2-4 March 2015, Tokyo, Japan [online available at http://kagakushi.org/iwhc2015/papers/01.JohnsonJeffrey.pdf, accessed: 7 March 2016].

Jonas, H.: 1984, *The Imperative of Responsibility*, Chicago: University of Chicago Press [German original: *Das Prinzip Verantwortung*, Frankfurt: Suhrkamp, 1979].

Karpenko, V.: 1998, 'Alchemy as donum dei', *Hyle: International Journal for Philosophy of Chemistry*, **4**, 63-80.

Lippmann, E.O.v.: 1933, Urzeugung und Lebenskraft: Zur Geschichte dieser Probleme von den ältesten Zeiten an bis zu den Anfängen des 20. Jahrhunderts, Berlin: Springer.

Newman, W.R.: 2004, Promethian ambitions: Alchemy and the quest to perfect nature, Chicago: University of Chicago Press.

Price, C.C.: 1965, 'The new era in science', *Chemical and Engineering News*, 27 September, 90-91.

Schmidt M.: 2009, 'Do I understand what I can create? Biosafety issues in Synthetic Biology', in: M. Schmidt; Kelle, A.; Ganguli-Mitra, A. & de Vriend, H. (eds.), *Synthetic Biology: The Technoscience and its Societal Consequences*, Dordrecht: Springer, pp. 81-100.

Schmidt M.; Kelle, A.; Ganguli-Mitra, A. & de Vriend, H. (eds.): 2009, *Synthetic Biology: The Technoscience and its Societal Consequences*, Dordrecht: Springer.

Schummer, J., 2003: 'The Notion of Nature in Chemistry', *Studies in History and Philosophy of Science*, **34**, 705-736.

Schummer, J., 2006: 'Historical Roots of the 'Mad Scientist': Chemists in 19th-century Literature', *Ambix*, **53**, 99-127.

Schummer, J.: 2009, 'The Creation of Life in Cultural Context: From Spontaneous Generation to Synthetic Biology', in: M.A. Bedau & E.C. Parke (eds.): *The Ethics of Protocells: Moral and Social Implications of Creating Life in the Laboratory*, Cambridge, MA: MIT-Press, pp. 125-142.

Schummer, J.: 2011, Das Gotteshandwerk: Die künstliche Herstellung von Leben im Labor, Frankfurt: Suhrkamp.

Thomsons, J.A.: 1922, *The Outline of Science*, New York: Putnam's Sons.

Whitman, W.B.; Coleman, D.C. & Wiebe, W.J.: 1998, 'Prokaryotes: the unseen majority', *Proceedings of the National Academy of Sciences*, **95** (12), 6578-6583.

van den Belt, H.: 2009, 'Playing god in Frankenstein's footsteps: Synthetic Biology and the meaning of life', *Nanoethics*, **3**, 257-268.

Chapter 17

The Normative Molecule: Patent Rights and DNA

Saurabh Vishnubhakat

Abstract: Throughout the biotechnology age, fears about the distortionary effects of property and other legal institutions upon the health and self-determination of individuals and societies have accompanied more popularly sensational fears about unscrupulous choices within the scientific community itself. Still, for most of that time the prevailing legal regime both in the United States and in Europe remained generally permissive of ownership of, and exclusionary power over, the fruits of much biomedical research, though this leniency took different forms and came about in different ways. In particular, the policy of the United States Patent and Trademark Office to grant patents on genetic compositions such as DNA sequences produced an extensive landscape of legal rights that would eventually provoke a backlash in both legal and popular opinion during the *Myriad Genetics* lawsuit. This case study examines the normative dimension of patent rights over isolated DNA sequences through the lens of the *Myriad* case, discussing the institutional context in which the case arose and identifying ethical lessons that the case offers.

1. Introduction

1.1 The magic microscope

In April 2011, the U.S. Department of Justice did something it had never done before, and it was a sign of things to come. The Solicitor General, the Justice Department lawyer who represents the U.S. government in lawsuits, came personally to argue the government's position in a patent case before the Federal Circuit Court of Appeals ('Federal Circuit', in the following). Ordinarily, the Solicitor General would have left a case like this to be argued by the solicitor of the specific agency that deals regularly with the subject matter of the lawsuit, especially in a complex field like patent law. Indeed, the Solicitor of the Patent Office regularly does this in Federal Circuit cases. At most, the Solicitor General might have sent a lawyer from his Justice Department staff if the case was more important than usual. Even then, agency lawyers from the Patent Office (who know the subject matter) and Justice Department lawyers (who coordinate legal policy across the executive branch) would have met and planned out the strategy together. What made this case special, and what did the Solicitor General's appearance portend?

The answer came from a confluence of three unusual problems, and all of them were related to the underlying science of DNA and to the question of expertise in genetic and genomic science. The lawsuit was about whether certain patents held by Myriad Genetics were valid. Myriad was founded in the 1990s by scientists at the University of Utah who, like others at the time, were trying to isolate the gene or genes associated with elevated risks of developing breast and ovarian cancer. The Myriad-Utah team, which also included researchers from the National Institutes of Health and elsewhere, were the first to isolate and publish the sequence of the BRCA1 and BRCA2 genes. The team then sought patents variously related to these gene sequences, and the Patent Office granted the patents. The Public Patent Foundation (PUBPAT) and the American Civil Liberties Union (ACLU), concerned about the effects of these kinds of patents, brought a lawsuit in 2009 to invalidate Myriad's patents. The lawsuit was assigned to Judge Robert Sweet of the U.S. District Court for the Southern District of New York. This was the case now pending on appeal in the Federal Circuit, following Judge Sweet's decision in the district court.

One problem was that the Patent Office continued to disagree with the Justice Department on a fundamental matter of patent law and policy. That was whether isolated DNA sequences themselves fall within the broad scope of patent-eligible subject matter – *i.e.*, whether isolated DNA sequences even belong in the patent system.[1] The disagreement

---

[1] The details of that patent policy matter are discussed in Section 2.2.

mattered. The Patent Office sits in the Department of Commerce and is an executive-branch agency. Ordinarily, the Patent Office has the power to grant or deny patent applications, and it has more expertise about science and technology than most other parts of the executive branch; certainly it has more expertise on patent law than any other agency. Meanwhile, the Solicitor General's office at the Justice Department sets legal policy across the executive branch. So when the Patent Office has one view about proper patent policy, the Solicitor General's office has a different view, and both sides have debated an impasse: who should prevail?

Who does prevail is simple: the Solicitor General. But is that a good and sensible rule? Perhaps the expertise of the Patent Office means that its view should be the official view of the U.S. government on all patent matters. Or perhaps the greater experience that the Solicitor General has coordinating across multiple specialized disciplines (like patent law, tax law, environmental protection, securities regulation, *etc.*) means that its view should be the official view of the government. The superior policy authority of the Solicitor General's office means that although it should, and does, listen to expert agencies and learn from them, it does not defer to them. The Solicitor General still makes an independent decision and, importantly, expects the agencies in the executive branch, like the Patent Office, to accept that decision and sign on to it.

That last expectation did not come to pass this time. The Solicitor of the Patent Office did not sign the government's brief, indicating continued and unreconciled disagreement with the Solicitor General about what the official position of the government ought to be, and indicating that disagreement publicly. It is not clear whether this played any role in the Solicitor General's decision to appear in the case. However, his unprecedented appearance while this intra-governmental disagreement was still a live issue highlighted the Patent Office's contrary position and made it an important sticking point. Oral arguments both at the Federal Circuit and later at the Supreme Court brought up this fundamental dispute between the expert opinion of the specialist agency and the official opinion of the generalist authority.

A second problem that complicated matters further was that the Patent Office was not the only executive-branch institution that had relevant scientific expertise, and even the experts disagreed. The position of the National Institutes of Health (NIH) diverged from that of the Patent Office. The NIH is an agency within the Department of the Health and Human Services and supports scientific research through extensive federally funded research grant programs, especially research in the life sciences. Thus, like the Patent Office, the NIH had significant claims to expertise at least in the underlying subject matter of Myriad's patents, even if it could not necessarily claim the same level of expertise as the Patent Office about patent policy. For this very reason, inter-agency discussions to consider policy questions are structured to consider all relevant positions across the executive branch before the government's official position is decided. And this, too, underscored the importance of the Solicitor General himself entering an appearance to argue in the Federal Circuit.

A third and more elusive problem was that the Federal Circuit itself was potentially at odds with the Supreme Court, either about the disputed patent policy matter in general, about the *Myriad* case in particular, or both. Federal judges are legal generalists, and the way they are selected – political nomination by the President and confirmation by the Senate – does nothing in particular to promote expertise in science and technology. The Federal Circuit is different. Created in 1982, the Federal Circuit Court of Appeals hears appeals from all patent lawsuits in the entire country. Other circuit courts in the federal system hear appeals only from their respective geographic regions. The Federal Circuit's jurisdiction was defined by subject matter rather than geography in order to promote uniformity and expertise on matters of patent law.

As a result, the Federal Circuit has extensive experience with patent law and receives hundreds of patent cases each year. Federal Circuit judges are steeped in patent doctrine, and of those who come to the bench, many have formal education in physical science or engineering disciplines. By contrast, the Supreme Court receives thousands of requests but chooses to hear fewer than a hundred cases each year, and of these, it is unusual that even one or two cases a year are about patent law. As for Supreme Court Justices themselves, if there is a field in which they develop expertise before coming to the bench, it is usually constitutional law or administrative law, certainly not patent law.

So just as the Patent Office came to the dispute with greater expertise but less authority than the Solicitor General, the Federal Circuit came to the case with greater technical and doctrinal expertise but less judicial authority than the Supreme Court. And unlike the executive branch, the two courts could not directly confer and debate the case. The Federal Circuit would have to decide the appeal for itself first, then wait to see whether the Supreme Court would take the case and, if so, how the Court would rule. So the direct involvement of the Solicitor General right at the Federal Circuit highlighted the case even more starkly for the Supreme Court's consideration.

So far, this may seem like a story about competing claims for power and claims of expertise, but not especially a story about genetic science itself. The clearest evidence against this came in the first minute of the argument of Neal Katyal, the Acting Solicitor General. The Myriad patents in question contained various DNA-related claims, but what eventually drove the lawsuit was a distinction between isolated fragments of genomic DNA (gDNA) and isolated fragments of complementary DNA (cDNA). The structure of these DNA fragments, especially their nucleotide sequences, were, at some basic level, the result of nature's handiwork. For this reason, the PUBPAT and ACLU challengers believed that both of these types of isolated DNA sequences, genomic as well as complementary, were not patent-eligible at all.[2]

The government took a more modest position. gDNA was not patent-eligible, agreed Katyal, because its nucleotide sequence does occur in nature – in that portion of the chromosome itself where the relevant gene is located – but cDNA was eligible because its nucleotide sequence does not occur naturally. When cDNA is synthesized from mature messenger RNA (mRNA), the introns have already been removed from the RNA transcript, leaving only exons. Thus, the removal of introns changes the nucleotide sequence from what occurs in nature to something different, akin to a magazine from which all the advertisement pages have been torn out.

---

2 In patent law, a 'product of nature' is not eligible for patenting, though a human-made invention can be patent-eligible even if derived from a natural product – if the invention is 'markedly different' from what occurs in nature.

To illuminate this distinction, Katyal invited the court to imagine a 'magic microscope' that could zoom in on nucleotide sequences. Using such a microscope, the sequence of an isolated gDNA fragment could be compared to the sequence of wild-type gDNA from the same gene *in vivo*. The comparison would reveal that the sequences were the same. Meanwhile, a comparison between an isolated cDNA fragment and any corresponding wild-type DNA or RNA would reveal that the sequences were not the same (because the introns had been removed).

This metaphor of the magic microscope was rhetorically powerful. It met the scientific expertise of both sides in the dispute with the government's own scientific expertise, but reflected shrewd choices about which scientific details should matter. The magic microscope argument pointed the court toward the sequence information contained in the BRCA1 and BRCA2 genes, and the normative implications of that information, rather than solely the chemical structure of the DNA that constituted the genes. It urged scientific and lay audiences: don't miss the forest for the trees. DNA is a chemical compound, yes, but that is not all it is.

1.2 Patent rights and the special case of DNA

Before going further, it is helpful to clarify the terms a bit. How does the intellectual property framework of patents work, and does DNA pose special problems for patent law? Intellectual property can be understood most simply by analogy to more familiar kinds of property. For example, a company that owns a car and a traveler who rents the car both hold property interests in the same underlying thing: the car. Where the underlying thing is some new and useful invention, the property interests people can hold over that invention are called patent rights, and they are one form of IP. (Other forms include copyrights and trademarks.)

What do these property interests entail? The interest of the company that owns the car includes various rights, like the right to remove unauthorized drivers from the car, the right to paint the car red, the right to sell the car, and so on. The interest of the traveler renting the car also includes rights, but understandably, these are fewer than – and, in some sense, inferior to – the owner's rights, as the rental agreement spells out.

Patent rights start out similarly. An inventor who holds a patent on her invention – say, a special wooden chair – can for a limited period of time exclude others from making the chair, using it, selling or offering to sell it, or importing it into the country. (After that time, the chair enters the public domain, and anyone can use it freely.) But what does it mean to think of the chair as an invention rather than just another tangible thing?

It means that if someone buys lumber, saws the lumber into parts, and assembles those parts into something that matches the patented chair, then he is infringing the inventor's patent right. The patent covers not just one specific physical chair, but all objects that shares the particular features of the chair. The infringer used his own materials, tools, and effort. Indeed, he has his own property interests in those things. Yet his rights do not extend to making the patented chair as he has done. This makes patents an extraordinary kind of property right, which are granted only as a reward for inventing something new and useful, something society would not have received (or would not have received as soon) without the inventor's contribution. Even with the patent's restrictions, society is still better off than if the new chair had never been invented.

Still, one can easily imagine how this setup can get messy where the invention has important social and dignitary implications, such as with DNA. It is one thing to grant patents as a reward for inventing a new organic solvent for degreasing engines. Is it the same to grant patents for discovering a gene sequence, or designing a new one? They are both chemical compounds, but subjecting them to private property rights can have dramatically different effects.

For several decades, the Patent Office saw no problem and granted gene-related patents routinely. The idea was that taking away the patent reward for DNA-related inventions would slow down investment in those inventions, harming society in the long run. This explanation assumed a lot about the nature and drivers of invention. Those assumptions are what had allowed Myriad to obtain their patents and what PUBPAT and the ACLU had to attack, at least as to DNA patents.

2. Ownership of DNA Through Patents

2.1 The Myriad genetics patents

Although PUBPAT and the ACLU were challenging fifteen claims across seven different Myriad patents, three claims are particularly relevant for the present purposes.[3] These are three of the composition claims from U.S. Patent No. 5,747,282 entitled '17Q-linked breast and ovarian cancer susceptibility gene'. The Federal Circuit and, later, the Supreme Court evaluated these three claims as being representative of all the composition claims in dispute.

By law, the claims in a patent can cover subject matter from any of four categories: compositions of matter, processes (which in patent law are synonymous with methods), machines, and manufactures (35 U.S.C. § 101). Here, the composition claims asserted ownership over identified isolated DNA sequence molecules themselves. The legal logic was that although a given DNA sequence may be quite large chemically, it is still ultimately a molecule and thus a composition of matter. The method claims, meanwhile, asserted ownership over ways of doing things related to the identified DNA sequences, such as making comparisons between the claimed DNA sequences and other DNA sequences or screening potential cancer therapies. These method claims were separately found invalid.

Considering the actual language of the composition claims that the Federal Circuit took as representative, the following are claims 1, 2, and 5 of Patent No. 5,747,282:

> 1. An isolated DNA coding for a BRCA1 polypeptide, said polypeptide having the amino acid sequence set forth in SEQ ID NO:2.
> 2. The isolated DNA of claim 1, wherein said DNA has the nucleotide sequence set forth in SEQ ID NO:1.
> 5. An isolated DNA having at least 15 nucleotides of the DNA of claim 1.

---

3 A patent is a stylized document with distinct sections and parts. The 'claims' of a patent are the statements at the end that specify what the inventor actually invented. Taken together, the claims in a patent must all point to a single invention, but each claim represents a slightly different embodiment of that invention.

The terms 'SEQ ID NO:1' and 'SEQ ID NO:2' refer to sequences that are fully spelled out later in the patent. For example, 'SEQ ID NO:2' refers to a particular amino acid sequence corresponding to a BRCA1 polypeptide; the sequence is Met-Asp-Leu-Ser-Ala-Leu-... and so on. Thus, claim 1 covers any nucleotide sequence that codes for the polypeptide having that amino acid sequence. As biochemists know, the genetic code is redundant such that the same amino acid can be coded by multiple nucleotide triplets. For example, the codons CGT, CGC, CGA, CGG, AGA, and AGG all code for arginine. The syntax of claim 1 – defining the DNA sequence by reference to the polypeptide sequence – ensures that all relevant redundancies are covered. Claim 2 then goes further by specifying one example of a nucleotide sequence, the one set forth in SEQ ID NO: 1, that codes for the amino acid sequence of SEQ ID NO:2. Claim 5 broadens the scope of the patent right even further by claiming all sub-sequences of claim 1 that are 15 nucleotides or longer.

To understand the scope of how broadly these claims sweep, consider just the first six amino acids that are listed above from SEQ ID NO:2, Met-Asp-Leu-Ser-Ala-Leu. The different nucleotide triplets that code for each amino acid are as follow:

Amino Acid	**Codons**	**Codons that Code for that Amino Acid**					
Met (Methionine)	1	ATG					
Asp (Aspartic acid)	2	GAC	GAT				
Leu (Leucine)	6	CTA	CTC	CTG	CTT	TTA	TTG
Ser (Serine)	6	AGC	AGT	TCA	TCC	TCG	TCT
Ala (Alanine)	4	GCA	GCC	GCG	GCT		
Leu (Leucine)	6	CTA	CTC	CTG	CTT	TTA	TTG

From these redundancies, this six-peptide fragment alone could be coded by 1,728 different DNA sequences ($1\times2\times6\times6\times4\times6$), each merely 18 nucleotides in length. For example, the following is but one of the 18-nucleotide sequences that would code for the Met-Asp-Leu-Ser-Ala-Leu polypeptide sequence. Moreover, each 18-nucleotide sequence would have 10 sub-sequences that were at least 15 nucleotides long.

Met			**Asp**			**Leu**			**Ser**			**Ala**			**Leu**		
A	T	G	G	A	T	C	T	A	T	C	T	G	C	T	C	T	A
15-nucleotide sequence																	
	15-nucleotide sequence																
		15-nucleotide sequence															
			15-nucleotide sequence														
16-nucleotide sequence																	
	16-nucleotide sequence																
		16-nucleotide sequence															
17-nucleotide sequence																	
	17-nucleotide sequence																
18-nucleotide sequence																	

This means that Claim 5 multiplies the 1,728 different DNA sequences to 17,280 different DNA sequences over which the claims assert ownership. As the length of the claimed polypeptide grows, the permutations of codons, the number of sub-sequences that are at least 15 nucleotides in length, and the overall number of isolated DNA sequences that are covered by the patent grows exponentially. The upshot of this broad patent scope is that composition claims directly covering isolated DNA sequences on the scale of a gene like BRCA1 or BRCA 2 are powerful legal instruments, and reflect the high stakes of the *Myriad* case.

In practical terms, the broad exclusionary power of patent rights over DNA as a composition means that even some new use for the patented DNA sequence or some new method involving the patented DNA sequence – a use or method that was unknown at the time of patenting – would still be encumbered by the patent rights that Myriad held. It would be as if someone walked around a piece of uncharted land, mapped it, and acquired a property right over it. Minerals in the land or oil below it would belong to the property holder, just as much as the land itself does, even if the owner had no knowledge of those additional resources. Indeed, this is a simplified example of how property rights in land actually work, though it is highly contested whether it is appropriate for the law to treat genetic resources in the same way.

2.2 The 'product of nature' doctrine in patent law

The *Myriad* case itself had already reached a significant milestone by the time it arrived at the Federal Circuit. The federal district court in Manhattan, where the case began, had given the challengers of the Myriad patents a clear victory. They had standing to file the lawsuit in the first place (a technical issue of court procedure but very important in public interest lawsuits like this). And according to the district court, all of the composition claims (as well as all of the method claims) of the patents in question were patent-ineligible. Myriad had come to the Federal Circuit to seek reversal of these judgments. Meanwhile, the direct and personal involvement of the Solicitor General, discussed above, meant that continued interest in the case – including review by the Supreme Court – was likely. The composition claims, whose scope swept so broadly, were the especially contentious element of the case. The Federal Circuit would have to both decide and explain which approach to take in evaluating the patent-eligibility of claims that covered isolated sequences of genomic DNA (gDNA), complementary DNA (cDNA), or both.

To do so, the Federal Circuit would have to apply the 'product of nature' doctrine. A longstanding rule of patent eligibility, the doctrine provides that natural products themselves are not allowed to be patented, though other things that build on products of nature may be eligible. How a decision maker might apply this rule depends to some extent on what the justifications for the rule are understood to be. One justification is that patent rights are an incentive and reward for those who invented the subject matter of the patent, but products of nature are not invented at all; they arise from natural processes without human design or intervention. Another is that patent rights exist to promote innovation, and because natural products are among the raw materials from which innovation is conducted, they should remain unencumbered for public use.

The first explanation is more analytical and emphasizes conditions and criteria for deciding whether something is a product or nature and therefore ineligible for patenting or is not a product of nature and so may, indeed, be eligible. The second explanation is more instrumental and emphasizes the desirability of the outcomes that are likely to come about based on whether something is deemed an ineligible product of nature or not. In practice, these explanations are not mutually exclusive. Courts rely in various ways and to varying degrees on both explanations as they apply the doctrine in accordance with traditional methods of legal analysis.

An important element of that analysis is respecting prior precedents. To grasp the product of nature doctrine, two precedents are especially helpful. One is the Supreme Court's decision in *Funk Brothers v. Kalo Inoculant* (Funk Brothers 1948). In the *Funk Brothers* case, the patent was directed toward a mixture of nitrogen-fixing bacteria from the genus *Rhizobium* that can infect and form nodules on the roots of leguminous plants. Before the patent, growing leguminous plants with the necessary capacity for nitrogen-fixation required using individual strains of *Rhizobium* for different kinds of plants. The different bacterial strains could not be mixed because they were believed to inhibit each other. The inventor discovered a combination of these bacteria that did not have this mutually inhibitive effect and allowed mixed strains and obtained a composition patent on the mixture of bacterial strains. The Supreme Court held, however, that the mixture was a product of nature, and thus ineligible, because each strain remained unchanged by the mixing and continued to perform its natural function.

The other helpful precedent is the Supreme Court's decision in *Diamond v. Chakrabarty* (Chakrabarty 1980). In that case, microbiologist Anand Chakrabarty had invented a transgenic bacterium from the genus *Pseudomonas* containing plasmids that coded for the ability to degrade hydrocarbons. The genetically engineered bacterium could break down elements of crude oil and was believed to be valuable in cleaning oil spills. Dr. Chakrabarty had sought method and composition claims, including a composition claim to the bacterium itself. The Supreme Court held that the bacterium was not a product of nature and was patent-eligible. The decision provided important doctrinal lessons. First, patent eligibility is broad, embracing 'anything under the sun that is made by man'.[4] Second, though products of nature are ineligible, an invention

4 This language is most closely associated in patent law with the *Chakrabarty* decision but was actually part of the legislative history of the 1952 Patent Act itself.

may distinguish itself from a natural product if it has 'markedly different characteristics from any found in nature'.

In its explanation, the Supreme Court drew a direct comparison to its decision in *Funk Brothers*. Whereas the root nodule bacteria were unchanged and did the same thing in combination that they did individually, Dr. Chakrabarty's engineered bacterium was a new organism that did not occur in nature. It is important to note that this conclusion is a matter of perspective. The *Pseudomonas* host bacterium did do what it had always done, ordinary cellular functions and all. Meanwhile, the plasmids that were spliced into the host bacterium occur naturally, though not in *Pseudomonas* itself, and also did what they had always done: code for hydrocarbon degradation, specifically of camphor and octane. Taken together, however, these two products of nature became one product that was 'markedly different' enough not to be considered natural. The mixture in *Funk Brothers* was a combination of bacterial strains, and the mixture in *Chakrabarty* was a combination of host bacterium and plasmids. The former was a product of nature, the latter was not, and the salient difference seems to have been that the latter resulted in a new organism. In other words, what the relevant unit of observation is can matter a lot.

2.3 The Myriad case

In light of these precedents, should isolated DNA sequences be considered products of nature? The district court's decision regarding Myriad's composition claims said yes (Myriad I 2010). Judge Sweet concluded that genes are carriers of biologically important information and that their legal status must therefore be evaluated in informational terms. In that evaluation, DNA sequences reflect the same information in their isolated form as they would in their wild-type form: the same set of nucleotides in the same order encoding the same amino acids. Thus, isolated DNA sequences are not only not 'markedly different' – in the informational sense, they are not different at all. This logic would later echo in the Supreme Court's decision as well.

The Federal Circuit's decision, meanwhile, said no (Myriad II 2011). Unlike the nucleotide chain of the wild-type BRCA1 or BRCA2 gene that occurs in nature, the nucleotide chain of the isolated gene is chemically disconnected from its adjacent nucleotides. In nature, covalent bonds would attach to the 3' and 5' ends of the gene. For the isolated gene, they end in a hydroxyl group and a phosphate group, respectively. As the Federal Circuit explained, this makes the isolated gene a "free-standing" molecule with a "distinctive chemical identity from that possessed by native DNA" (*ibid.*). This may seem like a technicality, but the Federal Circuit underscored that "a covalent bond is the defining boundary between one molecule and another". As for why this was different enough to be 'markedly different', the court further explained that "genes are in fact materials having a chemical nature and, as such, are best described in patents by their structures rather than their functions" (*ibid.*) notwithstanding that some may think of molecules, even macromolecules such as DNA, in terms of their functions and uses.

A fortiori, this discussion of why gDNA sequences like the isolated genes of Myriad's patents were not products of nature reflected a similar discussion of why cDNA was not a product of nature, either. cDNA not only had unbonded 3' and 5' ends (just as gDNA did) but also contained only exons: the non-coding introns had been removed. As a result, cDNA did not even correspond in terms of its nucleotide sequence to any naturally occurring nucleic acid. So if gDNA was not a product of nature, cDNA was certainly not a product of nature. Both were patent-eligible because, in drawing a comparison between the patented invention and the corresponding product of nature, the relevant unit of observation was chemical structure.

However, the Supreme Court saw things differently. In comparing isolated gDNA sequences for BRCA1 and BRCA2 to the wild-type gDNA sequences of those genes, it was not chemical structure alone that was relevant. Instead, the information contained in the nucleotide sequence was the relevant unit of observation. The reason for this, said the Supreme Court, was that the language of Myriad's own patent claims was expressed not in terms of chemical composition but rather in terms of sequence information. Moreover, the invention over which the patent claims purported to assert ownership did not rely on any chemical change that might have resulted from the cleaving of covalent bonds to produce an isolated section of gDNA. The claims focused on the genetic code reflected in the BRCA1 and BRCA2 gene sequences, and that would be the way in which their status as products of nature would be evaluated.

This approach prioritized genetic information over chemical structure. By this approach, claims to cDNA would still survive. Structural difference at the 3' and 5' end was the only basis for the patent-eligibility of gDNA, but cDNA differed in both structure and nucleotide sequence. Thus, the removal of introns to produce a new sequence meant that, even under the Supreme Court's more stringent approach, cDNA remained markedly different from the underlying DNA out of which it was transcribed. Indeed, this was essentially the result as well as the approach that the Office of the Solicitor General had counseled from the start.

The decision was highly anticipated and immediately controversial. PUBPAT, the ACLU, and a host of advocacy groups that were aligned with them had pushed for a broader conclusion that DNA sequences that correspond to naturally occurring genes, whether those sequences take the form of gDNA or cDNA, were patent-ineligible products of nature because their nucleotide sequences were determined by nature rather than by human design. The Supreme Court's *Myriad* decision fell short of that goal by leaving cDNA eligible. Meanwhile, Myriad itself and biotechnology companies that held DNA-related composition patents similar to Myriad's suffered a setback from the patent-ineligibility of claims to isolated gDNA. Those rights were no longer valid, and a great many others could now use those gDNA sequences freely.

The government's position, which prevailed, was widely seen as a compromise to ensure that the building blocks of genetic and genomic innovation were unencumbered and widely available, while the opportunity to recoup investments in developing libraries of cDNA sequences would remain viable for the biotechnology industry. Whether or not this pragmatic policy choice was what actually motivated the Supreme Court's decision, it was the result that the decision brought.

2.4 Myriad overseas: Revisiting the European Biotechnology Directive

The controversy was not limited to the United States. After the Supreme Court in *Chakrabarty* announced a broad purview of eligibility for patents in biotechnology, the European Commission in 1982 proposed its own legislative reform in order to make biotechnology-related inventions eligible for patents issued by the European Patent Office (EPO). These efforts culminated in the Directive on the Legal Protection of Biotechnological Inventions –commonly known as the European Biotechnology Directive – and were aimed at keeping the European Union competitive with the United States as to incentives for innovation, such as in biotechnology, requiring long-term capital investment. The EPO implemented the European Biotechnology Directive as part of the European Patent Convention, the multilateral treaty that authorizes the EPO to consolidate the patent application process in member countries across Europe.

Over its decade-and-a-half-long deliberations, the Parliament of the European Union had worked out a fairly specific position about whether and when patents can be issued over inventions related to DNA and other genetic materials. This was a different posture than that of the United States. The apparently broad mandate of *Chakrabarty* had come from a politically insulated Supreme Court, and so its stability over time depended on different institutional forces than did the legislative consensus that undergirded the European Biotechnology Directive.

As a result, the dramatic reversal in the Supreme Court's *Myriad* decision raised the possibility that the greater specificity and greater predictability and clarity of European patent laws on biotechnology made the EU more attractive for investment. For example, Article 5(1) of the European Biotechnology Directive provides as follows:

> The human body, at the various stages of its formation and development, and the simple discovery of one of its elements, including the sequence or partial sequence of a gene, cannot constitute patentable inventions.

By contrast, Article 5(2) and 5(3) provide as follow:

> An element isolated from the human body or otherwise produced by means of a technical process, including the sequence or partial sequence of a gene, may constitute a patentable invention, even if the structure of that element is identical to that of a natural element.
>
> The industrial application of a sequence or a partial sequence of a gene must be disclosed in the patent application.

The circumstances in the United States were more nebulous, requiring predictions about what might or might not be considered a product of nature in the future. Guessing wrong meant that large sums of today's investment could be lost tomorrow when others see a desirable but expensive patented technology and try to make it more widely accessible by removing its patent protections. These sorts of potential costs had to be weighed, of course, against benefits such as improved research access to natural products unfettered by exclusive rights. The real uncertainty was whether, on balance, the cost-benefit calculus would enhance net social welfare.

Analytically, the approach of the European Biotechnology Directive was to avoid making patent eligibility dependent on vague constructs such as what counts as a 'product of nature' (or an 'abstract idea' or 'natural phenomenon' – which are also ineligible under U.S. patent law). Instead, the Directive does much line-drawing between what is eligible and what is not. For example, human beings at various phases of development are not eligible for patenting. Other forms of life may be eligible, though they must not be caused to suffer unless there is substantial countervailing medical benefit. Discoveries, scientific theories, and mathematical methods are not considered inventions at all and so are not eligible. And in general, inventions must be of a technical nature and have a technical effect or industrial application in order to be eligible under the European approach. In all, the clarity of the European Biotechnology Directive and of its implementing regulations does not necessarily correspond to broader or more generous patent scope. In fact, much is excluded, but the boundary lines are easier to identify.

For Myriad, this meant that its corresponding European patents on BRCA1 and BRCA2 were adjudicated more straightforwardly on the basis of the relevant European patent law when those patents were challenged in the European Patent Office, rather than lurching from one judicial approach to another regarding the vague product of nature doctrine. Commentator Jessica Lai framed the point in the first major post-*Myriad* scholarly analysis of the BRCA patents in Europe this way: "neither BRCA1 nor BRCA2 were strongly patented in Europe," but for firms in the biotechnology industry, it is hard to ignore that the European

BRCA-related patents largely survived where the U.S. patents did not (Lai 2015).

The flip side of this dynamic is that although enactment of the European Biotechnology Directive reflected sufficient consensus in the European Parliament twenty years ago, popular unease with the patenting of human genes existed then and has only grown since the turn of the century. As Dr. Lai points out, important U.S. Supreme Court decisions can and do influence law and policy in other jurisdictions, especially given the preeminent role of the United States in shaping the international patent harmonization agenda. The growing disfavor, even in parts of Europe, for DNA-related patents may find ample support in both the *Myriad* decision and the subsequent decisions of the lower U.S. federal courts interpreting *Myriad*.

Taken together, this body of recent American jurisprudence reflects a discernible trend against allowing such patents where, for example, their application is in diagnostic medicine (Eisenberg 2015). For biomedical patents more generally, however, this trend may be somewhat more attenuated in magnitude or at least mixed in its applicability (Rai 2013). In either case, the finite resilience of *Chakrabarty* and similar judicial precedents is a reminder that legislative milestones such as the European Biotechnology Directive are not immune from reconsideration merely because of their longevity.

3. Ethical Implications

Until now, the discussion has traced the *legal* question of whether exclusive rights can be asserted over DNA sequences through patent law, and the biochemistry and genetics that underlie that legal debate have been scientifically straightforward. In fact, what was really contested in this legal debate with regard to the science was what scientific facts were relevant, and who was best suited to decide what was relevant. The key choice of the Supreme Court's decision in *Myriad* was to consider DNA not merely as a large molecule that has chemical structure and function but, more profoundly, as a vehicle for important genetic information. This section now evaluates the answer to that legal question in ethical terms. The themes of contested relevance and the power to decide remain important here as well, and are developed further.

3.1 Expertise and generalism

At the start of this decade, law professor Peter Lee published an important and wide-ranging paper in the *Yale Law Journal* applying the 'Two Cultures' thesis first advanced by C.P. Snow in postwar Britain. Snow's argument was that intellectual specialization in the liberal arts and the sciences, respectively, threatened a "gulf of mutual incomprehension" between these two broad segments of academic thought (Snow 1959). Professor Lee's insight, in turn, was that the 'Two Cultures' are a helpful lens through which to understand and evaluate U.S. patent law. The major premise of Professor Lee's paper is that "no matter how elegantly policymakers craft patent law, if generalist judges lack the capacity to administer it, the patent system cannot fulfill its objectives" (Lee 2010). This foreshadows, of course, the epistemic debate that would arise in the *Myriad* case between the more expert but narrowly specialized postures of the Patent Office and the Federal Circuit, on the one hand, and the more authoritative but generalist postures of the Solicitor General and the Supreme Court, on the other.

Professor Lee's thesis suggests a preference that generalism should enjoy primacy. The expert institution has its say, and the generalist institution listens carefully, but the generalist is who decides. So the expert had better be able to translate complex information persuasively. Indeed, this is how such decisions are actually made in practice. Generalist decision makers might choose at various times to give more or less deference to experts further down the chain of authority, but the need to translate complexity cannot be eliminated. The ethical dimension of this translational challenge lies in an interaction between specialists and generalists that remains to be considered: the dialogues between inventors and patent lawyers and among patent lawyers themselves.

Patent lawyers who represent inventors before the Patent Office to secure patent rights over inventions grapple most directly with the 'Two Cultures' problem. As lawyers, they must complete traditional legal training, including reasoning by analogy, inducing general principles from specific cases, and synthesizing conflicting propositions. However, permission to practice before the Patent Office also requires education or experience in a science or engineering discipline. Thus, patent lawyers must also demonstrate proficiency in the scientific method, including developing and testing hypotheses, reasoning quantitatively, and drawing or rejecting causal inferences in accordance with statistical norms.

As Snow and Lee recognized, these two epistemic orientations are far from mutually exclusive, but they do prioritize different modes of thinking, knowing, and – what is especially important – making arguments. Notably, representing clients in the Patent Office is not the only kind of patent lawyering. A great many lawyers negotiate business deals involving patent rights, file or defend patent-related lawsuits in the courts, and engage in a range of patent policy and other patent-related endeavors, all without having any training or epistemic orientation in science or engineering. These all may still reasonably be considered patent lawyers.

Ultimately, all patent lawyers must work in that system of social, political, and legal institutions where legal rights over complex scientific and technological innovation must be reconciled with the more broadly held values from which patent law draws its legitimacy in the first place. Because a patent legally empowers its owner to exclude others from making, selling (or even offering for sale), importing, or using the invention, the most commonly cited broader value underlying patent law is innovation in the technological and economic sense. On this view, what matters is whether patents on gene sequences or other DNA-related innovations promote or hinder research and whether they promote or hinder the development of industries to bring that research to market (Contreras & Deshmukh 2017; Cook-Deegan *et al.* 2012). This is not the only narrative that bears on gene patenting, however.

Apart from the view of patents as incentives to innovation, additional and often competing views exist as well of broadly held social values that patent law should accommodate (Contreras 2016). One view is that science should be concerned more with fostering collaborative and path-breaking discovery than with pursuing individual profit for incremental advances. Another view is that legal institutions should avoid changing course suddenly or dramatically because a stable institution that respects reliance interests can more effectively encourage those who undertake long-term research investments to view the institution's commitments as credible. Still another view is that broader distributive effects, especially disparate limitations on access, should be taken into account when evaluating patent laws and patent rights because even a system that does everything it should to promote innovation may still be defective if, for example, it prices out vulnerable or otherwise marginalized segments of society from the innovation's benefits.

Among these and other important narratives, some balance must be struck, and the recurring question of who decides comes again to the fore. However, beyond simply pointing to who decides in contemporary society, it is helpful to consider, in accordance with a well-developed system of philosophical legal ethics, how to decide who decides.

3.2 How to decide who decides

In their recent and highly readable intellectual history of legal ethics, David Luban and Bradley Wendel describe that well-developed system as an ongoing succession (Luban & Wendel 2017). The First Wave of legal ethics was rooted in moral philosophy and concerned primarily with the problem that a lawyer's zealous, one-sided advocacy to the client might make the lawyer complicit in the client's wrongs, inflict harms upon third parties, and reinforce structures of social power that reward unethical or otherwise undesirable actions. The Kantian view of people as ends in themselves, influential in Western thought, seemed to be at odds with the lawyer's apparent duty to treat the client as the end and the lawyer as the means. In response, First Wave legal ethics scholars attacked the 'neutral partisanship' conception that a lawyers should remain agnostic to the morality of the client's positions or demands, should dispassionately show partisanship toward the client's interests, and should be exempt from moral criticism for their role as mere intermediaries (Simon 1978).

Thus, when a lawyer who majored in biochemistry during college, graduated from law school, and now works at a patent law firm receives a request from a client biotechnology company to file applications in the Patent Office on isolated gene fragments that the client has sequenced, an ethical choice confronts the lawyer. Should she draft the patent application and the claims to maximize her client's potential future revenues through licensing and, if needed, litigation? Should she try to balance her client's desire for revenue with the interests of academic research, *e.g.*, by drafting claims that would cover industrial applications but not necessarily university research? Should she go even further and, either through claim drafting and the Patent Office process or by vigorous advice to the client, try to ensure that as broad a share of society as possible has access to the diagnostic or other benefits of the gene sequence? A First Wave legal ethicist would likely say that the answers to some of these questions are yes; that the primacy of the traditional lawyer-client relationship must yield in difficult cases to more general ethical requirements that apply to all individuals alike; and that the reason for this reconception is essentially moral in nature.

The Second Wave of legal ethics, by comparison, was rooted more in political philosophy. Rather than situating the lawyer in an individual moral stance, it was concerned primarily with the lawyer's unavoidably institutional role. The lawyer is not just another person with universal ethical obligations but is "part of a scheme of political institutions and practices that has the governance of the community as its end" (Luban & Wendel 2017). This institutional role is especially important in pluralist societies where the whole assumption of homogenous moral norms is inapt. Such societies require what influential legal theorists Henry Hart and Albert Sacks called institutional settlement, a reconciliation of competing or conflicting normative judgments (Eskridge & Frickey 1993). For this reconciliation to have meaning, lawyers – indeed, all members of society – must respect the results of the settlement process even if those results are unpleasant in individual cases. Conversely, for the process of securing institutional settlement to have legitimacy, it must consider a broad enough range of normative perspectives that the result can reasonably be considered deliberative and just.

Thus, for our patent lawyer whose client seeks patents on isolated gene fragments, ethical choices remain. The Second Wave legal ethicist would point to the responsibility of the Patent Office and Solicitor General to take the concerns of impoverished patients into account when formulating policy. This consideration could not take place in a vacuum, nor does it have to. Government agencies with policy authority over access to healthcare, such as the Centers for Medicare & Medicaid Services within the Department of Health and Human Services, might properly be included in an executive-branch discussion with the Patent Office about the broader social effects of gene-related patents. So might the Food and Drug Administration, which regulates clinical testing of pharmaceuticals and whose lawyers and policy advisors are familiar with the interactions of patent law with drug safety and efficacy.

Importantly, the institutionally political (rather than individually moral) stance of Second Wave legal ethics does not let our patent lawyer off the hook. In addressing arguments to various audiences such as executive agencies or federal courts, she remains obligated to educate herself and to educate and advise her client about various pluralist perspectives that are likely to bear on the client's interests. This obligation was crystallized starkly during the *Myriad* case in how the Federal Circuit and the Supreme Court worked to fulfill their respective responsibilities not only to determine which view of the product of nature doctrine would promote innovation or hinder university research or distribute access to breast cancer diagnostics one way or the other, but also to determine which of these objectives were worth pursuing at the expense of which others.

3.3 The dichotomy in Myriad

Recognizing these obligations recalls our starting point: the Justice Department and its magic microscope in April 2011. By this point, we can identify all the relevant actors as well as the epistemic orientations that underpin their respective arguments and decisions. Myriad Genetics, the innovator seeking to protect its patents and revenue streams, had solved a scientific problem and cast its solution in scientific terms. Myriad's orientation was more technical in its nature and more narrowly specialized in its expertise. The Patent Office had taken a similarly technical and specialized posture in granting Myriad's patents. This was true of the agency leadership who had previously deemed genetic sequences patentable and would later defend that decision in executive-branch deliberations. It was also true of the patent examiners who had evaluated Myriad's actual applications under then-existing laws and found them to be deserving of patents.

By contrast, PUBPAT and the ACLU were recasting Myriad's scientific solution in broader terms. In their view, the ability of molecular biologists to correlate genetic mutations with elevated risks of developing breast and ovarian cancer was only the beginning. Unless and until that scientific contribution could be met with legal and regulatory policy that ensured widespread usage of the invention as well as economic and fiscal policy that ensured widespread access by patients across the socioeconomic spectrum, the justifications for conferring exclusive patent rights on life-saving diagnostic methods were not satisfied. Thus, PUBPAT and the ACLU's orientation was much more broadly generalist.

Judge Sweet in the Southern District of New York was similarly generalist, as federal district judges tend to be. The Federal Circuit is a high-profile exception to this rule given its more specialized patent-related responsibilities, but the Supreme Court again restores the norm of favoring generalism. Executive branch orientation follows a similar upward trend toward generalism. Subordinate agencies are specialized according to their respective missions whereas high-level officials such as the Solicitor General are broadly generalist. In the middle are agency heads, who must translate between political superiors and technical employees.

Thus, at each step in a dispute, one may expect different chances of success depending on one's own orientation and that of the audience. PUBPAT and the ACLU found a favorable generalist audience in the district court, but Myriad and the Patent Office found a favorable specialist audience at the Federal Circuit. The Solicitor General and the Supreme Court had no easy task in disentangling the perspectives that had percolated up toward them. The Solicitor General's office had to come to grips with its lack of relevant scientific expertise relative to the Patent Office (and other agencies such as the NIH), just as the Patent Office had to come to grips with its lack of relevant authority. A similar dynamic described the generalist but authoritative Supreme Court and the expert but subordinate Federal Circuit.

In this kind of complex and multi-faceted dispute, it is not enough for our hypothetical patent lawyer simply to say, 'I have studied biochemistry and law, and because this isolated DNA sequence is a molecule, patent law's familiar rules on chemical inventions govern'. That may turn out to be the conclusion, as it was in the Federal Circuit's decision. However, our brief ethical analysis, the Solicitor General's position, and the Supreme Court's decision suggest that this insular view is not a proper starting point. As Myriad's lawyers as well as PUBPAT and the ACLU's lawyers discovered as the dispute progressed, this insular view excludes competing perspectives that may lack scientific or doctrinal expertise but nevertheless have authority from broader social values that the patent system should accommodate. These considerations also underscore the importance of Professor Lee's overall argument, that the patent system cannot fulfill its objectives unless the expertise of those inside the system can be translated effectively to the generalists who must administer it.

4. Conclusion

This brief and selective glimpse into one of the more contentious and far-reaching disputes over the ethical, legal, and social implications of genetic and genomic innovation has identified a few recurring themes that merit further study. One is the ongoing tension between experts and non-expert generalists in the often zero-sum sphere of political and social decision making. For reasons of institutional settlement and broader reasons of democratic legitimacy that are outside the scope of this discussion, generalists largely hold ultimate authority in the patent system.

However, the limits on expert input should not be taken as a sign that expertise is unimportant. If anything, the vesting of decisional power in generalists means that scientists, engineers, and others who have cultivated technological expertise bear a correspondingly greater responsibility to engage, educate, and persuade those outside their respective disciplines about what the facts are and how those facts matter.

Another recurring theme, related to the translational obligation of experts, is that the ethical stance of those with training in the natural sciences should extend beyond individual demands of moral philosophy, important though they are. It should further consider institutional demands of political philosophy as well. As with generalism, the salient lesson is one of engagement.

Such engagement may take the form of seeking advanced training in interdisciplinary problems. It may also take the form of being purposeful about the kinds of financial, social, professional, or other incentives to which to respond. Because the science of biochemistry, and the particularly compelling potential of genetics and genomics, are so intimately tied to basic questions of human health and self-determination, these calls to engagement apply with special force to those in the chemistry-related disciplines.

Further Reading

The following papers offer an overview of important issues that remain after the Supreme Court's *Myriad Genetics* decision: Cook-Deegan 2012, Rai & Cook-Deegan 2013, Sherkow & Greely 2015. These issues include the growing usage of trade secrecy over clinical data, the legal outlook for whole-genome sequencing, and historical lessons about how genes first came under the purview of the patent system.

References

Contreras, J.L.: 2016, 'Narratives of Gene Patenting', *Florida State University Law Review*, **43** (4), 1133-1200 [available online: ssrn.com/abstract=2485681, accessed 8 May 2020].

Contreras, J.L. & Deshmukh, V.G.: 2017, 'Development of the Personal Genomics Industry', in: S. Bouregy, *et al.* (eds.), *Genetics, Ethics and Education,* Cambridge: Cambridge Univ. Press [available online: ssrn.com/abstract=3092313, accessed 8 May 2020].

Cook-Deegan, R.M.: 2012, 'Law and Science Collide Over Human Gene Patents', *Science*, **338** (6108), 745-747, [available online: science.sciencemag.org/content/338/6108/745, accessed 8 May 2020].

Cook-Deegan, R.M.; Conley, J.M.; Evans, J.P. & Vorhaus, D.B.: 2012, 'The Next Controversy in Genetic Testing: Clinical Data as Trade Secrets?', *European Journal of Human Genetics*, **21** (6), 585-588 [available online: www.ncbi.nlm.nih.gov/pubmed/23150081, accessed 8 May 2020].

Eisenberg, R.S.: 2015, 'Diagnostics Need Not Apply', *Boston University Journal of Science & Technology Law*, **21** (2), 256-286 [available online: ssrn.com/abstract =2631679, accessed 8 May 2020].

Eskridge Jr., W.N. & Frickey, P.P.: 1994, 'The Making of the Legal Process', *Harvard Law Review*, **107**, 2013-2055 [available online: digitalcommons.law.yale.edu/ fss_papers/3843/, accessed 8 May 2020.

Lai, J.C.: 2015, 'Myriad Genetics and the BRCA Patents in Europe', *UC Irvine Law Review*, **5** (5), 1041-1076 [available online: scholarship.law.uci.edu/ucilr/vol5/ iss5/5/.

Lee, P.: 2010, 'Patent Law and the Two Cultures', *Yale Law Journal*, **120** (1), 1-83 [available online: digitalcommons.law.yale.edu/ylj/vol120/iss1/1/, accessed 8 May 2020].

Luban, D. & Wendel, W. B.: 2017, 'Philosophical Legal Ethics: An Affectionate History', *The Georgetown Journal of Legal Ethics*, **30** (3), 337-364 [available online: scholarship.law.cornell.edu/cgi/viewcontent.cgi?article=2686&context=facpub, accessed 8 May 2020].

Rai, A.K.: 2013, 'Biomedical Patents at the Supreme Court: A Path Forward', *Stanford Law Review Online*, **66**, 111-117 [available online: scholarship.law.duke.edu/ faculty_scholarship/3142/, accessed 8 May 2020].

Rai, A.K. & Cook-Deegan, R.M.: 2013, 'Moving Beyond Isolated Gene Patents', *Science*, **341** (6142), 137-138 [available online: science.sciencemag.org/content/ 341/6142/137, accessed 8 May 2020].

Sherkow, J.S. & Greely, H.T.: 2015, 'The History of Patenting Genetic Material', *Annual Review of Genetics*, **49**, 161-182 [available online: ssrn.com/abstract=2713761, accessed 8 May 2020].

Simon, W.H.: 1978, 'The Ideology of Advocacy: Procedural Justice and Professional Ethics', *Wisconsin Law Review*, **1978**, 29-144.

Snow, C.P.: 1959, *The Two Cultures* (London: Cambridge University Press) [available online: s-f-walker.org.uk/pubsebooks/2cultures/Rede-lecture-2-cultures.pdf, accessed 8 May 2020].

Legal Documents

European Parliament and the Council of the European Union: 1998, *Directive 98/44/EC* (European Biotechnology Directive) [bit.ly/2C4uiuM].

Supreme Court of the United States: 1948, *Funk Bros. Seed Co.. v. Kalo Inoculant Co.* (Funk Bros.), 333 U.S. 127 [www.leagle.com/decision/1948460333us1271450].

Supreme Court of the United States: 1980, *Diamond v. Chakrabarty* (Chakrabarty), 447 U.S. 303 [www.leagle.com/decision/1980750447us3031737].

Supreme Court of the United States: 2013, *Ass'n for Molecular Pathology v. Myriad Genetics, Inc.* (Myriad IV), 133 S. Ct. 2107 [www.leagle.com/decision/insco20130613e08].

U.S. Court of Appeals for the Federal Circuit: 2011, *Ass'n for Molecular Pathology v. Myriad Genetics, Inc.* (Myriad II), 653 F.3d 1329 [www.leagle.com/decision/infco20110729000t].

U.S. Court of Appeals for the Federal Circuit: 2012, *Ass'n for Molecular Pathology v. Myriad Genetics, Inc.* (Myriad III), 689 F.3d 1303 [www.leagle.com/decision/infco20120816212].

U.S. District Court for the Southern District of New York: 2010, *Ass'n for Molecular Pathology v. Myriad Genetics, Inc.* (Myriad I) [www.leagle.com/decision/infdco20100330948].

U.S. Patent No. 5,693,473: 1997, 'Linked breast and ovarian cancer susceptibility gene' [patents.google.com/patent/US5693473].

U.S. Patent No. 5,709,999: 1998, 'Linked breast and ovarian cancer susceptibility gene' [patents.google.com/patent/US5709999].

U.S. Patent No. 5,710,001: 1998, '17q-linked breast and ovarian cancer susceptibility gene' [patents.google.com/patent/US5710001].

U.S. Patent No. 5,747,282: 1998, '17Q-linked breast and ovarian cancer susceptibility gene' [patents.google.com/patent/US5747282].

U.S. Patent No. 5,753,441: 1998, '170-linked breast and ovarian cancer susceptibility gene' [patents.google.com/patent/US5753441].

U.S. Patent No. 5,837,492: 1998, 'Chromosome 13-linked breast cancer susceptibility gene' [patents.google.com/patent/US5837492].

U.S. Patent No. 6,033,857: 2000, 'Chromosome 13-linked breast cancer susceptibility gene' [patents.google.com/patent/US6033857].

Chapter 18

American Chemical Society Codes of Ethics: Past, Present, and Future

Jeffrey Kovac

Abstract: This chapter traces the evolution of the American Chemical Society (ACS) Code of Ethics from its inception to the present. It also discusses various supporting documents issued by the ACS which have some ethical content. The similarities and differences between the ACS code and several representative national and international codes are then explored. Finally, the strengths and weaknesses of the ACS code in providing guidance in finding solutions to current ethical problems are discussed.

1. Introduction

Codes of ethics reveal much about a profession; they tell us what it values and how it wants to be seen by the world. The code also helps to define the relationships within the chemical community, how professionals should ideally interact with each other, and shapes the relationship between the profession and society (Frankel 2009). A code of ethics formalizes the informal bargains that professionals make with themselves and with society and can be used as a guide in ethical decision making.

Tracing the evolution of a code shows how the profession responds to changes in circumstances, both internal and external. Chemistry has always been a science in the middle between the theoretical and the practical, between philosophy and craft. Chemists have developed substances that have greatly improved human life from pharmaceuticals to materials, but they have also been responsible for some of the worst environmental pollution. Chemists are employed in a wide variety of institutions from universities to production plants, so any code of ethics for chemists must account for the different responsibilities of these jobs. Crafting an appropriate code is a complex challenge.

After a brief history of codes of ethics, I will trace the development of the American Chemical Society (ACS) code of ethics from its origin in 1965 as the Chemist's Creed to the present to show how the code has adapted to changing circumstances. The ACS code is supplemented by several documents concerned with more specialized topics and their ethical content is described and analyzed. There are many other codes of ethics for chemists developed by national chemical societies and by international organizations so it is instructive to compare the ACS code with a few representative codes to see the similarities and differences. The final two sections are an assessment of the ACS code that suggests some changes to better deal with contemporary ethical issues.

2. Brief History of Codes of Ethics

The Hippocratic Oath, historically taken by physicians, dates to the fifth through third centuries BCE, although the earliest surviving written version is from about 275 CE. The word profession is Latin for 'bound by an oath'. In ancient Rome one's profession was the occupation declared under oath to a tax collector. About 100 CE, the Hippocratic Oath became the required oath for physicians. The earliest code of medical ethics was proposed by Thomas Percival of Manchester, England, in 1794 which led to codes of ethics for the practice of medicine beginning about 1850 (Baker 1999). The word 'code' also comes from Latin where it originally referred to any wooden board. Eventually the word came to refer to any paged book, or codex, and finally to a systematization of rules or laws (Davis 1999).

Although there is surprisingly little written about the history of codes of ethics, empirically we can see that codes began to proliferate in the mid-20th century as researchers struggled to respond to the scientific and medical atrocities of the Nazis (Metcalf 2018). Before World War II, the German Medical Society had promulgated a set of ethical guidelines for therapy and human experimentation but those had been negated by Adolf Hitler. At the war crimes trials in Nuremberg, the judges developed the ten-point Nuremberg Code which provides a basis for subsequent codes related to human and animal experimentation. Scandals such as the infamous Tuskegee experiment have led to further developments in biomedical ethics (Jones 1993). Professions other than medicine and law, including chemistry, began to write codes of ethics to formalize relationships within the profession and between the profession and society.

3. Evolution of the American Chemical Society Code of Ethics

Although the Federal Charter of the American Chemical Society, issued in 1937, lists among its objectives, "the improvement of the qualifications and usefulness of chemists through high standards of professional ethics, education, and attainments" (ACS 2016a), the Society did not adopt a formal code of ethics until 1965 when it issued 'The Chemist's Creed' (ACS 1965). The ACS Council Committee on Professional Relations and Status had been considering various versions of a code of ethics for several years, responding in part to a survey conducted in 1960 which showed that a majority of members thought having a code of ethics was a good idea. The August 1964 meeting of that committee considered a code of ethics that had been adopted by the Dayton Section in 1949. That code had originally been drafted by P. K. Rothemund who had advocated for registration for chemists after World War II. The document was further refined by Austin M. Patterson who was an editor of *Chemical Abstracts* (ACS News 1964). 'The Chemist's Creed' was adopted by the ACS Council a year later (ACS News 1965). The brief article in *Chemical and Engineering News* which reports the adoption of the document does not say why the title was changed from code of ethics to creed. Perhaps the committee and the council felt that a creed, which is a personal statement of values, was more appropriate than a code which is a set of rules. Although the vote was strongly positive, there was opposition. Those opposed to the new code argued that the provisions were too obvious and that anyone worthy of the label 'professional' automatically did what the statements promoted. It is interesting that 'The Chemist's Creed' is not mentioned in the Centennial History of the ACS (Reese 1976).

That relatively brief statement, 286 words in length, was revised and expanded in 1994 as 'The Chemist's Code of Conduct' (ACS 1994). The

1994 version underwent more minor revisions in 2007 and was given a new title, 'The Chemical Professional's Code of Conduct'. This latter code has been regularly revised, most recently in 2016. This section will examine the evolution of the formal statements on ethics as reflected in the three codes. Quotations in the rest of this section come from the three versions of the code.

'The Chemist's Creed' succinctly specifies the responsibilities of a chemist in eight areas: to the public, to the science, to the profession, to an employer, to the chemist him or herself, to employees, to students and associates, and to clients. Most of the statements are based on solid moral values such as truth telling, not cheating, and maintaining trust among professionals. For example, the chemist is admonished to "search for [chemistry's] truths by use of the scientific method, and to enrich it for the good of humanity". Further, the chemist should "maintain my professional integrity as an individual" and "hold the highest ideals of personal honor". The rest of the Creed consists of similar broad statements of moral and professional ideals. One interesting provision, which came directly from the original Dayton code, is that the chemist should "live an active, well-rounded and useful life", a refreshingly broad perspective on the professional life.

A large fraction of chemists are employed in the chemical industry, so there are statements regarding obligations to employers, employees, and clients. Although chemists employed in universities or research laboratories do have employment issues, these statements are primarily directed at the industrial sector. The sections on employees and clients are quite reasonable; they should be treated with respect and dignity, both well-accepted moral values. The responsibility to employers, on the other hand, demands complete loyalty. The chemist must "serve him undividedly and zealously in mutual interest, guarding his concerns and dealing with them as I would my own". This aspect of the code is, at best, morally questionable because chemists must further the employer's objectives even if they find those goals to be illegitimate for practical, legal, or ethical reasons. This should be contrasted with the Code of Ethics of the National Society of Professional Engineers (NSPE 2017). Although engineers are expected to "act in professional matter for each employer or client as faithful agents or trustees", the primary duty of engineers is to the health and welfare of the public, and they are not to work with “any person or firm which they have reason to believe is engaging in fraudulent or dishonest business or professional practices”.

‘The Chemist’s Code of Conduct’, issued in 1994, expands the provisions contained in ‘The Chemist’s Creed’. It contains nine sections. The section on “Responsibilities to Myself” was eliminated although some of the provisions were moved to other sections. The section on “Students and Associates” was divided into two separate headings and a section on “The Environment” was added.

The section on responsibilities to the public was strengthened. It opens by stating, “chemists have a professional responsibility to serve the public interest and welfare”. Further, “chemists should be actively concerned with the health and welfare of co-workers, consumers and the community”. These proactive statements reflect an increased concern with public health and safety compared to the earlier negative statement, “discourage enterprises or practices inimical to the public interest or welfare”.

The two sections on the “Science of Chemistry” and the “Profession” reflect core principles of professional ethics, including respect for the truth and ensuring that scientific contributions are thorough, accurate, and unbiased. There is also a nice statement of humility, “understand the limitations of their knowledge”. There are additional important provisions on the responsible conduct of research, including keeping accurate laboratory records, maintaining integrity, giving credit where it is due, and avoiding conflicts of interest. Finally, the document states that “scientific misconduct, such as fabrication, falsification, and plagiarism are incompatible with this Code”. These are the three categories of scientific misconduct identified by the Office of Research Integrity (Steneck 2004).

The section on the “Employer” is significantly different from the corresponding section in ‘The Chemist’s Creed’. Instead of complete loyalty, chemists are required only to “promote and protect the legitimate interests of their employers”, leaving the ethical judgment of what is legitimate to the individual. This statement is still problematic because the key term, ‘legitimate’, is not defined. For example, one of the legitimate interests of a corporation is making a profit. Are chemists obligated to protect that interest even if they judge that the company is selling products that are dangerous to humans or the environment, or using processes that generate more pollution than necessary? An additional problem is that there is no corresponding provision regarding protections for chemists if there is a disagreement as to what is a legitimate interest. This would vary depending on the employer, but the fate of many whistle blowers is usually not good (Glazer & Glazer 1989, Lubalin & Matheson 1999).

The fundamental message regarding employees, associates, and students is that they should be treated with respect. This is essentially the same as the responsibilities in 'The Chemist's Creed', but stated more clearly in the revised code. Two important additions to the section on students are statements that the tutelage of students is a trust conferred by society and that students should not be exploited. Because graduate education in chemistry is a kind of apprenticeship, this is an important protection. A senior graduate student can be very productive and is much cheaper to support than a postdoctoral research associate; there is a temptation to delay that student's graduation so that he or she can produce a few more articles.

The publication of *Silent Spring* by Rachel Carson in 1962 is often identified as the beginning of the environmental movement in the U.S. (Carson 1962). Events that followed include the founding of the Environmental Protection Agency in 1970. The toxic effects of chemical wastes at Love Canal in New York came to light beginning in 1976. Such incidents led to the passage of the so-called Superfund or The Comprehensive Environmental Response, Compensation, and Liability Act (CERCLA) in 1980. The Bhopal disaster in India, in which 500,000 people were exposed to methyl isocyanate gas and other chemicals and about 4000 died, occurred in 1984. The Responsible Care program of the American Chemistry Council was adopted in 1988 (American Chemistry Council 2017). The Code finally reflected this concern with a section on the Environment which tells chemists to "understand and anticipate the environmental consequences of their work", and "to avoid pollution and to protect the environment". This was an important advance that formally recognized the darker aspects of the chemical industry, although it is relegated to the end of the code suggesting to some that it is less important than the other provisions (Bensaude-Vincent & Simon 2008, p. 236).

The 2007 revision changed the name of the document to 'The Chemical Professional's Code of Conduct'. Most of the sections were not changed. The name of one section was changed from "Associates" to "Colleagues". That section and the section on employees were amended by adding language on avoiding bias based on grounds such as race, gender, ethnicity, disability, or other personal attributes. These additions reflected the increasing concern with bias in American society and were long overdue. The section on the environment was expanded to strengthen the language and to add provisions related to sustainable development. By 2007 the problem of limited resources, particularly of petroleum, and the need to consider the fate of future generations were broadly recognized and the Code reflected these concerns.

4. Supporting Documents

The ACS has developed several documents that supplement the code of ethics. Perhaps the most important of these is the 'Ethical Guidelines to Publication of Chemical Research' (ACS Publications 2015). In 1982 the editors of the ACS journals established a sub-committee to draft a set of ethical guidelines because they felt that the informal teaching by mentors or research groups, and the experience in preparing manuscripts was not systematic. The two core principles that guided the sub-committee were honesty and fairness (Bunnett 1983). The initial draft, presented to the Board of Editors in mid-1983, has been regularly updated and is a comprehensive statement of ethical publication practice which includes detailed guidelines for editors, authors, and reviewers and a brief section for authors publishing in the popular literature.

Probably the most important guideline for editors is that they give each manuscript unbiased consideration, judging it solely on its scientific merit and the quality of the presentation. Editors also need to avoid conflicts of interest and should allow another responsible person, such as an associate editor or a member of the editorial advisory board, to handle a manuscript from a close collaborator or former student or a manuscript on a topic that is close to the editor's own area of research. The same principles apply in the choice of reviewers. All of these provisions follow from the core principle of fairness. Finally, the whole process must be confidential.

For authors, the standard is completeness and accuracy in reporting the procedures, data, and conclusions as concisely as possible. Any professional reading the article should be able to understand what was done and to reproduce the work. If images are included, they must be free from misleading manipulation, something that is currently much easier to do with modern digital technology. The author must also cite the relevant prior work and identify other sources of information such as personal communications. It is also important to identify any unusual hazards in the reported procedures and to indicate that appropriate standards for experimentation on animal or human subjects have been followed, if necessary. Occasionally, an author will criticize previous work, which is acceptable, but this criticism must not be personal.

Fragmentation, dividing a larger article into smaller pieces, is to be avoided, although it is permissible to publish a preliminary short communication followed by a detailed article. Submitting the essentially same article to two different journals at the same time is a breach of ethics. Plagiarism is an even more serious offense. Authorship guidelines are described in detail. A co-author is someone who has made a significant scientific contribution to the work and shares responsibility for the results. Other, more minor, contributors should be acknowledged. It is also essential that financial and other conflicts of interest be disclosed. Chemists and other scientists are increasingly involved in commercial ventures and the publication of a positive result related to that venture might be financially lucrative.

The peer review system is an essential part of the publication process and all chemists have a professional obligation to participate. As with editors, the core ethical principle for reviewers is fairness. They should only agree to review manuscripts they are scientifically qualified to judge and they should review them objectively respecting the intellectual independence of the author. Reviews should be completed in a timely manner and judgments about the manuscript must be properly supported. If the reviewer needs to criticize a manuscript, that criticism should never become personal. Conflicts of interest should be avoided and the review process must be held confidential.

The guidelines for both authors and reviewers contain an interesting provision. Both are asked to inform the editor of "concerns with respect to manuscripts that report research that, based on current understanding, can be reasonably expected to provide knowledge, products, or technologies that could be directly misapplied by others to pose a threat to public health and safety, agricultural crops and other plants, animals, the environment, or materiel". There is, however, no corresponding provision in the guidelines for editors as to what they should to do with such concerns. The 'The Chemical Professional's Code of Conduct' requires that chemists be "actively concerned with the health and welfare of co-workers, consumers, and the community", and to "protect the environment", but it, too, does not provide any guidance as to what chemists should do if they have concerns that a substance or a process is dangerous.

The final section provides a few guidelines for chemists writing for the popular literature. First, they are admonished to be accurate and unbiased. The guidelines recognize that in writing for a lay audience, it will be necessary to simplify and to use more common words rather than precise technical language, but it is still important to strive for accuracy. The final provision is quite important. Chemists should not announce a discovery to the public unless it is secure enough to publish it in the scientific literature and that the discovery should be submitted to a journal as soon as possible so that it can be subjected to peer review. This provision is there, in part, to protect the public, but also to discourage such practices as 'publication by press conference'. A good example of this was the announcement of cold fusion in March 1989, a discovery that has yet to be verified (Close 1991).

The 'Professional Employment Guidelines', first issued in 1975 and revised regularly since then, are not primarily concerned with ethics, per se, but rather with good employment practices (ACS 2004). They do include statements opposing discrimination in employment and physical and verbal harassment. Some of this is prohibited by law, but the ACS guidelines go beyond the law and make a strong moral statement. They also make contact with the Code of Conduct in urging the chemical professional and the employer to minimize risks to the environment.

Similarly, the 'Academic Professional Guidelines' establish standards for academic institutions for the education of the next generation of chemists (ACS 2016b). There are sections concerning faculty, students and postdoctoral associates, departments and institutions. These guidelines also include a statement opposing discrimination as well as a statement that chemical scientists should maintain "high standards of honesty, integrity, ethics, and diligence in the conduct of teaching, research, and professional activities". There is also an emphasis on developing a culture of safety in the department. Although faculty members and departments are expected to develop an atmosphere in which students can learn and mature into working professionals, there is no explicit statement on the importance of ethics. For example, faculty are expected to serve as mentors, but the importance of modeling ethical professional behavior is not mentioned.

The final document is 'Scientific Integrity in Public Policy', which concerns the interaction between science and government (ACS 2014-2017). It includes sensible recommendations regarding the ideal way that Federal agencies and Congress should solicit and use scientific information. Scientists and engineers have an obligation to provide accurate and unbiased information and should avoid, or at least disclose, conflicts of interest.

5. Comparison with Other Codes

It is instructive to compare the ACS Chemical Professional's Code of Conduct with similar codes adopted by other organizations, both national chemical societies and international organizations. There are hundreds of such codes (OPCW 2015a) so I have to be selective. I will compare and contrast ACS code with those of two other chemical societies, the Royal Society of Chemistry and the German Chemical Society, two countries with important chemical industries, and two recent international codes, the 'Hague Ethical Guidelines', developed under the guidance of the Organization for the Prohibition of Chemical Weapons, and the recent 'Global Chemists' Code of Ethics'.

The Royal Society of Chemistry of the United Kingdom (RSC) has developed a detailed document entitled 'Professional Practice and Code of Conduct' (RSC 2013). It begins by listing three overarching behaviors, broad categories that are then used to organize the rest of the code. They include: (1) inclusivity – respect, (2) integrity – rigor, and (3) leadership – responsibility. These categories certainly correspond to important moral principles including trust, respect for other humans, truth telling, and also to the need for strong leadership within the chemical community to ensure that the ethical principles are upheld.

The first section of the code presents general ethical considerations, such as chemists "should never engage in an action that conflicts with their integrity or that of the Royal Society of Chemistry" and "have a duty to serve the public interest, and maintain and enhance the reputation of the profession". This is language quite similar to that found in the ACS Code. The remaining sections primarily concern industrial chemists. Among the topics are employer responsibilities, self-employment and consultancy, trade union membership, presenting legal evidence, and tribunals and inquiries. The latter three topics are not discussed at all in the ACS code. There are sections that cover education, environment, health and safety, and other legislation and communications. The guidelines for industrial chemists are broadly similar to those in the ACS code, but little is said explicitly about research ethics, which is a more prominent part of the ACS code.

Chemists in the UK, however, can turn to the 'Code of Practice for Research', issued by the UK Research Integrity Office, an independent agency offering guidance to universities and other research organizations (UKRIO 2009). This is a detailed guide covering all aspects of research. There is also a statement on research integrity of the Royal Society which sets out standards for all scientists in the UK. It articulates four key principles for ethical conduct: excellence, accountability, transparency, and responsiveness. For each principle there are guidelines for both researchers and institutions. For example, researchers are encouraged to strive for excellence and institutions are expected to create an environment where "the honest and ethical conduct of science is an expected norm". This is one of the strengths of this document. As I have previously argued, developing a culture of ethical behavior in a research group and an academic department is crucial to the cultivation of virtue (Kovac 2013a).

In contrast to the long and detailed codes of the RSC and the ACS, the Code of Conduct of the German Chemical Society is brief, less than a page in length (OPCW 2015). It is a statement of principles emphasizing the responsibility of chemists to society, the economy, the environment, and particularly to future generations. There is also a statement regarding the responsibility of chemists to fight against the misuse of chemistry, including the production of chemical weapons. Finally, there is an explicit statement that the code is binding on all members of the society. The code of the Royal Society of Chemistry contains a similar statement, something that is lacking in the ACS Code.

In 2015 representatives from various national chemical societies developed the 'Hague Ethical Guidelines' under the guidance of the Organization for the Prohibition of Chemical Weapons (OPCW), the international organization that oversees the provisions of the Chemical Weapons Convention (OPCW 2015b). These guidelines are written from the perspective of the OPCW, emphasizing the need to prevent the misuse of chemicals, particularly as weapons, but they also provide a strong statement on the relationship between chemistry and society. The core element of the 'Hague Ethical Guidelines' is, "achievements in the field of chemistry should be used to benefit humankind and protect the environment". This is followed by a statement regarding the importance of sustainability so that the needs of future generations are not compromised. The remaining sections of the guidelines address education, awareness and engagement, ethics, safety and accountability, oversight, and exchange of information. The statements on accountability and oversight are particularly concerned with ensuring that chemicals do not fall into the hands of those who would misuse them for illegal, harmful, or destructive purposes.

In 2016, a group of scientists from 18 countries gathered in Malaysia to draft the 'Global Chemists' Code of Ethics' (ACS 2016c). The workshop was convened by the ACS Office of International Activities. This idealistic statement urges chemists to be "role models, mentors and advocates of the safe and secure application of chemistry to benefit humankind and preserve the environment for future generations", echoing the language of the 'Hague Ethical Guidelines'. The statement on the environment emphasizes the importance of environmental sustainability and the proper use and disposal of chemicals and instruments. The sections on research and scientific writing and publishing continue the theme of benefitting humankind and protecting the environment while maintaining the highest standards of integrity. Finally, there are statements regarding safety and security that elaborate the principles put forth in the 'Hague Ethical Guidelines'.

The two international codes are concise statements of moral ideals. They reflect increased contemporary concerns with protecting the environment and the possible misuse of chemicals. They also attempt to portray chemistry in the best possible light, as a science committed to benefitting society. Chemists have long been concerned with the public perception of their science (Bensaude-Vincent & Simon 2008; Schummer *et al.* 2007). Although chemists have produced substances that have improved the human condition, these substances were often dismissed as 'artificial' and inferior to the 'natural' substances they replaced. They were often produced by processes that generated waste products that polluted the environment, in plants that occasionally had accidents resulting in many deaths.

Two important differences between the two international codes and that of the German Chemical Society, on the one hand, and those of the ACS and the RSC, on the other, are the emphasis on preventing the misuse of chemicals and on protecting the environment. Both the ACS and RSC codes are silent on the issue of preventing the misuse of chemicals. Both contain statements on the environment but they are much less prominent.

6. Discussion

Mark S. Frankel has identified three types of codes of ethics: aspirational, educational, and regulatory (Frankel 2009). All of the codes discussed in this chapter are largely aspirational, statements of the ideals to which chemists should aspire. Some of the codes, that of the German Chemical Society, the Hague Ethical Guidelines, and the Global Chemist's Code of Ethics, stop there. The ACS and the RSC codes add some educational and regulatory aspects, although neither is as detailed as the Code of Ethics of the National Society of Professional Engineers (NSPE, 2017). The ACS and RSC codes explicitly recognize the different roles that chemists play: although some of the provisions apply to all chemists, others, such as the responsibility to clients, apply only to those involved in consulting.

All professional codes are embedded in a larger moral landscape because all professionals simultaneously belong to several communities each with its own set of responsibilities (Kovac 2013b). Some of these responsibilities are explicitly recognized in the ACS code but others are not. We are all citizens of a national society with a history and with goals and ideals. With citizenship comes obligations. Second, almost all chemists are employed by an institution, a college or university, a government or private research laboratory, a government agency, or a corporation. Each of these has its own culture and expectations. Because a large fraction of chemists are employed by industry, the influence of the institution is very important, and both the ACS and RSC codes have several provisions that are primarily directed at industrial chemists. A further complication is that the chemist might be a manager, or even a company president, with responsibilities to a board of directors or the company shareholders. Many academic chemists are also entrepreneurs who are involved in start-up companies so there is a potential conflict of interest between their academic and commercial obligations. Another possible responsibility is the source of funding for the research being conducted which might impose constraints. Fourth, all chemists are members of the human community and have the same moral obligations as all other people. Simultaneous membership in these different communities can certainly give rise to moral dilemmas. For example, when does chemist's moral responsibility as a member of the larger human community take precedence over obligations to an institution or country? The moral landscape might be further complicated by the scientists religious beliefs and practices. Because chemistry is a secular pursuit, I will not consider the moral demands of particular faith traditions, but it is important to remember that religious beliefs can strongly influence certain moral decisions.

Some of these complications are recognized, at least implicitly in the ACS Code and the supporting documents. As noted, obligations to employers are included and the first provision of the code states the obligations of chemists to serve the public interest, although it is left to individuals to decide exactly what the public interest is and how best to serve it. Similarly, the section on the environment only tells chemists to "understand" the impacts of their work and to "recognize" the need to develop sustainable processes. The ACS statement on the environment has been strongly criticized by Bensaude-Vincent and Simon (2009, chap. 14) who point out that it is the last provision in the code and therefore seems to be the least important. They contrast its 'prudence' with the more 'precautionary' stance of the European Union's REACH (Registration, Evaluation, Authorization, and Restriction of Chemicals) which establishes a central database of information about chemicals to ensure that particularly hazardous materials are properly controlled or even banned (European Commission 2006). They call for a new culture of chemistry in which chemists consider all the consequences of their research. Roald Hoffmann has made a similar plea (Hoffmann 1997). It is not enough to synthesize a molecule to solve a particular problem. One must also consider the whole life of that molecule and try to determine what negative effects it might have in other contexts.

An example of the tensions between prudence and precaution and between commercial interests and health and environmental safety is the pesticide chlorpyrifos, marketed by Dow Chemical under the trade name Lorsban. Chlorpyrifos is an organophosphate closely related to nerve agents which is why it is so effective as an insecticide. It was banned by the EPA for most residential use in 2000 and environmental groups in the US have been trying to have the compound completely banned since 2007 (Pesticide Action Network 2006). Research has shown that chlorpyrifos can harm the developing brains of fetuses and children who eat food from plants treated with this compound. Dow Chemical has published an extensive report in which it throws doubt on the various studies that show adverse effects (Dow 2017). Farmers who find the compound useful are also opposed to a ban. A precautionary attitude would suggest that the compound be taken off the market because of the suggestion of harm and also because of its similarity to chemical weapons. On the other hand, a risk-benefit analysis might conclude that the needs of agriculture were great enough to compensate for a small number of health problems. Although several EPA reports have documented the potentially serious health effects of chlorpyrifos, it has yet to order a complete ban on its use.

If a Dow or an EPA chemist looked to the ACS Code for guidance on this question, what help would he or she get? The Code tells chemists that they should "promote and protect the legitimate interests of their employers", while understanding "the health, safety, and environmental aspects of their work". The ACS Code does not display a moral 'red flag'. Selling chlorpyrifos is a legitimate interest of Dow, at least from the perspective of company management and the shareholders. The individual chemist might disagree that selling a potentially dangerous product is a legitimate interest. The chemist might also decide that as a human being, he or she has a responsibility to oppose the marketing of the product either within the company or more publicly. As noted earlier, the ACS code does not provide any ethical protection for the whistle blower. Based on a strict reading of the code, as long as the chemist understands that there are potential health or environmental effects, nothing further need be done. This ambiguity seems to me to be at least morally problematic. The 'Hague Ethical Guidelines' are a bit stronger because of the similarity of chlorpyrifos to chemical weapons, but Dow certainly argues that the agricultural use of the compound is not a misuse of its product, so invoking the Hague Guidelines does not seem to solve this potential moral problem. There are many other examples of this tension, such as flame retardant chemicals in clothing and furniture (Slater 2012), where the economic interests of chemical companies are in conflict with health or environmental concerns.

7. Conclusion

It is unreasonable to expect any code of ethics to solve complex moral problems, particularly those that involve the different moral communities that chemists inhabit. Each situation requires a careful ethical analysis. Because ACS members have a wide variety of roles, it is also important to have the kinds of provisions that are in the current code. On the other hand, anyone reading the ACS code will come away with the impression that the most important ethical issues for chemists are interpersonal relationships such as those between employer and employees. Both the ACS code and the RSC code are mainly inward looking, concerned with the image of chemistry and how chemists treat each other. Although these issues are important, I would argue that they need to be subsidiary to the important ethical issues of our time such as integrity in research and the impact of chemistry on society.

It is important for the ACS Code to put more emphasis on the problems of today's world and the need to protect future generations. A revised ACS code would be both more relevant and useful if it communicated to the chemistry community and to society that chemists recognize the ethical issues concerning the relationship between science and society as crucial in today's world (Mehlich *et. al.* 2017). The code also needs to come to terms with the pressures of the contemporary research environment which seem to lead to more and more breaches of research integrity (Kovac 2015). Both the 'Hague Ethical Guidelines' and the 'Global Chemists' Code of Ethics' put the responsibilities to benefit humankind and to protect the environment up front, emphasizing their importance. A revised ACS Code should use these statements as a model for a provision regarding the relationship between chemistry and society. Although research integrity is mentioned in the current code, that statement needs to be strengthened to emphasize the centrality of responsible conduct of research to the trust that is the central value of science.

As Don Gotterbarn notes, the process of writing a code of ethics and then getting it approved is always political (Gotterbarn 1999). Within the ACS there are a variety of constituencies that are likely to have strong opinions about the language of the code. For example, representatives of the chemical industry will have a different view of provisions regarding the environment than chemists who work for regulatory agencies like the EPA. Another tension is how much specificity to put into a code. The most recent version of the ACS code contains a clause that cautions chemists to avoid bias and lists several possible sources of bias. When a code contains a list of specific groups, there is the possibility that a particular group will feel left out. This is where aspirational codes containing only broad principles have an advantage. The disadvantage, of course, is that broad principles are open to interpretation.

The ACS has revised its code of ethics several times since 1965 to respond to changes in both the chemical community and society. The last major revision was in 2007. In light of the problems of today's world, it seems that it is time for the society to once again take a careful look at its ethical standards. Any revision effort should take advantage of the growing research literature on ethics in chemistry which has confronted many important contemporary ethical questions. The code also needs to be supplemented by educational materials. The Guidelines of the ACS Committee on Professional Training explicitly call for instruction in ethics as part of an undergraduate education (ACS CPT 2015). Federal funding agencies also require ethics education for graduate students. Such education would be facilitated by the development of high-quality, easy-to-use educational materials that focused on the ethical problems faced by chemists and used the ACS Code of Ethics as a resource. The combination of a revised code with a strong philosophical basis and supporting educational materials would make the ACS a leader in professional ethics world-wide.

References

ACS: 1965, 'The Chemist's Creed' [available online: http://ethics.iit.edu/ecodes/node/6227, accessed 15 December 2017].

ACS: 1994, 'The Chemists' Code of Conduct' [available online: http://ethics.iit.edu/ecodes/node/3426, accessed 15 December 2017].

ACS: 2004, 'Professional Employment Guidelines' [available online: https://www.acs.org/content/dam/acsorg/careers/profdev/ethics/professional-employment-guidelines.pdf, accessed 15 December 2017].

ACS: 2014-2017, 'Scientific Integrity in Public Policy' [available online: https://www.acs.org/content/dam/acsorg/policy/publicpolicies/science-policy/scientific-integrity.pdf, accessed 15 December 2017].

ACS: 2016a, 'The Chemical Professional's Code of Conduct' [available online: https://www.acs.org/content/acs/en/careers/career-services/ethics/the-chemical-professionals-code-of-conduct.html, accessed 15 December 2017].

ACS: 2016b, 'Academic Professional Guidelines' [available online: https://www.acs.org/content/acs/en/careers/career-services/ethics/academic-professional-guidelines.html, accessed 15 December 2017].

ACS: 2016c, 'The Global Chemists' Code of Ethics' [available online: https://www.acs.org/content/acs/en/global/international/regional/eventsglobal/global-chemists-code-of-ethics.html, accessed 15 December 2017].

ACS CPT: 2015, 'ACS Guidelines and Evaluation Procedures for Bachelor's Degree Programs' [available online: https://www.acs.org/content/dam/acsorg/about/governance/committees/training/2015-acs-guidelines-for-bachelors-degree-programs.pdf, accessed 15 December 2017].

ACS News: 1964, 'PR&S Committee Studies Code of Ethics', *Chemical and Engineering News*, **42** (28), 102-3.

ACS News: 1965, 'Chemist's Creed', *Chemical and Engineering News*, **43** (39), 86-87.

ACS Publications: 2015, 'Ethical Guidelines to Publication of Chemical Research' [available online: https://pubs.acs.org/userimages/ContentEditor/1218054468605/ethics.pdf, accessed 15 December 2017].

American Chemistry Council: 2017, 'Responsible Care' [available online: https://responsiblecare.americanchemistry.com/, accessed 15 December 2017].

Baker, R.: 1999, 'Codes of Ethics: Some History', *Perspectives on the Professions*, **19** (1), 3-5.

Bensaude-Vincent, B. & Simon, J.: 2008, *Chemistry: The Impure Science*, London: Imperial College Press.

Bunnett, J.F.: 1983, 'Ethics in Publication', *Accounts of Chemical Research*, **16** (3), 73.

Carson, R.: 1962, *Silent Spring*, New York: Houghton Mifflin.

Close, F.: 1991, *To Hot to Handle: The Race for Cold Fusion*, Princeton: Princeton University Press.

Davis, M.: 1999, 'Writing a Code of Ethics', *Perspectives on the Professions*, **19** (1), 1-3.

Dow: 2017, 'Chlorpyrifos Studies: The Debate is in the Details' [available online: http://www.chlorpyrifos.com/pdf/chlorpyrifos_science-analysis.pdf, accessed 15 December 2017].

European Commission: 2006, 'REACH' [available online: http://ec.europa.eu/environment/chemicals/reach/reach_en.htm, accessed 15 December 2017].

Frankel, M.S.: 2009, 'Professional Codes: Why How, and with What Impact?', *Journal of Business Ethics*, **8**, 109-115.

Glazer, M.P. & Glazer, P.M.: 1989, *The Whistleblowers*, New York: Basic Books.

Gotterbarn, D. 1999, 'Two Computer-Related Codes', *Perspectives on the Professions*, **19** (1), 5-7.

Hoffmann, R.: 1997, 'Mind the Shade', *Chemical and Engineering News* **75** (45), 3.

Jones, J.H.: 1993, Bad Blood: The Tuskegee Syphilis Experiment, New and Expanded Edition, New York: Free Press.

Kovac, J.: 2013a, 'Reverence and Ethics in Science', *Science and Engineering Ethics*, **19**, 745-56.

Kovac, J.: 2013b, 'Science, Ethics and War: A Pacifist's Perspective', *Science and Engineering Ethics*, **19**, 449-60.

Kovac, J.: 2015, 'Ethics in Science: The Unique Consequences of Chemistry', *Accountability in Research*, **22**, 312-29.

Lubalin, J.S. & Matheson, J.L.: 1999, 'The Fallout: What Happens to Whistleblowers and Those Accused But Exonerated of Scientific Miscoduct', *Science and Engineering Ethics*, **5**, 229-50.

Mehlich, J.; Moser, F.; Van Tiggelen, B.; Campanella, L. & Hopf, H.: 2017 'The Ethical and Social Dimensions of Chemistry: Reflections, Considerations, and Clarifications', *Chemistry A European Journal*, **23**, 1210-18.

Metcalf, J.: 2018. 'Ethics Codes: History, Context, and Challenges', Council for Big Data, Ethics, and Society [available online: https://bdes.datasociety.net/council-output/ethics-codes-history-context-and-challenges/, accessed 14 May 2018].

NSPE: 2017, 'Code of Ethics' [available online: https://www.nspe.org/resources/ethics/code-ethics, accessed 15 December 2017].

OPCW: 2015a, 'Compilation of Codes of Ethics and Conduct' [available online: https://www.opcw.org/fileadmin/OPCW/SAB/en/2015_Compilation_of_Chemistry_Codes.pdf, accessed 15 December 2017].

OPCW: 2015b, 'The Hague Ethical Guidelines' [available online: https://www.opcw.org/special-sections/science-technology/the-hague-ethical-guidelines/, accessed 15 December 2017].

Pesticide Action Network: 2006, 'Chlorpyrifos Factsheet' [available online: http://www.panna.org/sites/default/files/ChlorpyrifosFactsheet2006.pdf, accessed 15 December 2017].

Reese, K.M. (ed.): 1976, *A Century of Chemistry*, Washington, DC: American Chemical Society.

RSC: 2013, 'Professional Practice and Code of Conduct' [available online: http://www.rsc.org/globalassets/03-membership-community/join-us/membership-regulations/code-of-conduct.pdf, accessed 15 December 2017].

Schummer, J.; Bensaude-Vincent, B. & Van Tigglen, B (eds.): 2007, *The Public Image of Chemistry*, Singapore: World Scientific.

Slater, D.: 2012, 'How Dangerous is Your Couch?', *New York Times Magazine*, September 6 [available online: http://www.nytimes.com/2012/09/09/magazine/arlene-blums-crusade-against-household-toxins.html, accessed 15 December 2017].

Steneck, N.H.: 2004, *Introduction to the Responsible Conduct of Research*, Washington, DC: Office of Research Integrity.

UKRIO: 2009, 'Code of Practice for Research: Promoting good practice and preventing misconduct' [available online: http://ukrio.org/wp-content/uploads/UKRIO-Code-of-Practice-for-Research.pdf, accessed 15 December 2017].

Chapter 19

Ethics and Chemical Regulation: The Case of REACH

Jean-Pierre Llored

Abstract: In this chapter I look at the way the Precautionary Principle shaped the rise of REACH, and how it is still exerting a directive and dynamic influence on the development and the evolution of this new European chemical regulation. In it I also query the extent to which REACH actually implements such an ethical principle, and I outline significant challenges that remain to be addressed for making ethical decision-making stronger in such regulation. The case of REACH is thus used in order to reflect upon the relationships between ethical considerations and chemistry.

1. Introduction

The use of chemicals has been increasing, mainly due to the economic development in various sectors including industry, agriculture, food processing and distribution, drugs, cosmetics, and transport (Pollak 2011). As a consequence, people are exposed to a large number of chemicals of natural and human-made origins. The trouble is that chemicals may have unintended and harmful effects both on human health and the environment. They may have immediate, acute effects, as well as chronic long term consequences.[1] Chronic, low-level exposure to various chemicals may result in a number of adverse outcomes, including damage to the nervous and immune systems, impairment of reproductive function and development, cancers, and organ-specific damages. In addition,

[1] Acute toxicity results from a single, short exposure; the effects usually appear briefly and are often reversible. Chronic toxicity results from repeated exposure over a long period of time.

environmental emissions arising from the use of chemicals vary in impact, depending on both the properties of the chemical at stake and the purposes and methods of its use.

The hazards associated with a chemical depend on its nature, the other chemicals with which it is mixed, and its relative proportion in the case of a mixture or a solution. Furthermore, the effects of exposure to a chemical are dependent on many factors. First, the amount of the particular chemical being present inside the body: the dose of a chemical that a person receives is dependent on the concentration of the chemical and on the frequency and duration of the exposure. Second, the type of exposure: the way the hazardous chemical enters the body determines how the material may travel through the body and affects organs or systems.[2] Third, the effect depends upon the physical and chemical interactions developed between the chemical and the body. Last but not least, the susceptibility of the individual receiving the dose: each person's body will react differently upon exposure – exposure to a hazardous material may affect one person more than others. In most situations, relatively safe chemicals may become toxic if the dose is high enough, and even potent, whereas highly toxic chemicals may be used safely if exposure is kept low enough. All chemicals are thus toxic at some dose and may produce harm if the exposure is sufficient, but all chemicals produce their harm under prescribed conditions of dose or use. The actual health risk of a chemical is thus a function of the toxicity of the chemical, and of the actual dose (or exposure) someone has to that chemical.

The 'regulation' of chemicals is the legislative intent of a variety of national laws and international initiatives such as agreements, strategies, or conventions. These international initiatives define the policy of further regulations to be implemented locally as well as exposure or emission limits. For instance, the Strategic Approach to International Chemicals Management (SAICM)[3] was adopted at the International Conference on Chemicals Management, which took place in February 2006 in Dubai,

---

2 Local injuries involve the area of the body in contact with the hazardous material; they are typically caused by reactive or corrosive chemicals, such as strong acids, alkalis, or oxidizing agents. Systemic injuries involve tissues or organs unrelated to, or removed from, the contact site when toxins have been transported through the bloodstream.

3 A presentation of SAICM is available at http://www.saicm.org.

gathering state governments and intergovernmental and non-governmental organizations. This regulation defines a policy framework which covers risk assessments of chemicals and harmonized labeling up to tackling obsolete and stockpiled products. Another recent international chemical regulation, namely the Globally Harmonized System of Classification and Labelling of Chemicals (GHS),[4] proposes harmonized hazard communication elements, including labels and safety data sheets. It was adopted by the United Nations Economic Commission for Europe in 2002. This system aims to ensure a better protection of human health and the environment during the handling of chemicals, including their transportation and use. The classification of chemicals is based on their hazard. While governments, regional institutions, and international organizations are the primary audiences for the GHS, it also contains sufficient context and guidance for those in industry who will ultimately implement the requirements which have been adopted.

Concerning regional regulations, I could refer, for instance, to the US Toxic Substances Control Act (TSCA) of 1976 which shaped the mandate of the Environmental Protection Agency (EPA) to protect the public from unreasonable risk of injury to health or the environment by regulating the manufacture and sale of chemicals.[5] This act does not address waste produced as byproducts of manufacturing, as did the Clean Water and Air Acts before it. Instead, TSCA attempts to exert direct government control over which types of chemicals could, or could not, be used in actual use and production. This regulation does not require chemical companies to perform risk assessments on new chemicals. By contrast, the new European chemical regulation, namely the 'REACH Regulation' – an acronym that I will clarify below – requires chemical companies that produce at or above the level of 1 metric ton per year to conduct a risk assessment, and demands the conduct of chemical safety assessments for *all* the chemicals produced from the companies that produce more than 10 tons or more per year. Requiring companies to propose risk assessments corresponds to an *unprecedented* type of policy for chemical

---

4 Online available at http://www.unece.org/trans/danger/publi/ghs/ghs_rev05/05files_e.html; accessed 22 August 2017.

5 A presentation of TSCA is available online at http://www.epa.gov/chemicals-under-tsca, accessed 22 August 2017.

regulation. This new policy has only been developed within the framework of the European Union hitherto, and reaches its culmination within the context of REACH. Indeed, the EU is the sole large region where something called the 'Precautionary Principle' (PP) is implemented and controlled by case law.

In this chapter, I look into the way this ethical principle shaped the rise of REACH, and how it is still exerting a directive and dynamic influence on the development and the evolution of this new European chemical regulation. I also query the extent to which REACH *actually* implements such an ethical principle. In brief, the REACH legislation is the case study analyzed here in order to reflect upon the relationships between ethical considerations and chemical regulation, and especially how ethics is involved and plays a role in this kind of regulation.

To do so, I first scrutinize chemical hazards and risks and highlight the ethical problems they raise. I then introduce the different types of environmental chemical policies that have existed so far while identifying their basic assumptions concerning nature, science, and the relationships between humankind and nature. Third, I narrate the historical developments of: (1) the integration of the Precautionary Principle in European environmental law; and (2) its consequences not only over the European chemical regulation implemented before REACH, but also over the emergence of the REACH regulation itself. Fourth, I introduce the basic characteristics of REACH before discussing, in the last part of this chapter, how REACH implements and differs from the Precautionary Principle. In doing so, I shall outline significant challenges that remain to be addressed for making ethical decision-making stronger in such regulation.

2. Ethical Implications of Chemical Hazards and Risks

Chemical hazards and risks are central when addressing the ethical issues in chemistry. The word 'risk' nevertheless encompasses different meanings depending on the context of its use. Most definitions of risk are based on a probability calculation. According to decision theory – the theory of rational decision making – a decision is said to be made 'under risk' if the relevant probabilities are available and 'under uncertainty' if they are unavailable or only partially available, *i.e.*, expressed with probability intervals, as it is the case, for instance, when a meteorologist says: ‘the probability of rain today is between 0.5 and 0.7’ (Peterson 2009). The EU regulation of food safety defines risk as a function of the probability of an adverse health effect and weighted by the severity of that effect (European Council 2002). Risk can also be defined by the stochastic nature of the consequences of an action,[6] with the stochastic range being expressed around a central value, whenever a random variable intervenes. By contrast, the ordinary meaning is ‘danger’, ‘possible damage’ or ‘threat of disaster’, and comprises no explicit consideration of probabilities.

In this chapter, I shall use the word ‘risk’ in its legal sense, the one that corresponds to the legal definition proposed by the European Agency for Safety and Health at Work in the framework of the Directive 98/24/EC of 7 April 1998 on the protection of the health and safety of workers from the risks related to chemical agents. The objective of this Directive is to lay down minimum requirements for the protection of workers from risk to their safety and health arising, or likely to arise, from the effects of chemical agents that are present at the workplace or as a result of any work activity involving ‘chemical agents’. The article 2 of this Directive provides many definitions closely related to REACH Regulation:[7]

> ‘Chemical agent’ means any chemical element or compound, on its own or admixed, as it occurs in the natural state or as produced, used or released, including release as waste, by any work activity, whether or not produced intentionally and whether or not placed on the market.
>
> ‘Hazard’ means the intrinsic property of a chemical agent with the potential to cause harm.
>
> ‘Risk’ means the likelihood that the potential for harm will be attained under the conditions of use and/or exposure.

According to this Directive, hazard is thus defined as an ‘intrinsic’ property of a chemical agent whereas risk is a relational notion related to the

---

6 A stochastic event or system is one that is unpredictable due to the influence of a random variable.

7 The text of REACH is available online at http://ec.europa.eu/environment/chemicals/reach/reach_en.htm, accessed 22 August 2017.

probability that the potential for harm will be attained *under the conditions of use and/or exposure*. In fact, risk is the *probability* that the exposure level is higher than the minimum level of toxicity.

In practice, most chemicals have not been duly tested for their environmental and health impacts beyond light toxicity tests. The 'unknown' is not only a feature of new molecules or materials recently invented by chemistry; it is the most common feature of chemicals that have been disseminated for several decades in the atmosphere, in water, and in the soil. As a matter of fact, chemical bodies are context-sensitive, and the ways in which they act upon the world always depend on the physical and chemical context in which they are. It is impossible for a chemist to give an *exhaustive* description of the chemical character and future behavior of chemicals. No one could have been able to predict that chlorofluorocarbons (CFCs),[8] such as chlorodifluoromethane, could cause ozone depletion from the basic knowledge of its composition, structure, and from what chemists already know about the chemical reactions involving this kind of substance.

I do not mean, of course, that it is impossible for chemists to describe chemical bodies using their composition and their internal structure only, as if they were in isolation. This strategy is often a very efficient way to produce new chemicals, or to explain a certain type of reactivity within a chemical reaction. I know how hard toxicologists and ecotoxicologists are working for setting up new methods to determine the toxicity of a mixture of chemicals. I am also aware that risk assessment of chemicals is a complex task which involves multidisciplinary teams, and that any model, as well as any biological test, is limited by its applicability domain, *i.e.*, can only answer the questions that have been asked. But the composition and the internal structure of a substance can change depending on what surrounds it, as it has been known for a long time by chemists in the case, for instance, of acid or oxidative properties, and as it is more than ever the case in nanochemistry. Chemical molecules and materials also need to be defined by their selective capacity to interact

[8] Chlorofluorocarbon (CFC) is a group of compounds that contain only carbon, chlorine, and fluorine. The most common representative is dichlorodifluoromethane (Freon-12). Many CFCs have been widely used as refrigerants, propellants (in aerosol applications), and solvents.

with one another within a precise context. As a consequence, the knowledge we have about them can only be open and provisional (Llored & Sarrade 2016). The philosopher of chemistry Joachim Schummer draws our attention to the fact that science does not create knowledge only; it also transforms the world and produces 'the unknown':

> [w]ith every production of a new substance, the scope of non-knowledge increases tremendously, by the number of undetermined properties of the new substance as well as by all chemical reactivities of the already existing substances with the new one. [Schummer 2001, p. 110]

Following this line of reasoning, Godard adds:

> New substances introduce new properties that are difficult to anticipate, with possible consequences that are difficult to fully comprehend [...]. Due to the massive number of new chemical substances that are being introduced into ecosystems, this creative process entails an increasing unpredictability of environmental changes. The creation of a new substance and putting it on the market generate a new unpredictable potential for harming the environment and public health, increasing the difficulties associated with the control of these harms. This is a legitimate source of concern: *Chemistry is a major factor in making our world unpredictable*. [Godard 2013, p. 87, my emphasis]

As a discipline, *chemistry is a permanent source of new unknowns*, which justifies our paying special attention to the risks it potentially raises for us, and for other forms of life. The chemists Paul Anastas and Tracy Williamson assert:

> With knowledge comes the burden of responsibility. Chemists do not have the luxury of ignorance and cannot turn a blind eye to the effects of the science in which we are engaged. Because we are able to develop new chemistries that are more benign, we are obligated to do so. [Anastas & Williamson 1996, p. 1]

Anastas and Williamson put emphasis on the ethical commitment to which chemists cannot but subscribe. In the same vein, Schummer adds:

> it is very likely that any new substance can be used to cause harm. Thus, we may expect that our chemist, while being unable to foresee the particular case of harm, knows well about the high probability of

> possible harm. Therefore, the knowledge argument turns to the contrary and does not help to excuse our chemist. [Schummer 2001, p. 112]

The incompleteness of our knowledge about chemicals and the actual possibility for chemists to produce a safer and greener chemistry thus engage the whole community of chemists from an ethical standpoint. Before studying the REACH Regulation, I would like to introduce the three kinds of chemical policies which have been proposed so far.

3. Kinds of Environmental Policies and their Basic Assumptions

For over a century industrial societies have considered nature to be both a rich reserve of resources and a dump for the waste produced by resource exploitation. Nature would always be able to eliminate production and consumption residues. In other words, the possibility of eventual regeneration has always been assumed. However, as Kleindorfer & Kunreuther (1987), Ruthenberg (2016), Fjelland (2016), and Eckerman & Børsen (2018), to quote but a few authors, have pointed it out, several human tragedies and environmental disasters – as for instance in the late 1950s and early 1960s in West Germany with thalidomide,[9] in Seveso (10 July 1976),[10] or in Bhopal (3 December 1984)[11] – have illustrated that nature, and human health, cannot be represented as 'something' able to endure unbridled developments.

In response to this situation, three main policies for environmental protection have been successively proposed. Each of them depends on particular representations of what nature and science are; representations

---

9 Thalidomide was originally introduced as a non-barbiturate hypnotic, but withdrawn from the market due to teratogenic effects. For more details, see Ruthenberg 2016.

10 Strange skin diseases suddenly appeared on people's faces one Saturday in 1976. Nobody knew at first that the nearby chemical plant had exploded and released a cloud of 2,3,7,8-tetrachlorodibenzo-p-dioxin (TCDD) (Kleindorfer & Kunreuther 1987, p. 5). The accident resulted in the creation of the European Community's Seveso Directive, called 'Seveso I', in June 1982, which was followed by the directive 'Seveso II', instructing common guidelines for the chemical industry and giving new standards for safety and public insight.

11 Union Carbide (India): between 3,000 and 10,000 people were killed and between 100,000 and 200,000 disabled, because of a gas leakage of methyl isocyanate used as an intermediate for the production of insecticides and herbicides. For more details, see Eckerman & Børsen 2018.

which, in turn, underpin the ways in which humankind *ought to* or *must* behave with respect to nature (de Sadeler 2002). In fact, the process which leads from one kind of policy to another could be termed 'superposition' rather than 'succession', since the stabilization of a new kind of policy does not necessarily imply the elimination of previous ones.

3.1 Curative environmental policy

According to the 'curative kind' of policy, nature can no longer cure itself: it should be helped to repair the damage inflicted upon it. Following this perspective, what has been polluted *can* be cleaned up; what has been destroyed *can* be restored; and what cannot be safeguarded *can* be replaced, be it by natural processes or through human intervention. Everything occurs *as if* the situation were totally under human control. Moreover, and for reasons of equity and feasibility, public authorities ought to apportion the economic cost of such intervention by requiring polluters to pay the cost of cleaning up after pollution and destruction. The 'polluter-pays principle' thus creates the economic conditions for reparation (de Sadeler 2002). This policy nevertheless implies *a posteriori responses* to a social problem: knowing that assigning clean-up costs to liable parties can become highly problematic, especially when environmental effects become too diffuse or reparation proves too costly, as is often the case with chemicals.

3.2. Preventive environmental policy

The situation changes from a curative to a 'preventive type' of policy as soon as damages become *irreversible*, such that reparation is simply impossible. This schema rests on the idea that science can determine with *certainty and precision* what level of damage will compromise the restoration of ecosystems and their species. But we can only prevent what we understand; it is difficult to prevent a problem that is not understood, and even more difficult to prevent the unknown, which is very often caused by the release of chemicals into the environment as we have previously pointed it out. As a consequence, a preventive policy assumes that science is able to find a precise solution to any kind of problem (de Sadeler

2002). This is the reason why prevention usually addresses *known risks* for which a full risk assessment, leading to quantitative estimates of exposure of various groups and estimates of expected damage, can be delivered. As such, prevention is highly demanding in terms of knowledge and information but, being based on rather precise estimates, offers a rational basis for policies, for coverage by social security institutions and for the business of insurance companies (Godard 2013).

3.3 Anticipatory environmental policy

The situation changes again with the emergence of a new type of risk, *i.e.*, *potential risks* or *hypothetical hazards or threats*. The destructive effects of chemical substances, as for instance PCBs[12] and DDT[13] on life and CFCs on the ozone layer, could not be understood before these substances had been produced and released into the environment. *In this case, the relevant question is not about how to prevent assessable, calculable, and certain risks, but rather about how to anticipate risks suggested by uncertainty, plausibility, and probability.* Uncertainties can be related to different factors, including: (1) the geographical scope of the potential for damage (*e.g.* chemical pollutants in the marine environment); (2) the temporal duration of chemicals (low levels of chemical contaminants exert impacts which are difficult to detect in short-term laboratory tests, but which can show up in the next generation because of the persistence of chemicals in the environment); (3) the duration of its manifestation (*e.g.* the impact of greenhouse gases on climate); and (4) the reversibility or irreversibility of this manifestation (*e.g.* the ozone layer depletion). This new kind of risk is characterized by the difficulty of identifying and quantifying causal links between a multitude of potential hazards (such as various types of emissions) and specific adverse effects (for example desertification). At this stage, scientists are largely

---

[12] PCBs (polychlorinated biphenyls) are human-made chemicals first produced in the late 1920s. They were used as cooling fluids in electrical equipment and machinery because of their durability and resistance to fire.

[13] DDT (dichlorodiphenyltrichloroethane) was developed as an insecticide in the 1940s, and was widely used during World War II to combat insect-borne diseases. For more details, see Børsen & Nielsen (2017).

dependent upon analogies or computer simulations to assess suspected risks (Godard *et al.* 2002, de Sadeler 2002).

This is the context in which a new 'anticipatory environmental policy', based on the Precautionary Principle, is emerging. Such a principle is about collective, potential, uncertain, and hypothetical threats. Not only has damage not yet occurred, but there is no irrefutable proof that it will occur at all. Notwithstanding this situation, a key-idea is that *uncertainty should no longer delay the adoption of measures intended to anticipate environmental degradation*. Precaution serves to prevent delay under the pretext that the 'true nature' of risks is not known and will be *fully* determined later. Conversely, it serves to brake hasty action, by urging delay in executing projects whose risks are considered to be insufficiently well-identified. Reaching this goal requires a change in our perception of time: today's choices must also reflect a still uncertain future (de Sadeler 2002). Recourse to the Precautionary Principle is therefore justified by considering the long term effects despite and beyond the presence of uncertainty. The REACH regulation, as we shall see, belongs to this third category. This type of policy is new in the context of chemical regulation, and seems to better address the specific situation of chemical hazards related to the 'unknown' implied by the release of chemicals into the environment. As Godard (2013, p. 87) states: "There is no better justification for submitting products derived from innovation in the field of chemistry to rigorous procedures of public control, and to place these procedures under the flag of the Precautionary Principle".

3.4 Precaution versus phronēsis

Notwithstanding the fact that this kind of policy is new, precaution and the PP are often considered to be just a contemporary form of prudence, synonymous with *phronēsis* in Aristotle's *Nicomachean Ethics* (Bourg & Schlegel 2001, Andorno 2004). But that is not the case. I claim that understanding their differences enables us to understand the specific ethical situation of our time. In his *Nicomachean Ethics*, and especially in the Sixth Book, Aristotle first deals with the knowledge of things whose originative causes are invariable – the principles of scientific knowledge, or *epistēmē* –, and focuses his attention on the fact that invariable causes can be replicated under similar circumstances, *i.e.*, satisfy the requirements of scientific stability and universality. He then points out how the knowledge that guides art and action differs from *epistēmē*. To do so, he refers to situations within which human beings have to make a decision when the causes are not stable and universal, but, are, by contrast, context-dependent and never fully known – as is typically the case with chemicals if we draw a parallel with our present study. Aristotle calls *phronēsis* the special type of wisdom relevant to practical decision of this kind (Dunne 1993, Birkholm 2016). This wisdom requires an ability to discern *how*, *when* – the opportune moment to act, *i.e.*, the *kairos* in Aristotle's terminology –, and *why* one may act despite the inderterminacy of the situation. *Phronēsis* is thus related to decision-making and action in cases of inderterminacy and uncertainty.

To better understand how *phronēsis* differs from precaution, we have to bear in mind the additional distinction drawn by Aristotle between *poiésis* and *praxis*. *Poiésis* encompasses art, technology, and the activity of production in the broadest sense of the term. It is related to the means we use in order to satisfy our needs and desires, independently of any moral reflection about the possible *bad* consequences that this use of means may have upon other people. To make this idea more concrete in the domain of chemistry, *poiésis* could be related to the production of chemicals in order to satisfy our need for transportation, independently of the consideration of the bad health impact of gasoline. That is the reason why, according to Aristotle, *poiésis* should be complemented with *praxis,* which is about the capacity we have, as human beings, to explore *with caution* not only ourselves, but also the city we live in. *Praxis* is thus related to political action within a particular community, and *phronēsis* means to take care both of ourselves *and* the *polis* – the city state in ancient Greece.

By contrast, Precaution and the PP are not about ourselves and the city only: they are related to decision-making and action in order to take care of the Earth and Humankind understood as a universal interrelated whole. They are not only about us or our relatives, or about the cities or countries to which we belong. They are also about future generations, and the right they have to live in good conditions. This idea is well captured by Jonas' famous sentence (1984, p. 11): "Act so that the effects of your action are compatible with the permanence of genuine human life", and is clearly related to the decision we have to make considering the long term consequences of our actions upon the Earth. The objects both of precaution and the PP are humankind – present and future –, other forms of life and future ecosystems, and the Earth. They are global, and not local only. Following this line of reasoning, the aim of precaution, as we shall see in the EU context, is related to the defense of a sustainable development thanks to which humankind both *preserves nature and creates 'new natures'*, almost in the ancient sense of birth and growth of a being – *phusis* and *perì phuseôs* in ancient Greek.

Precaution and the PP involve collective agents such as States, institutions, and 'the public', and call for a *deliberative form of democracy* in which all stakeholders have *the same right* to take part in the decision-making about science and technology. Deliberations on science and technology involve a whole set of ethical values concerning humankind, the Earth, ecosystems, democracy, honesty in science, fair trade, animal suffering, human well-being, which co-exists with a pragmatic representation of science according to which uncertainty does not mean the defeat of science and truth, but does mean the opportunity for the sciences and technologies to articulate different kinds of expertise and interests. It is from within this growing network of ethical values that the third kind of policy has emerged. And it is also in this context that the PP has been integrated into the European legislation. Having grasped some crucial differences between prudence and precaution, I can now study how and why the gradual integration first of precautionary measures and then of the PP into the European legislation fosters the transition from the preventive to the anticipatory type of environmental policy in the EU context, and why the PP has become a crucial principle around which REACH revolves.

4. The Precautionary Principle in European Environmental Law and European Chemical Regulation before REACH

In 1972, The United Nations Conference on the Human Environment held in Stockholm, Sweden, defended the 'ALARA Principle' – the acronym means 'As Low As Reasonably Achievable'. This radiation safety principle is based on the minimization of radiation doses and aims to limit the release of radioactive materials into the environment.

In 1976, the government of West Germany published the paper 'Vorsorgende Umweltpolitik' with the view to framing German Environmental Policy in the long run. This paper paved the way for the anticipatory type of environmental policy. It referred to land degradation caused by acid rain, and called for *precautionary actions* for protecting and taking care of natural resources. It first defended the idea that *waiting for scientific certainty before undertaking preventive action is all but acceptable*. It also pleaded for a long-term, continuous, and adaptive approach to environmental measures. The basic idea was to remain opportunistic in the use of technological progress to drive an ecological modernization of industrial processes (von Moltke 1987).

In the 1980s, North Sea Conferences called for 'precautionary approaches'. The Precautionary *Principle* appeared, for the first time, in the field of marine pollution, for instance within the 1992 Helsinki Conventions on the Protection and Use of Transboundary Watercourses and International Lakes, and in another dedicated to the Protection of the Marine Environment of the Baltic Sea Area. In line with the German *Vorsorgeprinzip*, the contracting Parties of the Helsinki Conventions aimed to take preventive measures when there is reason to assume that a substance or energy introduced, directly or indirectly, into the marine environment may cause harm to human health and marine ecosystems, even when there is no conclusive evidence of a causal relationship between inputs and their alleged effects (article 3(2)). North Sea Conferences encompass preventive and anticipatory types of environmental measures, and constitute a step towards the implementation of the new anticipatory type of policy. In the same vein, and still in 1992, The Rio de Janeiro Earth Summit gave the Precautionary Principle a worldwide public audience while the writers of the Maastricht Treaty focused their attention both on this principle and on preventive actions.

In France, the 1995 Barnier Law for the strengthening of environmental protection defined the Precautionary Principle in the following terms:[14]

> Absence of certainty, taking account of current scientific and technical knowledge, should not lead to postponing the adoption of effective and proportionate measures aimed at averting the risk of serious and irreversible damage to the environment, at an economically acceptable cost.

This law, which clearly belongs to the anticipatory environmental policy, focused on the concepts of proportionality, coherence, and regular revision of measures, as well as on the need for public authorities to organize an independent, competent, multi-disciplinary, transparent, and adversarial expertise (Godard 2013).

In May 1998, the European Court of Justice issued a judgment on the 'Mad Cow Disease' case between the UK government and the Commission. The judgment stated that the authorities were right in taking health measures without waiting to have full scientific certainty about causal links and the extent of damage. This decision gave the Precautionary Principle an autonomous *legal force* in an area different from that of the environment, *i.e.*, food and health safety (European Court of Justice 1998). In this respect, it epitomized an 'anticipatory turn', which includes both environmental care and sanitary safety.

Two main texts have formalized and clarified this 'anticipatory turn', *i.e.*, the way the Precautionary Principle is understood and used at the European Union level: (1) the Communication presented by the Commission in February 2000, and (2) the Resolution adopted by European heads of state at the Nice Summit in December 2000.[15] This Resolution clearly states that *measures taken on the basis of the Precautionary Principle should be continuously re-examined in the light of the development of scientific knowledge*. To this end, follow-up of the effects of decisions should be implemented and further research should be carried out *to reduce the level of uncertainty*. According to Rogers (2011), five

---

[14] Law 95-101, article 1, my translation [available online at https://www.legifrance.gouv.fr/eli/loi/1995/2/2/ENVX9400049L/jo/texte, accessed August 23 2017].

[15] European Commission 2000, European Council 2000.

crucial points have to come into focus in order to understand how the EU gives sense to precautionary actions and implements the anticipatory environmental policy:

(1) The proportionality to the chosen level of protection;
(2) The non-discrimination of the procedure, in particular in regard to imported products;
(3) The consistency with similar measures previously taken for known risks, but taking account of scientific progress and change of concerns in the society;
(4) The choice of measures based on the consideration of the potential benefits and costs or various possible actions, including the no-action option; and
(5) The periodic review of measures in the light of new scientific results.

In addition, Klinke *et al.* (2006) emphasize three additional points in order to understand how the Precautionary Principle is understood and used in Europe:

(6) The principle is implemented within a *sustainable development perspective* in line with the Brundtland Report (World Commission on Environment and Development 1987);[16]
(7) Public authorities are responsible for organizing risk assessment, which should be conducted *independently* and *transparently* on a *multidisciplinary basis*; and
(8) Civil society should be implicated and particular attention should be paid to *consulting all interested parties at the earliest possible stage*.

We have now understood how the PP has been integrated into the EU legislation as an active open-ended process of decision-making, how it has been defined in this cultural context, what it contains in terms of aims, commitments, and ethical values, and in what types of purpose it takes part. The situation is now clearer on how and why the gradual implementation of the PP has fostered the transition from the preventive type of environmental policy to the anticipatory one. I can now briefly

---

[16] A sustainable development is a development that meets the needs of the present without compromising the ability of future generations to meet their own needs. (World Commission on Environment and Development 1987).

describe the European Chemical Regulation that was in force prior to REACH, and the role played by the Precautionary Principle in the implementation of this later regulation.

Recognizing the risk posed by persistent organic pollutants (POPs) to human health and the environment, the 2001 Stockholm Convention on POPs laid down a precautionary approach as its main objective:[17]

> Mindful of the precautionary approach as set forth in Principle 15 of the Rio Declaration on Environment and Development, the objective of this Convention is to protect human health and the environment from persistent organic pollutants.

As a consequence, Precaution guided the listing procedure for new POPs. In addition, the 2001 London International Maritime Organization Convention on the Control of Harmful Anti-Fouling Systems on Ships, which prohibited the use of harmful organotins in anti-fouling paints used on ships, established a precautionary mechanism to prevent the potential future use of other harmful substances in anti-fouling systems (Heyvaert 1999). The sixth amendment to the Council Directive on the classification, packaging, and labeling of dangerous substances, which established an EU-wide notification procedure for substances introduced on the market since 1981, was intended to increase knowledge of the effects of substances and thereby to facilitate subsequent decision-making. Since 1982, producers and importers of new substances have been obliged to notify the competent national authority about, and provide full information on, any substance that has been introduced on the market. Moreover, the Seveso Directives already put the onus of continuously collecting and updating safety information on operators of dangerous industrial plants, leaving national public authorities with the role of assessing the performance of those private assessors (Fleurke & Somsen 2011).

Because procedures apply to new substances *only*, most chemicals have never been assessed in terms of their harmful effects on health and the environment (Bro-Rasmussen 1998). In order to fill the information gap concerning chemicals introduced on the market before 1981, the

---

[17] Article 1, p. 4 [online available at http://www.wipo.int/edocs/lexdocs/treaties/en/unep-pop/trt_unep_pop_2.pdf, accessed 22 August 2017].

Council Regulation envisaged a system of evaluation and control of the risks posed by *existing substances*: any community importer or producer of an existing substance in quantities exceeding 1,000 t/year must submit data on the ecotoxicity and environmental fate and pathways of that chemical to the Commission. In June 1999, the Council nevertheless stated that, because risk assessments had only been drafted for a very small number of existing substances pursuant to EC legislation, and none had been adopted, the current approach is unlikely to achieve an appropriate limitation of all risks posed by these substances to health and the environment (Winter 2000).

To remedy this situation, the European Commission adopted, on 13 February 2001, a *White Paper* setting out the strategy for a future Union Policy for Chemicals (Rogers 2003). The main objective of the new strategy was to ensure a high level of protection for human health and the environment in the light of the Precautionary Principle, while ensuring the efficient functioning of the internal market and stimulating innovation and competitiveness in the chemical industry (Santillo *et al.* 2000). This *White Paper* then led to the REACH Regulation.

5. REACH Basic Characteristics

Following the Chemicals White Paper agenda, the REACH regulation entered into force in spring 2007 and will be gradually implemented until 2018, when approximately 30.000 substances are expected to have been included in the whole procedure. REACH targets chemical substances which have not previously been covered by existing regulations. The acronym of this regulation introduces new obligations and procedures aiming at registering (R), evaluating (E) and authorizing (A) – or restricting the production of, and even forbidding – chemical substances (CH). According to its website:[18]

> REACH is a regulation of the European Union, adopted to improve the protection of human health and the environment from the risks that can be posed by chemicals, while enhancing the competitiveness of the EU

[18] Quoted from https://echa.europa.eu/regulations/reach/understanding-reach, accessed 24 August 2017.

chemicals industry. It also promotes alternative methods for the hazard assessment of substances in order to reduce the number of tests on animals.

Ethical and economic objectives are thus put forward and intertwined within this regulation; they follow the often called 'No data, no market' line.

Special attention is given to chemicals classified as carcinogenic, mutagenic or reprotoxic (CMR), and to POPs. A second central plank of REACH is the principle of substitution: if safer alternatives exist, certain dangerous substances – the 'Substances of Very High Concern' (SVHC) – must be phased out. Whereas previously chemicals could only be banned if proven to be dangerous, REACH requires EU industry and importers to prove that each substance intended for the market is safe for human health and the environment. This is what is sometimes referred to as 'reversal of the burden of proof'. Those objectives are included in the different steps of the procedure.

5.1 Registration

Registration concerns chemicals (substances and products) intended for the market, provided that their level of production exceeds one ton per year and producer. Many specific products are not registered under REACH. For instance, chemicals used in biocides, agriculture, and cosmetics are excluded from REACH, because they are covered by other legislation (Heyvaert 1999, 2008). Quantities below the threshold are exempted from the registration requirements, as are substances used for research and development only.

The information requirements depend on, and vary greatly with, tonnage level. The standard information that has to be submitted by each registrant consists in a 'technical dossier' made up of information pertaining to the identity, classification, intended use(s), produced or imported quantities, physical properties, and toxicological and ecotoxicological information of the substances. For substances produced or imported in quantities of more than 10 tons per year and producer, proponents should present a chemical safety report giving the results of toxicity tests and *defining appropriate management measures apt to guarantee a safe use*. For persistent, bio-accumulating and toxic (PBT) characteristics of substances and products, an exposure assessment and a risk characterization is required.

Registration is based on the 'one substance, one registration' principle. This means that manufacturers and importers of the same substance have to submit their registration *jointly*. The requirement to share information is a fundamental aspect of REACH. In so doing, registrants of the same substance can *reduce registration costs* and *avoid unnecessary testing, especially on vertebrate animals*. There are two mechanisms for data sharing: (1) substance information exchange forums used for existing substances that have been pre-registered; and (2) inquiries for new substances or existing ones that have not been pre-registered.

5.2 Evaluation

Evaluation is conducted by member states according to guidelines and criteria elaborated by the new European Chemical Agency (ECHA).[19] Evaluation comprises three different steps:

(1) *The compliance checking* during which ECHA examines any registration dossier to verify if the information submitted by registrants is compliant with the legal requirements.
(2) *The report evaluation* during which ECHA assesses the registrants' proposals concerning the animal tests they envisage in order to prevent unnecessary animal testing. To do so, ECHA invites third parties – *public* consultation – to submit scientifically valid information or studies addressing the substance and hazard endpoints in question on its website.
(3) *The substance evaluation*, undertaken by national competent authorities, on substances that have been prioritized for potential regulatory action because of concerns about their hazardous properties. Member States evaluate certain substances to clarify whether their use poses a risk to human health or the environment and request further information from the registrants to verify the suspected concern, if necessary.

---

[19] See https://www.echa.europa.eu/regulations/reach, accessed 22 August 2017.

In the case of a request for further information, registrants can comment on it within 30 days and update their dossiers with information relevant to the concern or fill the data gaps detailed within the draft decision. The evaluating Member State or ECHA re-examines the comments and updated dossiers and may amend the draft decision accordingly. The other Member States and ECHA, in the case of substance evaluation, then review the updated draft decision and the registrants' comments, and have 30 days to propose further amendments. This iterative procedure thus involves all the actors at the EU scale in order to increase the chance of making a good decision. As an illustration of the procedure, I refer to a document produced by ECHA in 2014 on octocrilene.[20] This substance, which is mainly used in cosmetics and personal care products, was suspected of causing long lasting harmful effects to aquatic life. The document was published so as to demand further research according to a precise methodological framework.

The selection of the substances to be assessed is the result of a preliminary identification made by Member states and ECHA. The list of these substances is used to define the Community Rolling Action Plan (CoRAP), which indicates substances for evaluation by the Member States in the next three years, and which is updated each year in March. It was set up for the first time in 2012, and the first decisions were given in the fall of 2013.

5.3 Authorization

Authorization of chemicals is based on evaluation. Depending on the level of danger and the quantities involved, a special authorization is needed. A key regulatory outcome of evaluation could be the imposition of restrictions on the manufacture, supply, or use of a substance. Substance evaluation may also lead to a substance being added to the priority list for authorization, or to a proposal to change the classification and labeling. Dangerous products are banned, unless it is demonstrated that

[20] 'Decision on substance evaluation pursuant to article 46(1) of regulation 46(1) of regulation NO 1907' [online availabe at: https://echa.europa.eu/documents/10162/13628/corap_sev1_228-250-8_dec_final_public_2796_en.pdf/, accessed 22 August 2017].

the benefits for society are higher than the possible harm to public health and the environment, as long as no viable alternative exists. In that case authorization with restrictions in scope and time can be delivered. In the case of SVHC, for instance, the obligation rests with firms to furnish proof that risks posed by this category of substances are either 'adequately controlled', or to show a 'socio-economic need for their continued use'.

The final decision is made by the Commission, taking into account the opinion of ECHA. If risks are shown to be 'adequately controlled', then the Commission must grant authorization. If it is impossible to contain the risks fully, the Commission, involving the European Parliament and Council, may grant authorization, depending on the severity of the risk and the viability of alternatives.

The following Section discusses to what extent REACH implements the PP, and reflects upon how ethics and this regulation could be further and better related to one another by referring to the eight criteria of Section 4.

6. REACH and the Implementation of the Precautionary Principle

Article 1(3) of REACH clearly claims the application of the PP:[21]

> This Regulation is based on the principle that it is for manufacturers, importers and downstream users to ensure that they manufacture, place on the market or use such substances that do not adversely affect human health or the environment. Its provisions are underpinned by the precautionary principle.

Precautionary elements that indeed underpin REACH include: (a) providing a continuous supply of data; (b) risk assessments for substances used in certain volumes; (c) shifts in the burden of proof; (d) the requirement to search for safer alternatives; and (e) provision concerning review and monitoring. Those five points conform to the criteria 2, 3 and 5, and are compatible with the criteria 1 and 6 of Section 4.

Moreover, Annex I contains general guidelines for assessing substances and preparing chemical safety reports, and reflects the Precau-

---

[21] See http://www.reachonline.eu/REACH/EN/REACH_EN/article1.html.

tionary Principle by insisting that information gaps be acknowledged, and that – in addition to scientifically established risks – 'potential effects' of substances have to be taken into account (criterion 3). A crucial point which remains in compliance with a precautionary approach is that authorizations are subject both to periodic review and conditions (criterion 5), including monitoring (Fleurke & Somsen 2011). The authorization list is also provisional and can be amended over time. Furthermore, the Commission eventually has the power to suspend the authorization pending the review. Nevertheless, it remains to be seen how compliance with criterion 1, related to proportionate measures, could be improved further within the evaluation step, and especially within the iterative procedure involving all the actors at the EU scale.

In addition, innovation, in the form of substituting hazardous chemicals by safer alternatives, is a crucial element of precautionary thinking in Europe, even if such a shift is costly and introduces heavy discrimination between industries, depending on their importance and domains of activity. The substitution requirement for substances that falls under the authorization procedure follows the Precautionary Principle objectives: REACH clearly states that the aim of REACH is to encourage, and in certain cases, to ensure that substances, technologies, and engineering processes of high concern are eventually replaced by less dangerous ones if suitable economically and technically viable and sustainable alternatives are available (criteria 4, 6).

However, while the 'safer alternatives option' is good in theory, it turns out to be much more difficult than was anticipated 10 years ago. For instance, all the efficient substitutes for brominated flame retardants have a similar risk of impacts. Unfortunately, the same technical properties often incur the same hazards. Thus, if REACH encourages chemists to search for safer alternatives, which is consistent with, if not a consequence of, its precautionary nature, the current results have to be discussed in a more balanced way.

In addition, making companies, which have a vested interest in marketing their products, responsible for producing the relevant risk data is a source of tension with the Precautionary Principle requirement to set up an independent and transparent expertise (Godard 2013). In this respect, the current form of REACH does not implement the PP fully, and additional improvements are necessary ethically speaking, and especially concerning the *full* respect of criterion 7.

Furthermore, stakeholders are not systematically involved in the different stages of registration, evaluation, and authorization; and the opportunity given to the public to comment on risk assessment and socio-economic analyses are not precisely articulated within the decision-making process (Hansen *et al.* 2007). That is, once again, a source of tension with the PP, particularly with criterion 8, and calls for improving the debate between citizens and scientists, and for changing the way we consider science, public opinion, and their relationships. The issue may be, at least partly, resolved by education. Deliberative democracy is not a reality, but an aim that remains to be reached.

7. Concluding Remarks and Perspectives

REACH reinforces the weight of expertise in political decision-making. It also requires these two areas to interact so as to face the challenges posed by the compounded, serious, and irreversible risks they must assess and manage under *scientific uncertainty*. It can do so because of the PP integration into the EU law. Even if efforts still need to be made in order for REACH to better implement the PP, especially concerning compliance with criteria 1, 7 and 8, we should bear in mind that the 'Community Rolling Action Plan' is recent, and that the road will remain long and difficult: (1) to carry out the safer option, (2) to speed the whole procedure, and (3) to increase the number of substances taken into account by REACH. However, the regulation is a guide for action, mainly underpinned by the PP.

In addition, REACH enhances research in toxicology and ecotoxicology, and poses challenges to the existing frameworks for chemical safety evaluation. It calls for further studies about long-term effects and prolonged exposure at very low concentrations. To address this situation, researchers discuss alternative procedures using inherent characteristics of substances, and amplify factors of damage or determinants of scale in order to identify filters, thresholds, and screening conditions (Klinke *et al.* 2006). They also use Quantitative Structure-Activity Relationship models (QSAR models)[22] for assessing the bioaccumulation of chemicals.

The recourse to vertebrate animals for testing has gradually been reduced thanks to *in vitro* and *in silico* assays, which, despite the huge efforts that remain to be made in this area, is, in itself, a very positive outcome from an ethical standpoint. It enables us to avoid animal suffering and to protect life. Further new kinds of methods are emerging. The 'Integrated Testing Strategy' (IST) is a combination of test batteries covering relevant mechanistic steps and organized in a logical, hypothesis-driven decision scheme, which makes efficient use of generated data and gains comprehensive information for making decisions on hazards or risks (Ahlers *et al.* 2008). Another new approach, the 'Adverse Outcome Pathway' (AOP), links events at the molecular level to adverse outcomes at the biological level of organization relevant to risk assessment (Ankley *et al.* 2010). In brief, the way we rationalize ecotoxicological assays is thoroughly changing. REACH is among the sources of this change because of the PP plea for a long-term, continuous, and adaptive approach to environmental measures. It demands to act in a proportionate, balanced, and pragmatic way before any disaster occurs.

Although REACH requires the protection of the environment, it does not clearly define what we should protect. The concept of environment is too loose and multifarious, and does not refer to a 'fixed reality', but, by contrast, to a 'time-evolving reality'. *This is the reason why, beside technical and scientific research, ethical reflection is necessary, and why it is so urgent.* There is an important need for interaction between society and science, such that *all citizens*, including scientists, define what they want to protect and how much they are willing to pay for it – weighing the costs of both use and non-use of a chemical/product – while science and technology offer solutions. It is also a matter for all citizens to define political, cultural, and economic priorities and values (Berthoud 2014).

In parallel with the growing role of the precautionary and anticipatory type of policy in chemical regulation, we are witnessing the ongoing

---

22 QSAR models relate physico-chemical properties or theoretical molecular descriptors of chemicals to a biological activity of the chemicals. They quantify a supposed relationship between chemical structures and biological activity in a data-set of chemicals.

recasting of the operational, symbolic, conceptual, technical, and normative frameworks of chemistry fostered and carried out by green chemists (Llored & Sarrade 2016). Sustainability and the Principle of Precaution are becoming the tenets of chemical innovation, especially in Europe. The widespread reference to eco-conception, waste recycling, and life cycle analysis, in publications is a clear indicator of this ethical trend in contemporary chemistry. It nevertheless remains to be seen whether such an ethical challenge could turn out to be compatible with an economic system based on consumption, competition, and individualism. This is one of the reasons why future chemists should keep their mind strongly open to ethics.

References

Ahlers, J.; Stock, F. & Werschkun, B.: 2008, 'Integrated testing and intelligent assessment: New challenges under REACH', *Environmental Science and Pollution Research*, **15**, 565-72.

Anastas, P.T. & Williamson, T.C.: 1996, 'Green chemistry: an overview', in: P. Anastas & T.C. Williamson (eds.), *Green Chemistry*, Washington: American Chemical Society, pp. 1-17.

Andorno, R.: 2004, 'The Precautionary Principle: A New Legal Standard for a Technological Age', *Journal of International Biotechnology Law*, **1** (1), 11-19.

Ankley, G.T.; Bennet, R.S. & Erickson, R.J.: 2010, 'Adverse Outcome Patways: A Conceptual Framework to Support Ecotoxicology Research and Risk Assessment', *Environmental Toxicology and Chemistry*, **29** (3), 730-41.

Aristotle: 1999, *Nicomachean Ethics*, English translation by T. Irwin, 2nd edition, Indianapolis: Hackett.

Berthoud, B.: 2014, The Precautionary Principle in EU Risk Regulation: A Matter of Priorities, Hamburg: Anchor Academic Publishing.

Birkholm, K.: 2016, 'The Ethical Judgment: Chemical Psychotropics', *Hyle: International Journal for Philosophy of Chemistry,* **22**, 127-48 [also in this volume, chapter 15].

Bourg, D. & Schlegel, J.-L.: 2001, *Parer aux risques de demain,* Paris: Seuil.

Børsen, T. & Nielsen, S. N.: 2017, 'Applying an Ethical Judgment Model to the Case of DDT', *Hyle: International Journal for Philosophy of Chemistry*, **23**, 5-27 [also in this volume, chapter 9].

Bro-Rasmussen, F.: 1998, 'The Precautionary Principle and Science-Based Limits in Regulatory Toxicology', in: D.M. Pugh & J.V. Tarazona (eds.), *Regulation for*

Chemical Safety in Europe: Analysis, Comment and Criticism, Dordrecht: Kluwer, pp. 97-110.

De Sadeler, N.: 2002, Environmental Principles. From Political Slogans to Legal Rules, New York: Oxford University Press.

Dunne, J.: 1993, Back to the rough ground: 'phronesis' and 'techne' in modern philosophy and in Aristotle, Notre Dame: University of Notre Dame Press.

Eckerman, I. & Børsen, T.: 2018, 'Corporate and Governmental Responsibilities for Preventing Chemical Disasters: Lessons from Bhopal', *Hyle: International Journal for Philosophy of Chemistry*, **24**, 29-53 [also in this volume, chapter 5].

European Commission: 2000, *Communication on the Precautionary Principle*, Brussels: European Commission (2 February).

European Council: 2000, 'Resolution of the Council on the Precautionary Principle', in: *Conclusions of the Presidency, Nice European Council*, Annex III, Nice: European Council (7-9 December 2000).

European Council.: 2002, 'Regulation (EC) No. 178/2002 of the European Parliament and of the Council of 28 January 2002 laying down the general principles and requirements of food law, establishing the European Food Safety Authority and laying down procedures in matters of food safety', *Official Journal of the European Communities*, L **31**, 1-24.

European Court of Justice: 1998, 'Case C-180/96: Agriculture – Animal health – Emergency measures against bovine spongiform encephalopathy – Mad cow disease', Brussels: European Court of Justice (Judgment of the Court of 5 May 1998).

Fjelland, R.: 2016, 'When Laypeople are Right and Experts are Wrong: Lessons from Love Canal', *Hyle: International Journal for Philosophy of Chemistry,* **22**, 105-25 [also in this volume, chapter 8].

Fleurke, F.M. & Somsen, H.: 2011, 'Precautionary regulation of chemical risk: How REACH confronts the regulatory challenges of scale, uncertainty, complexity and innovation', *Common Market Law Review*, **48** (2), 357-93.

Godard, O.: 2013, 'The Precautionary Principle and Chemical Risks', in: J.-P. Llored (ed.), *The Philosophy of Chemistry: Practices, Methodologies and Concepts*, Newcastle upon Tyne: Cambridge Scholars Publishing, pp. 65-96.

Godard, O.; Henry, C.; Lagadec, P. & Michel-Kerjan, E.: 2002, *Traité des nouveaux risques*, Paris: Gallimard.

Hansen, S.F.; Carlsen, L. & Tickner, J.A.: 2007, 'Chemicals regulation and precaution: Does REACH really incorporate the Precautionary Principle', *Environmental Science and Policy*, **10**, 395-404.

Heyvaert, V.: 2008, 'The EU Chemicals Policy: Towards Inclusive Governance?', in: E. Vos (ed.), *European Risk Governance: Its Science, Its Inclusiveness and Its Effectiveness* (CONNEX Report Series, vol. 6) Mannheim: MZES, pp. 185-221.

Heyvaert, V.: 1999, *Coping with Uncertainty: The Regulation of Chemicals in the European Union*, PhD Dissertation, Florence (Italy): European University Institute.

Jonas, H.: 1984, *The Imperative of Responsibility: In Search of an Ethics for the Technological Age*, translation by H. Jonas & D. Herr, Chicago: University of Chicago Press [first published 1979 in German as *Das Prinzip Verantwortung*].

Kleindorfer, P.R. & Kunreuther, H.C. (eds.): 1987, *Insuring and Managing Hazardous Risks: From Seveso to Bhopal and Beyond*, Berlin and Heidelberg: Springer.

Klinke, A.; Dreyer, M.; Renn, O.; Stirling, A. & Van Zwanenberg, P.: 2006, 'Precautionary Risk Regulation in European Governance', *Journal of Risk Research*, **9** (4), 373-92.

Llored, J.-P. & Sarrade, S.: 2016, 'Connecting the Philosophy of Chemistry, Green Chemistry and Moral Philosophy', *Foundations of Chemistry*, **18** (2), 125-52.

Peterson, M.: 2009, *An Introduction to Decision Theory*, 1st edition, New York: Cambridge University Press.

Pollak, P.: 2011, *Fine Chemicals: The Industry and the Business*, 2nd edition, Hoboken, NJ: Wiley-Blackwell.

Rogers, M.D.: 2011, 'Risk management and the record of the Precautionary Principle in EU case law', *Journal of Risk Research*, **14** (4), 467-84.

Rogers, M.D.: 2003, 'Risk analysis under uncertainty, the Precautionary Principle, and the new EU chemicals strategy', *Regulatory Toxicology and Pharmacology*, **37**, 370-81.

Ruthenberg, K.: 2016, 'About the Futile Dream of an Entirely Riskless and Fully Effective Remedy: Thalidomide', *Hyle: International Journal for Philosophy of Chemistry*, **22**, 55-77 [also in this volume, chapter 7].

Santillo, D.; Johnston, P.; Singhofen, A. & Krautter, M.: 2000, 'Hazard Based Risk Assessment and Management', in: G. Winter (ed.), *Risk Assessment and Risk Management of Toxic Chemicals in the EC*, Baden-Baden: Nomos, pp. 98-114.

Schummer, J.: 2001, 'Ethics of Chemical Synthesis', *Hyle: International Journal for Philosophy of Chemistry*, **7** (2), 103-24.

von Moltke, K.: 1987, *The Vorsorgeprinzip in West German Environmental Policy*, London: Institute for Environmental Policy.

Winter, G. (ed.): 2000, Risk Assessment and Risk Management of Toxic Chemicals in the EC, Baden-Baden: Nomos.

World Commission on Environment and Development: 1987, *Our Common Future*, New York: Oxford University Press [available online at: http://www.un-documents.net/our-common-future.pdf; accessed 22 August 2017].

Biographical Notes on the Authors

Klavs Birkholm is founder and director of the Danish think-tank TeknoEtik and honorary professor (ass.) of technoanthropology at Aalborg University, Denmark. He was 8 years appointed member of the Danish Council of Ethics, heading a number of investigative reports on, among others, Brain Computer Interfaces, Human Optimization, and new technologies for prenatal screening. For many years a well-known media figure, Birkholm was editor and host at DR (National Danish Broadcasting Company), his programmes featuring issues of philosophy, social sciences, new technologies, etc. Before that, he was editor at the Danish daily newspaper Information. His most recent books are *Politiske Smuler* (Political Crumbs), 2018; *De skjulte Algoritmer* (The Hidden Algoritms), 2018 (with Niels Frølich); and *Efter Mennesket* (After Humans), 2015. For the moment, Birkholm is taking part in a research project on the professional and ethical consequences of digitalization of social work in Denmark. Email: .birkholm@soundsense.dk

Tom Børsen holds a Master's degree in chemistry, a BA minor in Philosophy, and a PhD in University Chemistry Education. He is currently an Associate Professor at Aalborg University, Copenhagen, Denmark, where he serves as director of the Study Board for Techno-Anthropology and Sustainable Design. He is a member of the research group in Techno-Anthropology and Participation (TAPAR) and a board member of Centre for Applied Ethics at Aalborg University, Centre for Bioscience and Techno-Anthropology, and the think tank Techno-Ethics. Tom's research clusters around four aspects of Responsible Research and Technological Innovation: Ethical Technology Assessment, Robust Technologies through Action Research, Teaching Responsibility in STEM, and Interdisciplinary Policy Advice on Technological Risks. He is the co-author of 28 journal articles and 24 book chapters. Tom Børsen is the co-editor of the special issues of *Hyle* on *Ethical Case Studies of Chemistry*, and the book *What is Techno-Anthropology?* Email: boersen@plan.aau.dk.

Stephen Contakes is Associate Professor of Chemistry at Westmont College in Santa Barbara, CA, U.S.A. At Westmont he teaches courses in inorganic, organic, analytical, physical, and materials chemistry, a liberal arts chemistry course for nonscience majors, a seminar in science and religion, and on occasion biochemistry. Prior to serving on the faculty at Westmont, he completed undergraduate degrees in chemistry and chemical engineering at Lehigh University, earned his doctorate at the University of Illinois at Urbana-Champaign based on work in organometallic chemistry, completed postdoctoral research in bioinorganic and biophysical chemistry at Caltech, and served on the faculty of Azusa Pacific University. His research involves the synthesis of molecular assemblies and light-driven electron transfer catalysts and the exploration of past and present connections between chemistry and the humanities. Email: scontakes@ westmont.edu.

Ingrid C. Eckerman, born 1942, graduated in 1972 from Karolinska Institute, Stockholm, Sweden, and specialized in family medicine in 1977. She was employed as a general practitioner (GP) by the Stockholm County Council during 1977-2008. From 1997, she worked as a 'population health doctor'. In 2001, she took a Master's degree in Public Health at the Nordic School of Public Health in Gothenburg. Since 2012 she has been the editor-in-chief of the journal *AllmänMedicin* (FamilyMedicine). In 1992, she joined the NGO Swedish Doctors for the Environment (LfM), and was its president 2009-2014. In 1994, she participated in the International Medical Commission on Bhopal (IMCB). Her book *The Bhopal Saga: Causes and Consequences of the World's Largest Industrial Disaster* was published in 2004. Her engagement in Bhopal ended in 2011 when she was not allowed entrance to India. Email: Ingrid@eckerman.nu.

Ragnar Fjelland is a philosopher and physicist, and is currently professor emeritus. He was from 1992 to 2014 professor of philosophy of science at the Centre for the Study of the Sciences and the Humanities and Department of Physics and Technology, University of Bergen, Norway. His topics of interest include the significance of technology for the acquisition of scientific knowledge, ethical problems raised by modern science and technology, and the challenge of environmental problems to science. He was director of the Center in the periods 1993-95 and 1999-2001, member of the National Research Ethics Committee for Natural Science and Technology (NENT) 1991-1999, and member of the Norwegian Technology Board 2008-2017. He has been a visiting scholar at Universität Konstanz, Universiti Malaya, Kuala Lumpur, University of Toronto, University of California at Berkeley, and Max Planck Institute for the History of Science, Berlin. Email: Ragnar.Fjelland@uib.no.

Alastair Iles is Associate Professor of environmental policy at the Department of Environmental Science, Policy & Management at the University of California, Berkeley, U.S.A. He holds LL.M and S.J.D degrees from Harvard University, and researches sustainability transitions, chemicals policies and politics, and sustainable food systems. Currently, Iles is co-editor-in-chief of the sustainability transitions domain of the *Elementa: Science of the Anthropocene* journal. Email: iles@berkeley.edu.

Claus Jacob is Professor of Bioorganic Chemistry at Saarland University in Germany. He studied Chemistry and Philosophy at Kaiserslautern, Leicester, Hagen, and Oxford, from where he obtained his DPhil in Bioinorganic Chemistry under the supervision of Allen Hill in 1997, supported by a stipend of the German Merit Foundation. After a postdoctoral stay as Feodor Lynen-Fellow of the Alexander von Humboldt Foundation at Harvard Medical School with Bert Vallee he joined the University of Exeter in 1999 as Lecturer / Senior Lecturer and Saarland University in 2005 as Junior Professor. Claus is interested in various aspects of Bioorganic Chemistry, from redox modulation and nutraceuticals to biovalorization and nanotechnology, currently taking part in European co-operations and projects and editing the journal Current *Nutraceuticals*. His interest in Philosophy of Chemistry centers around aspects of scientific explanation, chemical impurities, symbolism and language, and the role of communication in scientific discovery. Email: c.jacob@mx.uni-saarland.de.

Taylor Jashinsky is a curriculum architect and writer for the Hollis Company, in Austin, Texas, U.S.A. With a bachelor's degree in Social Sciences from Westmont College, Jashinsky's passion for global development, storytelling, and travel, led her to Southeast Asia, where she worked in cross-cultural communications and corporate public relations for three years. Jashinsky served as the development agency, World Concern's, Communication Liaison in Southeast Asia, and later worked as the Corporate Affairs Account Manager for Myanmar's top PR agency, ERA Myanmar. Under her mentorship, she led a campaign that won the SABRE Award for Cause-Related Marketing in 2019. The chapter republished in this volume was Jashinsky's first publication. Email: Taylorjashinsky@gmail.com.

Jeffrey Kovac is Professor of Chemistry and Director of College Scholars, Emeritus, at the University of Tennessee, Knoxville, U.S.A. He was educated at Reed College (B.A. 1970) and Yale University (M.Phil 1972, Ph.D. 1974). After two years as a postdoctoral research associate at MIT, he served on the the faculty of the University of Tennessee from 1976 until his retirement in 2018. An interdisciplinary scholar, Dr. Kovac has published extensively in statistical mechanics and thermodynamics, polymer science, computer simulation, chemical education, history, and philosophy. His most recent book is *The Ethical Chemist: Professionalism and Ethics in Science*, Second Edition (Oxford 2018). His honors and awards include Phi Beta Kappa, the Woodrow Wilson Fellowship, Fellow of the American Association of Science, as well as numerous awards from the University of Tennessee for teaching, advising, and service. Email: jkovac@utk.edu.

Jean-Pierre Llored is Associate Professor in humanities and social sciences at the Ecole Centrale Casablanca, Morocco, and is attached to the Department of Humanities and Social Sciences, Ecole Centrale-Supélec, Paris, France. He was first trained as a chemist (chemical engineer and Professor 'agrégé' of chemistry) before receiving a PhD from the Free University of Brussels, Belgium, and becoming a Doctor in epistemology and history of science at the Ecole Polytechnique, France. His papers deal with philosophy of chemistry, philosophy of technology (Artificial Intelligence), metametaphysics, mereology, and ethics. He edited the volume *Philosophy of Chemistry: Practices, Methodologies, and Concepts* (Cambridge Scholars Publishing, 2013) and *La mésologie, un nouveau paradigme pour l'anthropocène* (Hermann, 2018). Email: JeanPierre.LLORED @centrale-casablanca.ma.

Abigail Martin is a Research Fellow at the University of Sussex School of Business in the Science Research Policy Unit (SPRU), U.K. She is an interdisciplinary social scientist working at the intersection of STS, political ecology, and environmental governance. She received her PhD in Environmental, Science Policy and Management from University of California-Berkeley and researches struggles over the development of biofuels, distributed solar and natural gas to mitigate climate change. She also maintains an active research agenda on innovation for sustainability and the circular economy. As a visiting professor, she has taught a variety of courses on environmental health, justice and policy at the University of California Berkeley School of Public Health, San Francisco State University, the University of California Davis, and the Mills College Lokey School of Business and Public Policy. Email: abigailmartin880@gmail.com.

Christine Meisner Rosen is an Associate Professor at the Haas School of Business at the University of California, Berkeley, U.S.A. She is one of the founders of the Berkeley Center for Green Chemistry, where she currently serves as an Associate Director. An historian of business and the environment, she received her PhD in American History at Harvard University in 1980. In addition to teaching courses on and conducting research related to green chemistry, corporate environmental strategy and management, and energy innovation and sustainability, she teaches and writes about American business history. She is finishing a book on the history of America's early struggles with industrial pollution between 1840 and 1920 that uncovers and explores parallels between those early struggles and our current struggles to mitigate and protect society from the emission of greenhouse gases and industrial toxins. Email: crosen-1900@berkeley.edu.

Søren Nors Nielsen is an Associate Professor of Sustainable Biotechnology as well as technoanthropology, at the University of Aalborg in Copenhagen, Denmark. He holds a Master of Biology from the University of Copenhagen, a PhD on the modeling of evolution in Danish shallow lakes from Risø National Laboratory and National Environmental Research Institute, and a Dr. agregado in Ecology, University of Coimbra. He has been teaching in more than 60 courses in systems analysis, environmental modeling and management, ecosystem theory, cleaner production, industrial ecology, and ecotoxicology at various universities in Denmark and many other countries. Since 1989 he has been working on ecosystem evolution and development mainly from a thermodynamic point of view, expanding the approach to society. He has co-authored a large number of papers in peer-reviewed journals, book chapters, and books. Email: nors@plan.aau.dk.

Klaus Ruthenberg studied chemistry and philosophy in Berlin and Göttingen and holds degrees in both disciplines. After some years in research institutes and industry he became a professor of chemistry at Coburg University of Applied Sciences in Bavaria, Germany. In the 1990s, he was a co-founder of the International Society for Philosophy of Chemistry (ISPC). Besides teaching general, analytical, and physical chemistry, he is lecturing in the history and philosophy of science and in bio- and medicinal ethics. He was a visiting scientist at various institutes abroad, among them Leuven (Philosophy), Buenos Aires and Montevideo (Philosophy), and Cambridge (History and Philosophy of Science). His main research interests are philosophizing chemists and chemical concepts. His publications appeared in paper collections and journals like *Hyle*, *Foundations of Chemistry*, *Kantstudien*, *Studies in History and Philosophy of Science*, and he has edited, together with Jaap van Brakel, the collection *Stuff: The Nature of Chemical Substances*. Email: klaus.ruthenberg@hs-coburg.de.

Joachim Schummer is an independent scholar and author based in Berlin, Germany, and on the island of La Palma, Spain, after having worked for many years at various universities around the globe. He graduated both in chemistry and philosophy and received his PhD and Habilitation (second doctorate) in philosophy from the University of Karlsruhe (now KIT). He is the founding editor-in-chief of *Hyle: International Journal for Philosophy of Chemistry* (since 1995) and has served on the boards of various journals and organizations, including UNESCO. His research interests include the philosophy, history, ethics, and sociology of sciences and technologies, broadly construed. He has published numerous books on chemistry, nanotechnology, synthetic biology, and general science, including *The Public Image of Chemistry* (2007) and *Nanotechnology Challenges* (2006) both with World Scientific Publishing. Email: js@hyle.org.

N. Dane Scott is Professor of Ethics in the Franke College of Forestry and Conservation at the University of Montana, U.S.A. He holds a Ph.D. in philosophy from Vanderbilt University and a B.S. in soil science from the University of California Riverside. Scott's primary research interests focus on ethical issues arising from emerging technologies and philosophy of technology. He has published numerous articles and book chapters examining the ethical and philosophical implications of climate engineering and biotechnology. In 2018 Scott published, *Food, Genetic Engineering and Philosophy of Technology: Magic Bullets, Technological Fixes and Responsibility to the Future*, Springer Series: The International Library of Environmental, Agricultural and Food Ethics. Email: dane.scott@mso.umt.edu.

Janet D. Stemwedel is a Professor of Philosophy at San Jose State University, U.S.A., where her research and teaching focus on issues in philosophy of science and the ethics of scientific practice. She earned PhDs in Chemistry (1994) and Philosophy (2001) from Stanford University, and has written on the philosophy and ethics of science for outlets including Forbes and Scientific American. Email: janet.stemwedel@sjsu.edu.

Saurabh Vishnubhakat is a Professor in the School of Law and Professor in the Dwight Look College of Engineering at Texas A&M University, U.S.A. His research explores intellectual property and administrative law, especially from an empirical perspective. His research has been cited in federal appellate and trial court opinions, agency reports and rulemaking, and over two dozen Supreme Court briefs. He has published in leading journals including the *Washington & Lee Law Review* and *Iowa Law Review*, the intellectual property and technology journals of the Harvard, Yale, Columbia, Berkeley, and Duke law schools, and the peer-reviewed *Journal of Economic Perspectives* and *Journal of Law and the Biosciences*. He previously advised the USPTO's first two chief economists and was a postdoctoral associate at Duke Law School. He holds a B.S. in chemistry from Georgia Tech and J.D. and LL.M. degrees from the University of New Hampshire Franklin Pierce School of Law. Email: sv10@law.tamu.edu.

Acknowledgement

The chapters of this volume are reproduced with permission from *Hyle: International Journal for Philosophy of Chemistry* (www.hyle.org). They first appeared as:

Birkholm, K.: 2016, 'The Ethical Judgment: Chemical Psychotropics', *Hyle: International Journal for Philosophy of Chemistry*, **22**(1), 127-148.

Børsen, T. & Nielsen, S.N.: 2017, 'Applying an Ethical Judgment Model to the case of DDT', *Hyle: International Journal for Philosophy of Chemistry*, **23**(1), 5-27.

Contakes, S.M. & Jashinsky, T.: 'Ethical Responsibilities in Military-Related Work: The Case of Napalm', *Hyle: International Journal for Philosophy of Chemistry*, **22**(1), 31-53.

Eckerman, I. & Børsen, T.: 2018, 'Corporate and Governmental Responsibilities for Preventing Chemical Disasters: Lessons from Bhopal', *Hyle: International Journal for Philosophy of Chemistry*, **24**(1), 29-53.

Fjelland, R.: 2016, 'When Laypeople are Right and Experts are Wrong: Lessons from Love Canal', *Hyle: International Journal for Philosophy of Chemistry*, **22**(1), 105-125.

Iles, A.; Martin, A. & Rosen, C.M.: 2017, 'Undoing Chemical Industry Lock-ins: Polyvinyl Chloride and Green Chemistry', *Hyle: International Journal for Philosophy of Chemistry*, **23**(1), 29-60.

Jacob, C. & Walters, A.: 2005, 'Risk and Responsibility in Chemical Research: The Case of Agent Orange', *Hyle: International Journal for Philosophy of Chemistry*, **11**(2), 147-169 (substantially revised for this volume).

Kovac, J.: 2018, 'American Chemical Society Codes of Conduct: Past, Present, and Future', *: International Journal for Philosophy of Chemistry*, **24**(1), 79-95.

Llored, J-P.: 2017, 'Ethics and Chemical Regulation: The Case of REACH', *Hyle: International Journal for Philosophy of Chemistry*, 23(1) ,81-104.

Martin, A. & Iles, A.: 2020, 'The Ethics of Rare Earth Elements Over Time and Space', *Hyle: International Journal for Philosophy of Chemistry*, **26**(1), 5-30.

Martin, A.; Iles, A. & Rosen, C.: 2016, 'Applying Utilitarianism and Deontology in Managing Bisphenol-A Risks in the United States', *Hyle: International Journal for Philosophy of Chemistry*, **22**(1), 79-103.

Ruthenberg, K.: 2016 'About the Futile Dream of an Entirely Riskless and Fully Effective Remedy: Thalidomide', *Hyle: International Journal for Philosophy of Chemistry*, **22**(1), 55-77.

Schummer, J.: 2016, '"Are You Playing God?": Synthetic Biology and the Chemical Ambition to Create Artificial Life', *Hyle: International Journal for Philosophy of Chemistry*, **22**(1), 149-172.

Schummer, J.: 2018, 'Ethics of Chemical Weapons Research: Poison Gas in World War One', *Hyle: International Journal for Philosophy of Chemistry*, **24**(1), 5-28.

Schummer, J.: 2020, 'The Chemical Prediction of Stratospheric Ozone Depletion: A Moral Model of Scientific Hazard Foresight', *Hyle: International Journal for Philosophy of Chemistry*, **26**(1), 31-54.

Scott, D.: 2018, 'Ethics of Climate Engineering: Chemical Capture of Carbon Dioxide from Air', *Hyle: International Journal for Philosophy of Chemistry*, **24**(1), 55-77.

Stemwedel, J.D.: 2016, 'The Case of the Finicky Reactions: A Case Study of Trust, Accountability, and Misconduct', *Hyle: International Journal for Philosophy of Chemistry*, **22**(1), 9-29.

Vishnubhakat, S., 2020: 'The Normative Molecule: Patent Rights and DNA', *Hyle: International Journal for Philosophy of Chemistry*, **26**(1), 55-78.

Name Index

V

W

Subject Index

E

F

G

H

I

Q

R

S

CPSIA information can be obtained
at www.ICGtesting.com
Printed in the USA
BVHW010038150621
609565BV00002B/3

9 789811 233531